Praise for <u>Heaven in a Chip</u>

"Highly recommended."—*Library Journal*

"Very readable. A provocative final chapter promotes the idea that digital networks will be able to hold our own (still-fuzzy) consciousnesses, putting an end to human death."—*Publishers Weekly*

"Remarkable things happen once you question that Aristotelian logic (A or not-A). Certain to appeal to readers curious about where leading-edge types are headed."—*Booklist*

"It's the creative other side of this renaissance scholar—the one that makes music and writes fiction—that shines in his latest project. When you finish the mind-boggling epic . . . you'll realize that those two sides of Kosko's or anyone else's personality—creative and scientific—are not so opposite."
—*Bookbytes*

"You have to admire a book that dares to ask big questions and return unorthodox answers. There's much to like about the way that Kosko brings together voices from far-flung fields of endeavor. Where else can you find pearls of wisdom from Bill Gates, Stonewall Jackson, and Albert Einstein?"—*Industry Standard*

"This timely book focuses on the constantly increasing level of complexity and fuzziness. This text demonstrates how no modern problem can be approached from a binary viewpoint: the constraints of 'yes' vs. 'no' have been split wide open. Throughout the text the author's arguments are clear, precise, and engaging."—*Fatbrain.com*

"Bart Kosko is the kind of guy you dream about having as a dinner guest. A true polymath, Kosko holds degrees in economics, mathematics, philosophy, and engineering, and his encyclopedic knowledge of each of these diverse fields is evidenced in his recent book. . . . A must read for futurists."—*Restoration Magazine*

"[*Heaven in a Chip*] well suits the new century. A truly original writer and thinker, Kosko combines a bold capacity for common sense with an equally bold optimism."—*Liberty*

"[*Heaven in a Chip*] is a book about science, technology, and a vision of where the world is headed in the next few years. For church leaders, it is a book about ethics, values, and the meaning of life. The Christian church will need to adopt some position in response to the technological change that is imminent."—*Quest*

"Kosko is not a deathist."—*Cryonics*

Heaven in a Chip

Fuzzy Visions of Society
and Science in the Digital Age

Previously published as *The Fuzzy Future*

Bart Kosko

THREE RIVERS PRESS • NEW YORK

Published by Three Rivers Press, New York, New York. Member of the Crown Publishing Group.

Random House, Inc. New York, Toronto, London, Sydney, Auckland
www.randomhouse.com

Three Rivers Press is a registered trademark and the Three Rivers Press colophon is a trademark of Random House, Inc.

Originally published in hardcover as *The Fuzzy Future* by Harmony Books in 1999.

Printed in the United States of America

Design by Maggie Hinders

Library of Congress Cataloging-in-Publication Data

Kosko, Bart.
    [Fuzzy future]
    Heaven in a chip : fuzzy visions of society and science in the digital age / Bart Kosko.—
1st pbk ed.
        Originally published: The fuzzy future. 1st ed. New York : Harmony Books, c1999.
        Includes bibliographical references and index.
        1. Fuzzy logic.   I. Title.
    QA9.64.K67   2000
    511.3—dc21   00-061563

ISBN 0-609-80567-3

10 9 8 7 6 5 4 3 2 1

First Paperback Edition

For Tony Kosko
1920–1970

who taught me to think for myself
and whose early death taught me to think by myself

Contents

The road to the future is always fuzzy.

ANONYMOUS

Preface

Suppose we replace your brain with a computer chip. How would you change? Would you still be you? What if we replace only two-thirds or half your brain with a chip? Would you think only black-and-white digital thoughts? Would your digital brain house a digital mind? Or would your mind use fuzzy logic? Would your thoughts be fuzzy or gray?

This book looks at such fuzzy or gray issues as they arise in the digital age. But just what do these terms mean? What is fuzzy? What is digital?

The term fuzzy means shades of gray between 0% and 100%. Most concepts are fuzzy because they have inexact borders. There are no hard lines between water that is warm and not warm or between sunsets that are red-orange or not red-orange or between front teeth that are crooked and not crooked. These concepts have opposites that shade into each other.

Hard lines occur most often in math and in politics. Circles and squares have black-and-white boundaries because for well over 2,000 years we have defined them that way with binary logic. But we can define an infinitude of fuzzy circles and squares that generalize the old binary circles and squares. A fuzzy circle might look like a jagged ideal circle. Its level of jaggedness would give a measure of its fuzziness or how much the circle looked like a non-circle. An ultra-fuzzy circle might look as if one painted a circle with a spray can rather than drew a circle with an ink pen.

Politics is all about drawing hard lines and backing them up with the force of the law. The line between a legal and illegal blood-alcohol level is a hard line because the state draws it as a line. A hard line in math depends on the fiat of definition. A hard line in politics depends on both fiat and force. That is why the same blood-alcohol level can land you in jail in one state or country and leave you free to drive at risk in another.

The state supplies much of society's demand for hard lines. A glance at any tax code shows that governments tend to over-supply this demand. The legal language of the tax code itself is shot through with fuzz of all types. This fuzz allows more choices but seldom for the taxpayer. It gives more license to those in power because it lets them draw lines through wider spheres of action. It gives them logical wiggle room. This is the fuzzy version of the Golden Rule: Those with the most power tend to draw the hard lines.

Binary logic has always been the logic of power.

The term digital age means an information age based on the binary units of information 1 and 0 or yes and no. These on–off bits are digital building blocks. They define what is true and what is false. We can use any two symbols and then mark them on paper or press them in mud or build them into the two states of an on–off switch or logic gate. A digital computer chip may contain millions of tiny on–off switches based on whether enough electrons flow across a gate or whether a well traps enough electrons.

A digital circuit in a chip may store a bit value as 1 if the input voltage exceeds three volts and store it as 0 if the voltage is less than two volts. The chip must have some way to round off voltages that fall between two and three volts in the fuzzy "noise margin." A digital system based on fiber optics assigns the bit value of 1 if a light is on in the fiber and assigns the bit value 0 if it is off. It too must round off the fuzzy middle ground of dim light values. The same holds for a digital polymer memory chip of the future. Such a plastic chip assigns the bit value of 1 if a molecule lies in one state and assigns the bit value of 0 if the molecule lies in some other state. The more futuristic digital quantum computers go further. They not only assign a quantum bit or "qubit" the value of 1 if an atom's nucleus spins in an "up" state and assign it the value of 0 if the nucleus spins in a "down" state. The statistical laws of quantum mechanics allow the same qubit to take on both the value 1 and the value 0 at the same time.

Computer chips have fueled the digital age but they did not invent it. Binary logic did. The "law" of either-or did. But binary logic does not work as well with how we think or reason. Our minds must work to define a circle as the locus of points equally far from a center point. We struggle to trace out the logic steps in a barroom argument or in a judicial decision or in a mathematical proof.

Minds are not digital processors. Our concepts are fuzzy to the core and our reasoning is approximate. The statement "Red apples taste good" holds for each of us only to some degree. The statement's fuzz or vagueness stems in part from the subjective nature of taste and from the fuzziness of taste and goodness. The statement's fuzz stems in large measure from the fuzziness of an objective "fact": A given red apple is not a pure red or 100% red apple. There is a continuum of exceptions between a pure red apple and a pure not-red apple.

There is always a continuum of exceptions.

We draw hard lines through fuzz to help manage the exceptions. This gives up accuracy for simplicity but favors action over description. It lets us paint the world with the 1s and 0s of the digital age. The technique is quick and dirty but it often gets things done. It helps us pick and sort and sell red apples as well as make claims about how good they taste.

Trouble occurs if we take these binary carvings too seriously. Then we can lose the balance between fuzz and precision. We can confuse the quick sketch of a mountain with the mountain itself. This has always fed our binary instinct to chop up society into logical groups or castes and to judge others as either in or out or with us or against us. We have all felt the sting of this binary instinct since the days of the playground. Perhaps none have felt it more than have the children born of interracial marriages. The designer children of tomorrow may feel it too. The binary instinct runs so deep in our thinking and in our institutions that it may have some basis in our genes. Natural selection may have tipped the balance in favor of those who were quick to round off and act over fuzzy thinkers who paused and questioned binary authority.

The irony of the digital age is that things are fuzzier than ever. Fuzz grows as the bits pour in. More precise "facts" have not made it easier to draw a line that decides whether a fetus is alive or whether a share of stock is a good buy or whether crashing a country's banking software is an act of war. Bit streams of 1s and 0s have changed the practice of law and art and science but bit streams have not sharpened the borders of these fields. A tele-present surgeon in France can cut out a brain tumor from a patient who lies awake on a bed in Canada. Such digital feats can save lives and challenge local monopolies on medical services. But they do not sharpen the border between brain tumors and non-tumors or the border between those who have the skill to supply our medical needs and those who do not.

The digital age has its own uncertainty principle: *Issues get fuzzier as their parts get more precise.* Lots of small blacks and whites add up to more gray. The overall uncertainty stays the same or maybe even grows. Digital precision does not make it go away. We now draw weather maps in exquisite digital detail while our concepts of light rain and cool breeze and partly cloudy skies remain as fuzzy as in the days before TV. Sonar and GPS satellite data let us track which boats pass through almost any square or cubic meter of ocean water. Yet most property rights in the ocean remain fuzzy and that is not good. We do not know who owns most of the ocean because it has few hard borders between mine and thine. So we suffer a growing "tragedy of the commons" of pollution and over-fishing.

The Internet gives the most striking example of how fuzz grows in a digital medium. Each state and country tries to draw hard legal lines through the fuzzy borders of Internet libel and obscenity and national security. Others try to tax Internet talk or to tax Internet commerce. Most of these efforts are in vain. They collapse both over the compound fuzziness of the competing legal concepts and most of all over the fuzziness of the governments' jurisdictions.

The Internet raises an old fuzzy question in a digital context: Where does one government end and another begin? And this gives rise to a vexing fuzzy question at a higher level: Who draws the lines on those who draw the lines?

Governments began as monopolies on force over a range of land. Their logic was the old logic of the sword. Maps drew the hard borders down to the last atom and swords backed up the maps. These binary force monopolies do not extend so well from atoms to bits. The borders blur a bit at a time. Governments may still issue their legal edicts in cyberspace but in most cases they lack the force to back them up. The most zealous state efforts to define and stamp out Internet gambling have only spurred innovations in on-line payment and encryption.

Competition among bits further increases fuzz. The Internet competes with the force monopolies and makes them compete with one another. In this way the world of bits now draws hard lines on the world of atoms. The financial markets were the first to pass through these fuzzy boundaries and give governments a very real run for their money. Central bankers may not like it but the humble Euro-dollar (not the new Euro currency) thrives as a stateless currency in cyberspace.

This book reviews fuzz in the first chapter and then looks at how fuzz lies at the core of both political power and scientific truth. The book ends with how fuzz lies at the essence of self: Is biology destiny? Are we more than our synapses? Will consciousness change when it shifts from the electrochemical reactions in three pounds of flesh to a bit stream of 1s and 0s in a chip? Will religion still hold a working monopoly on the concepts of soul and afterlife and heaven?

Science has not killed God but it has dethroned Him. That subtle power play has put all ideas up for grabs. The digital age will further challenge and shape our ideas as we shift our focus more from the old world of atoms to the new world of bits. The shift from atoms to bits and from chimp brain to chip brain may not be a smooth one. But much of it will be fuzzy.

Introduction: Creeping Fuzziness

I hear fuzzy logic has hit the big time.
They say it's number one in Tokyo.
Well I'm the first to say I ain't no Einstein.
But I had fuzzy logic long ago.

I had fuzzy logic before fuzzy logic was cool.
Fuzzy thinking helped me get through twelve long years of school.
Fuzzy logic helps me keep my married life content.
And fuzzy thinking made me a success in government.

> BOB HIRSHON
> "The Fuzzy Logic Song"
> (Reprinted with permission of the author)
> Broadcast on National Public Radio: 8 October 1993

When the prophet, a complacent fat man,
Arrived at the mountain top
He cried: "Woe to my knowledge!
I intended to see good white lands
And bad black lands.
But the scene is gray."

> STEPHEN CRANE

Gray is the color of truth.

> MCGEORGE BUNDY
> Former National Security Advisor: 1967

It's funny how when we're the recipient of pain we're clear that it's
black and white. But when we've got something to gain there are
shades of gray.

> DR. LAURA SCHLESSINGER

FUZZ CREEPS INTO A PROCESS BY DEGREES.
Suppose we cut out a small chunk of your brain. We replace that brain chunk with a tiny computer chip that acts the same as your fleshy brain chunk did. You are still you. But your mind's housing is now slightly less a brain and slightly more a non-brain.

Fuzz has crept into the process.

Now suppose we cut out a new brain chunk and replace it with a new chiplet. Then your mind's housing is still less a brain and still more a non-brain. We can keep chipping away at your brain like this until your mind's housing has changed fully but smoothly from brain to non-brain. You would think in a chip rather than in three pounds of meat. And you would still be you.

Or suppose the state lets you have some say in how it spends your tax money. Right now your say is not fuzzy but binary: You have zero say. Elected and appointed officials spend your money for you. You have only your vote and its very tiny say in who gets elected.

But suppose this year the state lets you say where you want it to spend just 1% of your tax bill. You might tell the state to spend that small tax chunk to pay down its debt or to fix the roads or to help cure cancer. You might like that small fuzzy taste of freedom of choice so much that you would next want to say where the state spends 2% or 10% or even 50% of your tax money.

Then fuzz would have crept into the legal structure of our social choices. It would have blurred some of our black–white lines of power. Chapter 4 looks at fuzzy tax forms as tools that can fuzzify our social choices.

Or suppose the man and woman next door like you so much that they want to have a child whose genes are the same as yours. They want a designer child and the design is you. Suppose you do not feel the same as they do and you want to stop them. You own yourself. So you feel that means that you own your genetic blueprint. The law may agree with you. But the law may not agree if the couple wants a child with only 95% or 50% or 33% of your genes.

Fuzz would have crept into our deepest rights of private property. And it does not stop there.

Creeping Fuzz in Science and Technology

Fuzz adds choices as it creeps into a process. It offers shades of gray between the extreme choices of black and white. This continuum of gray choices can challenge simple either–or world views that range from a child's claim that you are either a friend or an enemy to a scientist's view that all "well-formed" statements of fact are either true or false.

Fuzz gives up such simple claims to certainty. Fuzz does not make us choose between the claims that the sky is blue *or* that it is not blue. Fuzz lets us say that the sky is both blue *and* not blue to some degree. It extends Aristotle's logic chop of *A or not-A* to allow the yin-yang option of *A and not-A* to some degree. This extension may seem minor but it defies over 2,000 years of formal logic and mathematics. It is also common sense.

Fuzz has crept into our notions of common sense since at least as long as we have had language. Our words describe fuzzy or vague patterns like *cool air* or *large tree* or *high price*. We use terms like *healthy tongue* or *pretty face* or *fair trial* even though we cannot define their borders with hard either-or lines. We combine these fuzzy patterns into sentences like "If the price is high then the demand is low" to form the basic building blocks of our commonsense knowledge. We reason and converse with these fuzzy terms even though no two people mean the exact same thing by them.

Modern science has not been so tolerant of fuzzy thinking. It has often ignored fuzz or tried to define it out of existence by allowing only black and white truth options.

The formal language of science is the black-white language of math. That language deals best with logical truths like "One plus one equals two" or "All bachelors are unmarried men" or "Blue is blue." Logical truths are indeed 100% true because we construct them that way in accord with the laws of binary logic. The statements are true because of their form and not their content. They do not describe the world. They are statements of the formal system we call logic. They are not statements of empirical science itself. Science deals with real or alleged facts of the world. It deals with what the old philosophers of metaphysics used to call cause and effect.

The formal language of science uses the same black-white truth structure to describe factual truths like "Grass is green" or "Gravity causes erosion" or "You cannot break a fluid in two." It is still digital talk. Each statement is in principle either true 100% or false 100% even if we do not know which. This still holds if we describe the tentative nature of such binary claims with a probability hedge like "probably" or "frequently" or "usually." The statement "Grass is probably green" just gambles on whether grass is green or not green. The probability hedge does not alter the binary status of whether grass is green. Science at this level looks a lot like a casino. Both gamble on the binary-truth status of claims.

But it often works. The math-like talk of science has proven so powerful a tool in how we model and control the world that we sometimes forget that the math talk itself only approximates the world it describes. The Earth is not a perfect oblate spheroid and does not orbit the Sun in a perfect ellipse. The energy of an object does not exactly equal its mass times the square of the speed of light. And the sky is not either blue or not blue. The truth or accuracy of these claims of fact are matters of degree.

So fuzz creeps into our very notions of truth.

Fuzz also reminds us that black-white truth lies beyond our grasp: No one has produced a pure truth of the world. We have measured some statements about the energy of pulsars to more than 14 decimal places of accuracy. We would have to get the science right to infinitely many decimal places both to produce a pure binary truth and to know that we had produced it. This point has an economic flavor: Fuzz reminds us that binary precision is not a free good. We would have to pay dearly to achieve it and we never have.

How we describe the world remains gray or fuzzy.

At least one school of modern physics boldly rejects this world view. It contends that the universe itself is nothing but a pile of binary information bits—a vast pile of 10^{120} or so 1s and 0s. Chapter 11 looks at this radical binary thesis of "it from bit." The it-from-bit thesis has a sense of the fantastic. It comes about when we ask what happens if we throw the whole universe into a black hole. But in the end it results from projecting binary logic onto the world and taking this assumption to its logical conclusion. You derive from an assumption what you put into it.

Fuzz has made much deeper inroads into engineering and the commercial world that depends on it. Thousands of people drive the Volkswagen New Beetle or the GM Saturn that have fuzzy automatic transmissions.[1] Many more own a camcorder or washing machine or microwave oven that uses fuzzy logic in its control chip. Canon introduced the first fuzzy camcorder in 1990 with its H800 model. It used fuzzy if-then rules to focus the lens based on the relative contrast and brightness of image segments. More advanced fuzzy camcorders use fuzzy rules to stabilize the image when the user's hand shakes.[2] The fuzzy rules can tell movement within the image frame from movement between image frames. All the points in the image move in the same direction if the user's hand shakes the camcorder.

Chips have put fuzzy logic in business.

Hundreds of consumer products use fuzzy logic in their control chips. And a new car may have over 100 chips in it. The firms that sell these fuzzy products often do not mention fuzzy logic in their ads. Many of the products come from Japan or Korea because these countries produce so many of the world's consumer electronics. These Eastern countries also had less philosophical resistance to the idea of using fuzz in science and engineering. Culture surely played some role in at least the start of the process. The yin-yang symbol is after all on the flags of South Korea and Mongolia. Fuzz made these products smarter when someone reprogrammed their control chips. The software changes may cost only a penny or two per device and seldom require new hardware.

Fuzzy engineering has also spread to Europe and other continents. Germany took the lead in fuzzy applications in Europe in the late 1990s just as Japan took the lead in the Far East a decade before. The German applications tend to be more in heavy industry and process control. Most Japanese applications still tend to be in consumer electronics though they also include applications in manufacturing and banking and information systems. The Germans followed the engineering lead of the Japanese into fuzzy systems. This involved its own mix of German culture and economics. Concepts of yin and yang were not part of it.

Fuzzy systems in Germany reduce sway in large industrial cranes and control the temperature in plastic molding machines. A fuzzy system in Bonn treats waste water to reduce phosphate. Fuzzy systems in Hamburg and Mannheim control the burning process at waste incineration plants.[3] Fuzzy engineers have tested a fuzzy system to control the power output of a Belgian nuclear reactor.[4] Fuzzy systems in Brazil help guide trains and search and drill for offshore oil.[5] A fuzzy system in South Africa helps verify typed passwords.[6]

Other fuzzy software systems now route elevators in large buildings and assist many business and investment decisions and even filter out junk or spam e-mail.[7] Engineers have reported thousands of fuzzy applications in the formal literature besides those mentioned here. Chapters 9 and 10 explore the formal structure and limits of fuzzy systems and how they can learn their own sets of fuzzy rules from training data.

Fuzzy ideas have spread well beyond the camcorder. You can now read about formal fuzzy logic in one of the new fuzzy science journals or fuzzy textbooks.[8] Or you can see it in action at one of the dozens of fuzzy workshops and conferences held each year around the world. Scientists who might once have claimed that "fuzzy logic is the cocaine of science"[9] opened their July 1993 issue of *Scientific American* and found my article on fuzzy logic.[10] The science establishment had absorbed a literal "fringe" technology. Fuzz has even crept into modern art and gets passing mention in science-fiction novels and films.[11]

Gray has become okay. It has almost become cool.

The rest of the book requires a deeper knowledge of fuzz and we need to pause here to review it. The reader with some feel for fuzzy logic can safely skip ahead. Each chapter uses only slightly more fuzzy concepts than the one before it. But each chapter is sufficiently self-contained so that the reader can still jump from chapter to chapter in random access.

Sometimes we use fuzz in the simple sense of shades of gray. The chapters in Part One do this as they explore the fuzzy structure of political systems and taxes and property rights and warfare. This type of fuzz occurs when we ask who owns you or who owns the moon. The chapters in Parts Two and Three combine shades of gray with formal fuzzy systems. Part Two focuses on the science involved in fuzzy systems and the fuzziness involved in scientific systems. Part Three applies all these fuzzy ideas in conjectures about the digital future and its balance of fuzzy and digital concepts.

The key to it all is the simple and expressive fuzzy set.

Fuzzy Logic for Beginners

Fuzzy logic is reasoning with vague concepts.

Fuzzy logic is Mr. Spock's worst nightmare: It lets us do science without math. It lets us compute with words and shades of gray rather than with brittle equations that someone guessed at. Fuzzy logic is a branch of machine intelligence that tries to make machines think as we think—and perhaps better than we think.

There is new math behind fuzzy logic and it took thousands of years to find it. That finding is the pure contribution of fuzzy logic. The good news is that you do not need to know this math to use fuzzy logic. You need only press a button to wash your clothes in a washing machine that uses fuzzy logic in its control chip. You can also program and reprogram many fuzzy systems in English or in other natural languages. I will from time to time hint at some of these new results in fuzzy math when it leads to a new idea or when it helps me argue a point. I will restrict the formal statement of these and other math results to the endnotes.

Even at the math level there is fuzzy irony: The math of fuzzy logic is not fuzzy. But that math lets us extend and fuzzify all of mathematics.

Fuzzy logic lies at the heart of the old tug of war between thought and action. We do not accept a jury verdict of 75% guilty even if the defendant is in fact only 75% guilty or if just 75% of the jurors think she is guilty. We like our actions black or white even if our thoughts are gray or fuzzy. We often round off our fuzzy thoughts to make a point or to take a stand or to cast a vote. The goal is to get computers to work with more than just our rounded-off black and white concepts.

Fuzzy logic models a vague concept like *cool air* with a fuzzy set. But before we define a fuzzy set it helps to first review what we mean by a set.

A set contains objects. Each object either belongs to the set or not. There is no middle ground. The set of even numbers contains the number 2 but not the number 3. A set is an abstract binary structure. It allows no partial membership. Set membership is an either-or affair. The number 2 belongs 100% to the set of even numbers while the number 3 belongs 0% to it.

So a set is a digital structure. We can write a 1 if an object belongs to the set and write a 0 if it does not. The converse is also true:[12] A digital bit list of 1s and 0s defines a set. An object belongs to the set if there is a 1 in that slot in the bit list. It does not belong to the set if there is a 0 in that slot.

A fuzzy set allows partial membership. An object can belong to a set to any degree or "shade of gray" between 0% and 100%. The statement "The sky is blue" may be true only 80% because the sky may be only 80% blue and 20% not blue. The value 80% is a fuzzy truth value. We can also view this in terms of a set with fuzzy borders. Each sky that we see belongs to the set of blue skies to some degree.

Now consider the concept of *cool air.* We all have some feel for this concept. It affects how we dress and design our shelters and choose our blankets. But what does the concept mean?

Fuzzy theory gives a subtle answer: *The meaning of a concept is the fuzzy set that defines it.*

This answer implies in one stroke that concepts are both vague and relative. Cool air is vague or fuzzy because all air is both cool and not cool to some degree. Cool air is relative because no two people mean the same thing by the term cool air. The meaning varies among speakers and even varies for the same speaker over time. What each of us means by the term cool air may change from season to season or sometimes from day to day. So cool air does not have just one meaning. It has an infinitude of meanings and yet these meanings are all of a piece. A fuzzy set captures this idea.

The following figure shows just one of infinitely many fuzzy sets that can define the concept of cool air:

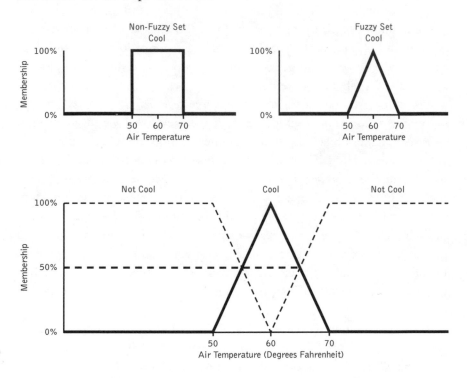

FUZZY SET AND ITS COMPLEMENT

This chapter's formal lesson in fuzzy logic depends on a good look at this figure.

The rectangle defines cool air as a binary set. All air temperatures are either cool 100% or not-cool 100%. A rectangle always stands for a black and white concept and vice versa. The binary set gives at best a crude idea of what we mean by cool air. Note the hard jump from none to all or from all to none at the border of cool and not cool. This means that an air temperature of 49.999999999° Fahrenheit is 0% cool but that an air temperature of 50° Fahrenheit is 100% cool. The binary jump is extreme and arbitrary. It makes most sense in a logical dictatorship.

The triangle defines a fuzzy set centered at 60° Fahrenheit. It allows only the lone air temperature 60° Fahrenheit to be 100% cool. All other air temperatures are cool to some degree less than that. The degree of coolness falls as the air temperature falls below 60° or as it rises above 60°. The two sides of the triangle define two continua of membership degrees. Rectangles cannot express this spectrum of partial or graded set membership.

The air temperatures 55° and 65° have a special fuzzy status in this fuzzy set of cool air. They are both cool 50% *and* not-cool 50%. The 50% factor prevents this from being the kind of logical contradiction (like "1 = 2") that would destroy mathematics. These two 50% values are pure fuzzy values or midpoint values like the cup that is both half empty and half full. They lie as much in the set of cool air as they lie out of it. The lower figure shows this. The fuzzy set of cool air intersects the fuzzy set of not-cool air at the 50% midpoint value. A fuzzy set always intersects its opposite at the "paradoxical" 50% midpoint value.[13] Binary sets never intersect their opposites.

Both the cool-air rectangle and triangle are relative because we can also draw them wider or thinner or shift them up or down the temperature axis. We change the meaning of cool air if we change the fuzzy set that stands for it. We might want to center the cool-air triangle at 60° in the winter and center it at 65° or 70° in the summer. And what is 90% cool for us might be only 70% cool for someone else. We might want to break the symmetry of the triangle and make the right side longer than the left side. The slope of these sides describes how fast the gray borders shade into black or white. The fuzz falls as the slope grows steeper. A binary rectangle has the steepest slope of all.

A simple but hard question arises at this point: Why draw the fuzzy set of cool air as a triangle? Why not draw it as a trapezoid or as a smooth bell curve or as any other curve? The age-old principle of sufficient reason demands a reason for this choice of curve rather than for any other.

The bad news is there is no good answer.

Triangles consist of line segments. We can easily draw them and compute with them. Fuzzy engineers often use them to get a quick first-cut system. But triangles give a simple type of fuzzy system (called "piecewise linear") that may be too crude to control or model a real process. So engineers must tune or adjust the triangles or try other curves to evolve a more complex fuzzy system.

This is also where fuzzy systems join with their sister technology of neural networks or "brain-like" learning systems. Neural systems are pattern computers. They can learn and store and recall the fuzzy pattern of a face or a tree or a tank even though no software programmer can define these patterns to the binary satisfaction of a digital computer.

A neural system learns fuzzy sets from experience. It learns these sets just as our brain learns fuzzy patterns like cool air or red apple or rough surface when the flux of experience rolls over our surface receptors and trains our brain. A neural system acts like the front end or sensory part of our brain. It learns patterns. A fuzzy system processes these fuzzy patterns and acts more like the cognitive part of a brain. It reasons with patterns.

We can teach the fuzzy pattern of cool air to a neural system much as we would teach the concept to a child. We show the neural system air samples and tell it whether and to what degree the air is cool. Each sample will slightly move the fuzzy-set curve or change its shape.

This is a key property of learning: Learning always changes something. A neural-fuzzy system changes the shape of its fuzzy sets. Watching TV or surfing the Net changes slightly the biochemistry of our brains. Each image leaves its footprint in our synapses. Think how many training samples change a teenager's brain to teach him the all-important teen concept of "cool." A triangle would be far too simple to model it.

So what is the best shape for a fuzzy set? We may never know. This question is much deeper than it sounds and fuzzy engineers will no doubt ask it for centuries. The math space of possible curves is just too vast to search and carve up with today's minds or today's computers. Neural learning remains our best search tool.

My graduate students and I have run extensive computer simulations looking for the best shape of fuzzy sets when we want to model a known test system as closely as we can. We let fuzzy sets of all shapes and sizes compete against one another. We also let neural systems tune the fuzzy-set curves to improve how well they model the test system. Triangles never do well in the contests. The best curves often have many bumps in them and defy a simple verbal description.[14]

This fuzzy analysis of cool air should give the reader cause for pause. Cool air is one of our simplest sensory concepts. Yet we open an infinite Pandora's box of fuzzy sets when we ask what the concept means.

The moral is that we all live in our own private conceptual worlds to a far greater degree than we may realize. We talk with the same sounds and write with the same symbols. But we do not mean the same things by these sounds and symbols. I call this the *conceptual anarchy*.[15]

Anarchy tends to be a good thing in the digital age.

User friendliness results from the conceptual anarchy. Smart machines can adapt how they perform to what we mean by our fuzzy concepts. Suppose you drive a car that has a fuzzy air conditioner. The fuzzy system has if-then rules in it like "If the air is cool then set the blower to slow" and "If the air is just right then set the blower to medium" and "If the air is warm then set the blower to fast." You might adjust the fuzzy sets that define cool air and warm air just as you have to readjust the seat and rearview mirror after someone else drives.

A voice system would let you tell the system which test blasts of air are cool and which are warm. Then the neural system would adjust the fuzzy sets. Other systems could try to estimate what you mean by cool and warm by how you set and reset the air blower. The rule structure would not change. Cool air would still map to a slow blower and warm air would still map to a fast blower. But the fuzzy sets that define cool and warm would change. The fuzzy system would adapt to help you find your niche in the conceptual anarchy.

A second look at the fuzzy-set triangle for cool air raises many other questions about the nature of fuzz and randomness. Why do we draw the fuzzy set as a thin line? Is that not artificially precise? Does anyone really think that 65° air is exactly 50% cool?

The answer is that this too is an approximation. We might really want to say that 65° air is "about" 50% cool or "nearly" 50% cool or "roughly" 50% cool. Then we draw the triangle with a spray can rather than with a fine-line pen. This gives an *ultra*-fuzzy set where a new fuzzy set models each fuzzy degree like 50%. The new fuzzy set can be a triangle or a curve of any shape. This can in theory go on forever. We can model each fuzzy degree in the new fuzzy set with another fuzzy set to create an ultra–ultra-fuzzy set and so on up the ladder of abstraction.

Others might wonder why we call the fuzzy-set triangle a fuzzy set rather than a probability distribution like the notorious bell curve of IQ scores. A technical answer is that the probability of any exact value like 65° being cool is always zero. A better answer is that standard probability theory just places odds on binary events. It works with sets that need a binary rectangle.

We can expand the probability math so that we can put odds on fuzzy-set curves like triangles. This gives the probability of a fuzzy event because it asks whether a vague event will occur. An example is when the weatherman says "There is a 10% chance of light rain tomorrow." The light–rain pattern is a fuzzy subset of all rain patterns because all rain is both light and not light to some degree. The light–rain concept bears the defining fuzzy stamp of having some overlap with its own opposite.

We can also turn this around and measure the fuzz or vagueness of a random event. Consider the statement "There is a slight chance of any rain at all tomorrow." It asserts that raining or not raining are both binary events and random events. It further asserts that whether the random raining event occurs is vague or inexact or fuzzy. A "slight" chance is a fuzzy subset of the set of all possible chances. We might describe this fuzz with a triangle or other fuzzy-set curve centered over the exact value of 10%.

We can unpack still more random structure from the triangle that stands for cool air. The fuzzy view sees a measurement of 65° air as an air temperature that is cool to the degree 50%. There is a roundabout way to see this in purely random terms and it is only fair to present it. This view does not see the triangle as a locus of degrees or fuzzy membership values. It instead sees them as a locus of conditional probabilities. Then we can equate the fuzzy statement "65° air is cool 50%" with the random statement "There is a 50% chance that the air is cool given that the air temperature is 65°." Far more people speak of air as cool or warm to some degree than speak of it in terms of conditional probabilities.[16]

The last question is the fuzzy bottom line: How fuzzy is one set or concept versus another?

The vagueness or fuzziness of fuzzy sets itself admits degrees. The fuzzy-set triangle in the figure is fuzzier than the binary-set rectangle because binary sets are all fuzzy to degree zero. But suppose we slightly move the left side of the rectangle and let it slope a little from left to right. Then the rectangle becomes a trapezoid and just barely a fuzzy set. The set gets fuzzier the more we move the leg to the left and slow the transition from black to white.

A theorem gives the answer: The fuzziness of a concept is the degree to which the concept equals its own opposite.[17] It is the degree to which the yin–yang symbol holds and a set or concept A equals *not-A*. The fuzz of cool air depends on how much a concept like cool air overlaps its opposite not–cool air. Maximum fuzziness occurs when a set or concept equals its own opposite.

The same theorem also shows a property unique to fuzzy theory. The fuzziness of a concept is the degree to which a part partially contains a whole. This strange whole-in-the-part relation occurs only to degree zero for binary sets and concepts. A related theorem shows that what we mean by probability in the gambling sense of a relative frequency is the same whole-in-the-part relation but among binary sets.[18] An example of relative frequency is the ratio of head flips to total coin flips or the ratio of the number of men who get prostate cancer at age 50 to the total number of 50-year-old men. This whole-in-the-part relation (called subsethood to reflect partial inclusion or containment) measures the degree to which one set or concept belongs to another. Cool air has some fuzzy overlap with warm air. So the two concepts or sets contain each other to some degree.

Fuzzy sets are the building blocks of fuzzy systems.

A fuzzy system builds a bridge from inputs to outputs. The bridge consists of rules like "If the price of gold is high then the demand is low" or "If the wash water is very dirty then add a lot more detergent" or "If the air is cool then set the blower to slow." The rules might map market facts to buy or sell commands. Or they might map sensor measurements to control actions. They always convert inputs into outputs.

A fuzzy system acts like a human expert who has a lot of if-then rules of thumb in his head. The expert somehow uses these commonsense rules to buy gold or wash clothes or adjust an air conditioner or perform any task. The expert can seldom explain how he performs as he does. He may cite some if-then rules but the non-expert can rarely match his performance by acting on those rules. Imagine trying to play the violin by merely asking a concert violinist how he plays.

No one wants to guess at the equations that will wash a load of clothes or grill a steak or focus a camera lens. Sometimes we get lucky and guess well and that is all for the good. But the bridge of rules may offer a simpler and better way to control a system or process. And it does so often enough to make fuzzy systems a multibillion-dollar success in consumer electronics and process control.

A fuzzy system fires all its rules to some degree for each new input. Suppose a sensor measures the air temperature of 65°. The measurement will fire the rule "If the air is cool then set the blower to slow" to degree 50% because the input is cool only 50%. The same input measurement fires the rule "If the air is cold then set the blower to stop" to a much lesser degree because the input of 65° is cold to a much lesser degree than it is cool. The same holds for the other rules that have warm air or very warm air or hot air as their if-part fuzzy sets.

The fuzzy system fires all rules in parallel and then adds the result and takes the average. That gives the final blower command as a type of weighted average. It converts measurement inputs into control outputs. This fire-and-sum process may occur many times per second when fuzzy software runs on a computer chip. The endnotes in chapter 9 give the complete details.

A GM Saturn fires its rules four times a second to update its six fuzzy inputs[19] and then makes the hard call of when to downshift as the car goes downhill. The fuzzy system acts as a super-fast expert who watches the slight changes in road conditions to decide when to shift the four gears: "If the car speed is slow and if the grade is negative and if the braking time is long then downshift one gear."

The weakest part of a fuzzy system is its rules. It may be hard to get good rules for a system if there are too many variables or too few experts or too few data. Stock picking is a good example. The system of share prices depends on thousands of variables. And there are very few investment experts whose long-term track records match either the market average or even perhaps their own claims of their stock-picking prowess. Such experts can quickly insert their knowledge and hunches in a fuzzy system. They can use neural systems to refine their systems. But the fuzzy rules will capture their mistakes as well as their wisdom.

The biggest problem with rules is that most fuzzy systems need too many of them. More variables make a system more realistic. We might add humidity and light intensity to a fuzzy air conditioner to better control the air blower. The good news is a fuzzy system can model any system with enough rules. That is a theorem. The bad news is *rule explosion*. In general the number of fuzzy rules blows up exponentially with the number of variables. That too is a theorem.

This is the "curse of dimensionality." It strikes all math systems that try to work with too many variables. Chapter 9 looks at some coping strategies. Faster chips will always eat slowly away at this computational block and allow us to use fuzzy systems with ever more rules and variables. But the problem is structural. There will always be a limit to the number of variables our technology can work with no matter how powerful our digital computers.

A digital chip may seem an odd host for the vague reasoning of fuzzy logic. How can the 1s and 0s of binary chip logic house the gray blobs of fuzzy logic?

Again the math of fuzzy logic is not fuzzy. It is as black and white as "Two plus two equals four." The chip does not even have to give a digital approximation to most fuzzy concepts as a compact disc does when it reconstructs the sound of music it has sampled 44,100 times per second. We have worked out the math so that chips can make exact fuzzy decisions just by adding and multiplying a few numbers. That is why there was a fuzzy revolution at the height of the digital age.

It was not always so.

A Brief History of Fuzzy Logic

The ancient Greek Zeno talked of how a sand heap became a non-heap a sand grain at a time. The concept A smoothly transitioned to its opposite *not-A*. The Buddha said the same thing of a chariot that comes apart a piece at a time. Logician Bertrand Russell returned to these and other "paradoxes" at the turn of the century when he and Alfred North Whitehead wrote the first major new book on logic since Aristotle.[20]

Russell observed that "everything is vague to a degree you do not realize till you have tried to make it precise."[21] A head full of a hundred thousand or so hairs passes from not-bald to bald a hair at a time. A table passes to a non-table a molecule at a time. Things pass smoothly from A to *not-A* and spend most of their time as a mix of both. The either-or logic of Aristotle applied to math but not to the world or the science that tried to describe that world.[22] The black-white world of symbols did not apply "to this terrestrial life but only to an imagined celestial one."[23]

Russell had opened the door to vague or fuzzy logic. Soon others explored it and laid its formal foundation. The Polish logician Jan Lukasiewicz first worked out fuzzy or multivalued logic in the 1920s as a direct extension of binary logic.[24] All statements were true and false to some degree. The only real constraint was that the truth percentages had to add up to 100%.[25] The claim that "Lemons are yellow" is 90% true implies that the claim that "Lemons are not yellow" is 10% true and vice versa.

The next advance came in 1937 when quantum philosopher Max Black published a journal paper on vagueness.[26] Black drew a graph of the first fuzzy set in this paper. Philosophers largely ignored the Black paper as they had largely ignored the vague work of Russell and Lukasiewicz. It was the heyday of "logical positivism."[27] This radical view held that only the formal statements of science or math or logic had "meaning." All other talk was "meaningless." Philosophers tried to get people to speak in the black-white logic of math and to define a table or a mountain down to the last molecule.[28]

Fuzzy logic crossed over into engineering in 1965 when Lotfi Zadeh published the landmark journal paper "Fuzzy Sets" and made *fuzzy* a new adjective in science and math.[29] Zadeh wrote the paper when he was still chairman of the department of electrical engineering at the University of California at Berkeley. He had come to the United States from Iran in 1944 to study at MIT and then at Columbia. He had helped lay the foundations of modern system theory in the 1950s. He soon followed his iconoclastic instincts and moved into vague logic.

Zadeh gave the field a new name and a new home and a whole new math framework. He never gave full credit to the fuzzy logicians who came before him but he largely made up for this by waging a one-man crusade for fuzzy logic in the information sciences. He wrote papers and lectured around the world at workshops and conferences as the decades passed. Critics and fans slowly took notice and in time so did the press.

The term *fuzzy* crept into the scientific underground. It became a code-word for those who doubted that either-or thinking was the hallmark of science and who searched in vain for that single binary statement of fact. Fuzzy researchers published papers in obscure journals and in time published them in their own journals.[30] They had to fight against the silly name *fuzzy* all the while. They still do.

The tide turned when fuzzy gadgets appeared on the market in Japan and Korea. The Japanese had paid attention when in 1980 the firm of F. L. Smidt & Company in Copenhagen first used a fuzzy system of rules to control a cement kiln.[31] Control of the kiln had often taxed human workers. The fuzzy system used rules of the form "If the oxygen percentage is high and if the temperature is high then slightly reduce the coal feed rate." The fuzzy system used less fuel and controlled the kiln as well as or better than humans did.

These and other applications grew out of the pioneering work of fuzzy engineer Ebrahim Mamdani of Queen Mary College in London. Mamdani first worked out and applied the direct ancestor of today's fuzzy rule-based systems.[32] Mamdani's work in the 1970s began modern fuzzy engineering and the shift in the field from philosophy and linguistics to business and engineering.

The Japanese soon embraced fuzzy logic in spirit and in business. Japan's Ministry of International Trade and Industry (MITI) joined with over 40 firms in March of 1989 to form the $70 million Laboratory for International Fuzzy Engineering Research (LIFE) in Yokohama.[33] The LIFE program ran from 1989 to the spring of 1995. The Japanese practice the rare art of not only ending government programs but ending them on schedule.

The breakthrough came in 1988 when Hitachi put a subway under fuzzy control in the city of Sendai north of Tokyo.[34] Then came the flood of smart ovens and car systems and robot graspers and TV tuners and hundreds of other gadgets.[35] The Sendai fuzzy subway system replaced human drivers along a 13.6 kilometer north–south track with 16 stations. It stopped more smoothly and used less fuel than did the best humans. The joke among fuzzy engineers was that the fuzzy field would crash if the subway ever did.

I once asked Lotfi Zadeh about patenting fuzzy sets before he became one of my Ph.D. advisors. He said the patent would have run out anyway before the field took off in the commercial world. But he joked that he would be a millionaire if in 1965 he had not titled his first paper "Fuzzy Sets" but had instead called it "Fuzzy Sex."

Zadeh never got his million dollars.

He did get something in June of 1995 that meant much more to him and that again told the worlds of science and engineering that fuzzy logic had arrived. The Institute for Electrical and Electronics Engineers (IEEE) gave Zadeh its 1995 medal of honor for his work in fuzzy sets.[36] This is the highest award in the IEEE and the closest thing that the information sciences have to a Nobel Prize.

Fuzzy logic had become mainstream.

Creeping Fuzz in the Digital Age

Fuzz took less than a century to creep from a logician's toy to an engineer's tool. We now turn to how fuzz can affect politics and the concepts that lie beneath much of modern social discourse and science and culture in the digital age.

The rest of the book has three large parts. The first part deals with politics in the broad sense of how we define and make social choices. Those choices range from which party or ideology we endorse and what taxes we pay to how we deal with neighbors who blast loud stereos or launch smart cruise missiles.

The second part deals with new results in the science and engineering of fuzzy systems. The formal ideas and problems of fuzzy logic have come a long way even in the 1990s and have stabilized around a few key concepts. These ideas are now creeping into a wide range of applications and fields that range from smart cars to financial analysis to the information structure of the physical universe. Fuzzy and neural systems have profound limits and yet these smart tools open a new door in science: They help us model the world without guessing at ever more complex equations. This is a quiet "paradigm shift" taking place at the computational frontier of statistics.

There is a fuzzy structure that runs through the first two sections. Each chapter deals with more fuzzy variables (and thus larger fuzzy cubes or "hypercubes" that represent all fuzzy choices).[37] This chapter has focused on fuzz in one dimension as in the spectrum of cool air that runs from 0% cool to 100% cool. This defines a line segment (a fuzzy "cube" of 1 dimension). Chapter 3 deals with politics and advances from the fuzzy line segment to the fuzzy square (a fuzzy cube of 2 dimensions). Chapter 4 adds more dimensions of social choice with a fuzzy tax form (which works with fuzzy cubes of say 10 or more dimensions). Chapter 11 takes this to the limit when it goes head to head with the radical claim that the world is just a vast pile of bits. It presents a fuzzy view of the world as a type of high-dimensional fuzzy fluid (or wave in a fuzzy cube of vast dimension where each point in the fuzzy cube is a possible world).

The third part of the book deals with culture in the digital age and how the concepts in the first two parts help pave the way for the smart digital worlds of tomorrow. These smart worlds converge on an answer to the age-old question of where will it all end.

Those old science-fiction movies are wrong. Humans like us won't run the digital future even if some of those future humans do wear shiny metallic suits. The movies are wrong because people will have fuzzed into something between what we now think of as human and non-human. They will have walked down the digital path of steady parts replacements and come to the end in a stream of 1s and 0s.

It ends where few would have thought it would end in a cold sparse universe made of a few atoms and a lot of void. It ends in a digital heaven on Earth or something close to it.

Part One

Fuzzy Politics

2

Fuzzy Politics

A new science of politics is needed for a new world.

ALEXIS DE TOCQUEVILLE
Democracy in America

Habit is the only thing which imparts certainty.

JOHN STUART MILL
Utilitarianism

There is a real tendency amongst people for degree of belief to approach to certainty. Doubt and skepticism are for most people unusual and, I think, generally unstable states of mind.

ROBERT H. THOULESS
"The Tendency to Certainty in Religious Belief"
British Journal of Psychology
Volume 26, number 1, 1935

POLITICS IS the daily example of how our fuzzy worlds clash with the black and white lines that run through them and through us. The lines tell us when it is legal to vote or drive or smoke or drink or leave a country. The rub is that someone else draws the lines.

Politics always comes down to someone or the state or the judge drawing hard lines through the gray fuzz of our right to act or speak or own or sit still. The hard lines give the borders that define our rights.

The hard lines not only define power in politics. The hard lines are power.

The five chapters in this first part look at this fuzzy conflict with an eye to how a fuzzy view can both blur the lines and offer more choices.

Chapter 3 starts with how we define ourselves on the political landscape. It explores the old labels of left and right in politics. The labels make up a fuzzy spectrum but one that has little meaning. The left–right line embeds in a fuzzy cube if the line connects two cube corners and passes through the cube center. A square gives the simplest form of a fuzzy cube of more than one dimension.

Social scientists have found that a fuzzy square better describes the fuzzy patterns of politics than does a simple left–right line. More complex analyses require more complex fuzzy cubes.

Chapter 4 looks at social choices and puts forth a fuzzy tax form as a means to help make them. A fuzzy tax form can work at the federal or state or local level. The lone taxpayer gets to tell the state where part of his taxes go and to what degree. This gives more say to those who pay and can help fund research bounties. You get more research breakthroughs if you pay for them. It could also help endow the state and over time reduce the total tax burden.

Chapter 5 explores to what degree you own yourself. You begin as a unique gene print or genome and grow by degrees into a sentient being with legal rights and complaints. The state draws hard lines through the fuzz of where your self ends and where the non-self world around you begins just as it draws a hard line of life and death through the legal fuzz of a growing fetus. But the fetus still grows smoothly from not-alive at conception to fully alive at birth. The issues get more complex if you have the power to change your genes and thus the power to change your self to some degree.

Chapter 6 shows how fuzzy property rights blur the lines between mine and thine. This fuzz tends to hurt more than help. A fuzzy theorem of sorts lies behind what happens when someone blows smoke in your face or blasts the stereo next door or drills for oil beneath your house. Behind these issues lies the still deeper issue of the degree to which a theory matches fact.

Future social experiments at sea and in space may someday test this matching. These experiments will require explicit defuzzification schemes.

Chapter 7 deals with war. Politics often ends in war. And in some sense it trades in shadow wars of state against state or state against man or any of hundreds of groups against one another. Information attacks will make it harder to draw the line between an act of war and not-war. Should Turkey declare war on Greece if Greek hackers in Athens bring down the Turkish stock exchange?

Future "smart" wars may differ in kind from the wars and battles of the past just as nuclear weapons changed the structure of war in the 20th century. Fuzzy and other smart systems will help boost the machine IQs of cruise missiles and other smart weapons that will become ever harder for a foe to shoot down. The result will be a shift in the structure of warfare: For the first time in military history it will be cheaper to attack than to defend.

What holds for politics holds for war. The battlefield of the future will be fuzzy as well as digital.

Left and Right and Neither: The Fuzzy Political Square

Power tends to corrupt and absolute power corrupts absolutely.

> LORD JOHN EMERICH EDWARD DALBERG ACTON
> letter to Bishop Mandell Creighton: 3 April 1883

The fundamental concept in social science is Power in the same sense in which Energy is the fundamental concept in physics.

> BERTRAND RUSSELL
> *Power*

Power wins not by being used but by being there.

> JOSEPH SCHUMPETER

Power never takes a step back—only in the face of more power.

> MALCOLM X

The Melians said to the Athenians: "We see that, although you may reason with us, you mean to be our judges; and that at the end of the discussion, if the justice of our cause prevail and we therefore refuse to yield, we may expect war; if we are convinced by you, slavery."

> THUCYDIDES
> *History of the Peloponnesian War*

The power of a man is his present means to obtain some future apparent good.

> THOMAS HOBBES
> *Leviathan*

Political power grows out of the barrel of a gun.

MAO ZEDONG

Government is the use of power to punish.

B. F. SKINNER
Science and Human Behavior

The only purpose for which power can be rightfully exercised over any member of a civilized community, against his will, is to prevent harm to others. His own good, either physical or moral, is not a sufficient warrant.

JOHN STUART MILL
On Liberty

The extent of my social or political freedom consists in the absence of obstacles not merely to my actual but to my potential choices to my acting in this or that way if I choose to do so.

ISAIAH BERLIN
Four Essays on Liberty

A quarter century of genetic studies has consistently found that for any given region of the genome, humans and chimpanzees share at least 98.5% of their DNA.

ANN GIBBONS
"Which of Our Genes Make Us Human?"
Nature: volume 281, 4 September 1998

The ultimate power is to set the agenda.

ANONYMOUS

WHAT IS POWER?

Power is the ability to make things happen. That is what 17th-century English philosopher Thomas Hobbes meant when he said that your power is your present means or ability to achieve future ends. That is what 20th-century English philosopher Bertrand Russell meant when he defined power in his book *Power* as "the production of intended effects." You have power if you can get what you want or intend. And that is what people mean when they claim that God is all-powerful or omnipotent. He has total or even infinite power. He can make anything happen.

Political power is more personal. It is the state's ability to get things done by force or coercion and get away with it. Political power lets those who wield it reward and punish with impunity. It is what Mao said grows from the barrel of a gun. This often takes the form of a legal system. Scholars even define the legal system as governmental social control.[1] So political power limits our choices. It creates a zero-sum game with social power or our freedom to act as we choose. Political power grows as social power falls and vice versa.[2]

Fuzz gives choices. We can pick and choose from the shades of gray that define fuzziness. More degrees mean more options. This can increase either social power or political power. It depends on who gets to make the choices. And more degrees mean we have more doubt about how we or someone else will choose. Some people want to use force to reduce other's choices and thus reduce the doubt about how they will choose.

The result is politics.

Politics has always been about using power to limit choices and set agendas. The extreme case is pure tyranny. We do just what the tyrant tells us to do. We obey each decree. Society chooses something if and only if the dictator chooses it.[3] There is no fuzz because there is just one choice.

Next comes the binary choice. You break the law or you don't. You sign the contract or you don't. You show your ID or you don't. Politics tends to lay force behind one of the two options. So they are power decrees after all. Only a fair vote gives you a real choice between two opposites. Members of democracies pride themselves on letting their members vote without pressure even if they can vote for only two candidates.

Fuzziness gives a type of anarchy of choice in the political world just as it gives an anarchy of concepts in the mental world. It is hard to pressure someone to pick any one degree like 83% or 20% when they can choose from the whole spectrum. So more often than not we simply don't let them indulge in fuzzy choices. We have the state draw a hard line for them and then force them to use it.

One famous case is the U.S. electoral college. Most Americans think they elect their president and vice president by a simple winner-take-all vote. John runs against Jane and the one who gets the most votes wins. This is not how it works. And Americans would have to amend their Constitution to change or abolish it.[4] Many have tried to do so for almost 200 years.

Each state has a fixed number of electoral votes. Each state holds in effect its own winner-take-all vote in the so-called general ticket system. If John gets more votes than Jane in Texas then John wins *all* the electoral votes in Texas. So we round off the vote in each state and ignore the voting choices of those who voted for the lesser candidate.

The next president need not be the one who gets the most votes. The winner is the one who gets the most electoral votes. John could in theory lose to Jane if John got more votes than Jane but got them in states where Jane got few votes. John would win these states but might lose the rest to Jane. Jane's binary round-offs could add up to more than John's. Such an outcome would no doubt lead to a quick end to the electoral system.

Politics contains many other conflicts between fuzzy choices and binary rules. Indeed the logic of politics is just that conflict: *The state draws public black-white lines through private gray choices.* And it backs them up with force.

This makes a good rule of thumb outside of science: If you find a binary line in your social life then the odds are a politician drew it. And of course parents draw lines for their children just as teachers draw them for their students. No one forgets the sting of falling short of the line between passing and failing. Parents and teachers act as local quasi-governments when they draw and enforce these lines.

The state draws the big lines for us. The state tells us when we have become an adult or a divorcée or a felon. You can pay a fine or lose a license or even go to jail over a mere matter of words and how the state defines them. Your car can wiggle only so much in its lane before a traffic cop draws a "discretionary" line and sees it as weaving or reckless driving. And you have to hope for the best each time you file your taxes.

The conflict between fuzzy and binary choices runs still deeper than this in politics. It affects the way we define ourselves on the political landscape. And most of all it affects how others define us and draw lines through us or through our beliefs and actions.

Here begins the theory of fuzzy cubes.

The Left-Right Spectrum: The 1-D Fuzzy Cube

Are you left or right? Are you liberal or conservative? These terms run through almost all modern political discourse. The left and the right define the poles of our political thought. They lie opposite the primal line that cuts political ideology into two pieces. And they give a simple debate format to TV talk shows and newspaper op-ed columns.

But what do they mean?

We learn to use the left-right terms by example. Ronald Reagan and Margaret Thatcher were right. François Mitterrand and Mikhail Gorbachev were left. Clint Eastwood and Arnold Schwarzenegger are right. Robert Redford and Warren Beatty are left.

Political scientists sometime say that the left wants change and the right does not.[5] That may have described many congresses and parliaments during the 1930s in the days of the New Deal and later in the 1960s during the days of Lyndon Johnson's Great Society. Today it may be the reverse in the United States and South America and elsewhere. The right tends to be more "radical" if not more "progressive" in its proposed market-based policies to privatize some state functions and abolish others.[6]

Legend has it that the left-right terms came from the period of the French Revolution. The radicals of 1789 sat to the left of the presiding officer of the French National Assembly. The name stuck and French intellectuals from Charles Fourier to Jean-Paul Sartre have claimed it ever since and have seen themselves as champions of the politically weak.

Conservatives often cite the 18th-century British statesman and writer Edmund Burke (1729–1797) as a founding father of the right even though he was something of a liberal in his day. Burke made his mark in conservative history when he questioned the goals and values of the French Revolution in his 1790 book *Reflections on the Revolution in France*. The American Thomas Paine wrote his 1791 book *The Rights of Man* to rebut Burke.[7]

The modern left can take much of the credit and blame for the welfare state. It can point to programs from Social Security to Medicare to food stamps as examples of the will of the left in action to help the poor. The modern right can take credit and blame for often opposing those programs and the increased spending and borrowing that tends to fund them.

The right can also point to its own legal gift to modern politics: the victimless crime. Some now call this a "consensual" crime.[8]

The right has outlawed private pleasures from drugs to gambling to prostitution to pornography in almost all countries that have an active right-wing party. These laws find their roots in the prohibitions of whichever religions the right supports or has supported in the past. The United States leads the modern world with well over 1 million persons in jail or prison. Over half of them are there because of drug prohibition. Six percent of the British use outlawed drugs compared with something like 12% of Americans or almost 6 million citizens who spend almost $49 billion a year on black-market drugs. Americans spend about $50 billion a year on legal tobacco products.[9]

There is still the question of what the left-right terms mean.

People may not be sure what the terms mean but they are sure the terms are fuzzy. You first learn the map of politics by learning who is more left or more right than you. The far left shades into socialism or communism. The far right shades into fascism or Hitler-style national socialism.[10] We address this strange claim below. Somewhere between these extremes lies the centrist or moderate creed. The left-right spectrum sets and limits the agenda for political debate and thought.

Fuzzy sets capture the segments of the left-right spectrum. Far-left socialists overlap to some degree with liberals. These overlap with moderate liberals. And these overlap with moderates and so on out to far-right fascists. The sets are fuzzy because people belong to them only to some degree. So they also do not belong to them to some degree. Few people are pure liberals or conservatives or pure moderates.

The next figure draws these fuzzy sets as overlapping triangles or trapezoids and hence as partial subsets of the left-right spectrum:

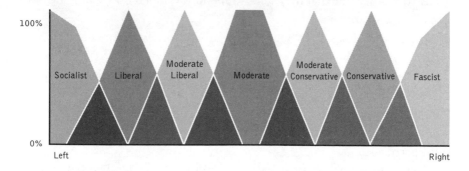

Each fuzzy set is vague and relative. People are liberal and not liberal to some degree. And different people draw the liberal fuzzy set in different ways. Some might want to draw the set wider or more narrow or more to the left or to the right. We could also add or delete some fuzzy concepts. We could add ultra-conservative to the right of conservative. Or we might widen the sets and limit them to only liberal and moderate and conservative.

Fuzzy concepts help flesh out the left-right spectrum but they still do not explain what the left and right terms mean. They just help show the circular nature of the concepts that underlie the left-right spectrum.

What is a liberal or a democrat? Someone who is left. Who are left? Liberals and the non-right are left. What is a conservative or a republican? Someone who is right and not left. And round and round goes the logic. To say that someone is a liberal because he is left-wing says no more than that he is liberal. It does not explain what the term means. The same holds for the terms moderate and conservative.[11] We define them in terms of one another and do not tie them to other concepts or variables. But you define a term if and only if you replace it with other terms (if and only if you replace the definiendum with the definiens). In the best case we replace the term with other terms or variables that we can measure.

The problem lies in the label of the left-right axis: It does not have one. What quantity or quality varies from left to right or from low to high? What does it measure? The left-right spectrum does not tell us. It does not give us a way to reduce or replace its terms in favor of simpler terms that define them.

Some have proposed the size of the state as a measure. Liberals want more government and conservatives want less. That may hold in some cases of social spending. It does not hold in many cases of defense spending or in granting farm subsidies or in waging war on victimless crimes. Both sides accuse the other side that they want to spend too much on some things and not enough on others.

So we arrive at a simple but startling conclusion. We often do not know what we mean when we talk about modern politics.

The left-right spectrum is fuzzy but its terms are circular. It favors name calling far more than it explains social behavior or helps group ideas into a logical taxonomy. It also reflects much of the sense of upheaval in the field of political science itself.[12] No wonder so many Americans want a third party.

Yet we do see a pattern in those we call left or right. The neural networks in our brain cluster these people and belief systems into groups. We cluster like with like and end up with at least two clusters. That much is empirical fact. We find political pattern A and political pattern B. People of pattern A tend to call for the same kinds of changes and oppose the same kinds of changes. This also holds for people of political pattern B. They tend to claim that some issues are the key ones and they use similar language and slogans to defend their claims.

The issue is whether the two patterns define logical opposites. We know only that the patterns A and B differ. This restates what we mean when we group them as distinct patterns A and B rather than as the same pattern. It takes much more to show that pattern B depends on A in the polar sense that B is *not-A* and vice versa. A black crow differs from a white duck but the two bird types are not polar opposites. Their feather colors may lie on opposite ends of a gray-scale spectrum but they share many other design features. A crow is not an anti-duck in the sense that day is the opposite of night or that on is the opposite of off.

The left and right have far too much in common to count as pure opposites. They both support markets and national defense and police protection to some degree. They differ on only a subset of issues or differ only in their degree of support for issues. Debate focuses on these issues and tends to ignore the common ground. TV screens and op-ed columns focus the debate still further on the points on which the two sides disagree. The media demand for sound bites that really bite gives both sides reason to use language that exaggerates their differences.

We need to map out the common ground and the contested ground. Politics is too complex to reduce to one variable. It needs at least two. It needs a bigger concept space.

The left–right line needs a plane.

The Fuzzy Political Square: Multidimensional Ideology

We can embed the left–right spectrum in a fuzzy square. A square consists of infinitely many line segments from left to right or from top to bottom. We won't use these lines. We will use instead one of the two long diagonals that crisscross each square.

The square itself is a type of fuzzy cube. Each point in the cube stands for how much the two objects or patterns A and B belong to it. The points in our case will stand for degrees of two types of freedom. The four corners stand for the four binary or all-or-none cases. The points have fuzzy values or percentages only if they lie inside the cube. It turns out that the theory of fuzzy sets is a theory of cubes.[13]

A line segment is the simplest "cube." The percentages from 0% to 100% define a cube of one dimension. It measures the values of one fuzzy variable. The left–right spectrum is such a 1-D fuzzy cube even though it does not state what the lone axis measures. Two fuzzy variables lead to a fuzzy square or 2-D fuzzy cube. Three lead to a solid cube or 3-D fuzzy cube and so on. The next chapter casts social choices as points in fuzzy cubes that may have dimension 10 or higher. We cannot picture these high-D fuzzy cubes in our mind's eye because we cannot picture how four distinct line segments can all be mutually perpendicular. But the notation of math lets us work just as easily with cubes that have a million dimensions as those that have just two or three. This becomes crucial in chapter 11 where a fuzzy cube that describes the universe as a point in it has more dimensions than there are atoms in the universe.

We focus now on just the two variables that define a fuzzy 2-D cube or fuzzy solid square. The variables measure two types of freedom and we now define them. That begins with how we define freedom itself.

What is freedom?

Pop singer Janis Joplin claimed in an iconic Baby-Boomer song that freedom means having "nothing left to lose." That slogan may describe the risk profile of a condemned man. And it points out that we may have more options for future action if we remove a constraint on our behavior like acting to preserve our wealth. But the slogan's focus on options is too broad. It also suggests that political freedom is any type of freedom like being free for the day or free to speak or free of a headache. The freedom we argue and sometimes fight about is political freedom. It deals with "legal" force.

Freedom is a negative concept in the sense that it is the absence of nonfreedom. It is the absence of restraint. And political freedom is a special case.

Political freedom is the absence of state restraint. You are free just to the degree that the state does not force you to act in some way or does not use force to keep you from acting in some way. Political freedom or liberty ends where state coercion begins.

Such freedom does not mean that you have the means or the power to act as you choose. Your power again is your ability to make things happen or your present means to achieve future ends. Janis Joplin's slogan of "nothing left to lose" is a good description of having no power. The bankrupt prisoner can have no power in this sense and still have little or no political liberty.

Robinson Crusoe is free to do as he wants on his island. He does not have the means or power to use some freedoms. He can fish and hunt and crack clamshells and swear all day at the top of his lungs. But he has no way to watch the news or to blast through rock or to fix an abscessed tooth.

Power gives more choices or options. Freedom lets you use what options you have. So more choices or options create more opportunities for coercion. The more freedom you have the more choices the state allows you to make without punishing you. And we can in theory avoid mental terms here and cast talk of choices and such in behavioral terms. You reveal your choices or your preferences by how you act or fail to act.[14] Actions in space-time speak louder than thoughts in neural circuitry.

Governments sometimes play with words to confuse the brain act of choice with the body act of choosing. Thus the U.S. government claims that the Social Security tax is a "voluntary contribution." It claims that the Internal Revenue Service depends on "voluntary compliance." Citizens can choose freely not to pay such taxes in their minds only. The state will punish them if they do not pay their taxes in the flesh. This is the same "voluntary compliance" that the classroom bully depends on when he asks a child to hand over his lunch money or his flip phone: Pay or else.

Freedom does *not* depend on whether the agent has a "free will."[15] This is less controversial than it sounds.

A robot is free to the extent that it can act on the desires that someone has programmed into its chip brain. We might add a random seed to its choice logic to make it harder for us to predict its choices and actions. We can add nonlinear dynamics that make it still harder for us to predict its behavior because then not only will similar inputs not always lead to similar outputs but the same input will often lead to different outputs.

A free robot can do as it pleases but it cannot please as it pleases.

The same is true of "bio-robots" such as ourselves. Other forces control what the robot pleases just as to some degree genes and the world outside our skin control us. That does not affect our political liberty. Modern philosophers have been careful to sort out these issues and avoid the apparent dichotomy between free will and determinism. Harvard philosopher Willard Van Orman Quine sums up hundreds of years of analysis on this issue:

> Like Spinoza, Hume, and so many others, I count an action as free insofar as the agent's motives or drives are a link in its causal chain. Those motives or drives may themselves be as rigidly determined as you please.[16]

Quine's "agent" can just as well be a self-programmed robot as a post-hominid political subject.

Freedom splits into at least two fuzzy sets of freedoms or free actions.

This means freedom is multidimensional. You can be free in one way to some degree and not free in some other way to some other degree.

Civil liberties make up the first broad fuzzy set of free actions. Most of these have to do with going to hell in your own fashion. They range from issues of what you can say or write to when the police can search you or tap your phone to whether you can buy or sell sex or drugs or alcohol or poker chips. Many deal with how far the state will allow you to pursue your religious beliefs. Most countries have few professed gay atheists in office. Some Muslim countries outlaw atheism and agnosticism as most Western countries did at one time. Saudi Arabia demands that even non-Muslim tourists fast during the day in the month of Ramadan.

Civil liberties also deal with what information the state will let you see or send. Sweden forbids adult TV before 9 P.M. and will not let firms run ads during kids' shows. Great Britain forbids adult TV before 9:30 P.M. France forbids adult TV from 6 A.M. to 10:30 P.M. while Germany forbids it before 11 P.M. Each state draws its own hard lines between the fuzz of the obscene and the non-obscene. The National Security Agency forbids American citizens and firms to encrypt some of their software with some math schemes. The NSA does this even though you can find the math schemes in the open literature and even though most other countries allow their use.[17]

Economic liberties make up the other broad fuzzy set of free actions. They fall with a rise in taxes or tariffs or regulations. They also fall in the face of forced rent controls or minimum wages or forced health care or just about any other intrusion of the state into the economy. They reach their lowest point in a command economy. The early Soviet Union tried to achieve such a state with its forced five-year plans but fell far short of the ideal of central planning. It allowed workers to switch jobs largely at will and allowed some to grow produce for market on small plots of land. Today the U.S. military is the closest thing the world has to a large pure command economy.

State spending gives a rough measure of economic freedom. The more it spends the more it takes in taxes and so the less the economic freedom. The state in Sweden spends about 68% of its gross domestic product (GDP) on social programs. Germany spends about 49% of its GDP. The U.S. federal government spends about 33% of its GDP. Singapore spends about 20% of its GDP.[18] State spending on defense also limits economic freedom to some degree. So does outlawing mail on Sunday.

The fuzzy political square has two axes. It lists civil liberties from 0% to

100% along the vertical axis. It lists economic liberties from 0% to 100% along the horizontal axis.

Each point in the square defines a simple fuzzy set. The point measures both the degree to which you are a civil libertarian and the degree to which you are an economic libertarian. To be a civil libertarian is to favor more civil liberty than not. To be an economic libertarian is to favor more economic liberty than not. A liberal is a civil libertarian but not an economic libertarian. A conservative is an economic libertarian but not a civil libertarian. A populist is neither. A libertarian is both.

The next figure shows a fuzzy political square with the names of the four patterns in its four quadrants. The quadrants include the names of some of their recent spokesmen in the U.S. media:

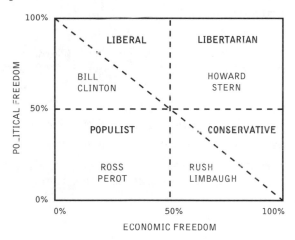

Note the dashed line that runs from the upper-left corner to the lower-right corner. It forms one of the square's two diagonals since it passes through the midpoint of the fuzzy square. It is a theorem that this diagonal connects the two binary vertices of pure civil libertarianism with pure economic libertarianism. And so we get as a corollary that the dashed line is the old left-right spectrum embedded in a 2-D fuzzy cube.

The left-right extremes are still opposites and the left and right still meet in the middle. Conservatives such as radio talk-show host Rush Limbaugh[19] still lock horns with liberals such as President Bill Clinton and disagree on most social issues much as the largely conservative states of Germany and Japan differ from the largely liberal states of Sweden and Canada. The fuzzy square charts their conflict in terms of how much of which freedoms they trade for state control. The advance here is that this left-right line falls out of the fuzzy square as a type of theorem. It follows from the geometric structure of the fuzzy political square.

And the fuzzy square's two variables avoid the circularity that plagues the lone left-right spectrum. Each point on the diagonal is a special weighted average of pure civil and economic libertarianism. If a liberal on this line is a civil libertarian to degree 60% then the liberal is also an economic libertarian to degree 40%. The two percentages sum to 100% if and only if the political position lies on the diagonal. This is the zero-sum structure that polarizes the left and right political positions. But we do not define liberal as simply non-conservative and we do not define left as non-right and then go on in a circle to define conservative as non-liberal and right as non-left. We instead define liberal and conservative by how they combine the two independent variables of civil and economic liberty.

The fuzzy square shows four fuzzy patterns and two pairs of opposites. Liberals oppose conservatives and a diagonal shows their opposition. Populists oppose libertarians and the other diagonal shows their opposition. The four patterns all oppose one another to some degree but not always over the same issues. Liberals and libertarians largely agree on issues of civil liberties. They disagree with populists and conservatives on these issues. Conservatives and libertarians largely agree on issues of economic liberties. They disagree with liberals and populists on these issues.

The fuzzy square shows how the left-right spectrum defines these third parties out of existence.

It further shows that the "third party" vote consists not just of one response to the left-right dogma but of the two opposing views of libertarianism and populism. This helps explain why no single third party has emerged that stands for the large number of voters who do not support the left-right choices of liberalism and conservatism. Libertarians and populists differ even more from each other than they differ from either liberals or conservatives.

The fuzzy square defines libertarians as both civil libertarians and economic libertarians. This is the formal definition of limited government. Many libertarians prefer the older term "classical liberal"[20] or the newer term "market liberal."[21] The term "liberal" still tends to mean libertarian outside the United States and inside scholarly journals where it refers to constitutional liberalism. Historian Francis Fukuyama gave a typical definition of political constitutional liberalism in his controversial book *The End of History:* "Political liberalism can be defined simply as a rule of law that recognizes certain individual rights or freedoms from government control." John Stuart Mill and other British radicals began the Liberal Party in England in the early 1880s.[22] They derived their platform of civil liberties from the British empiricist philosophers Thomas Hobbes and John Locke and from the common-law tradition in England that stretches back even before the barons in the plain of Runnymede forced King John to sign the Magna Carta on 15 June 1215. The Liberals derived their platform of laissez-faire economics from the market economics of Adam Smith and David Ricardo.

A 1994 Gallup poll found that 22% of Americans are libertarian.[23] Most vote as Democrats or Republicans.

The most famous libertarian American was Thomas Jefferson or rather his philosophy that "that government is best which governs least."[24] Jefferson was hardly the champion of liberty as a slave owner. He clearly said one thing and did another. DNA evidence has shown that he likely had at least one child with his quadroon slave Sally Hemings.[25] Hemings was also the half sister of Jefferson's wife. Jefferson may also have fathered other of her seven children. He was at home at Monticello nine months before all but one of them was born. Jefferson's hypocrisy was brutal and studied. That does not affect the strength of his arguments for severely limited government. But it does make their espousal look like special pleading: How can a champion of liberty own and sell other men and let his wife and children live as slaves? Benjamin Franklin also owned slaves for many years but at least in his old age he became an abolitionist.

The world has changed so much in 200 years that the great statesman and writer Jefferson now finds his modern libertarian successor in the millionaire radio shock-jock Howard Stern. Stern ran for governor of New York on the libertarian ticket in 1994. He dropped out of the race over something of a libertarian issue of privacy. The state required that he reveal his finances and he refused to do so.

Howard Stern's place in the fuzzy square reflects what *Rolling Stone* and other news sources see as a libertarian trend among the 130 million or so Americans under the age 35 and the still younger Generation X.[26] Political scientists have long observed that a person's ideology tends to form when she comes of legal age. Today's young tend to see government as a tool of the old and graying. Many if not most young people doubt that they will ever get a Social Security check. One 1994 study found that the state pays a 72-year-old in her lifetime $98,600 more in benefits than she ever paid to the state in taxes. But a 27-year-old will pay $203,000 more in taxes in his lifetime than he will get in state benefits.[27] Your place in the fuzzy square may depend to a large degree on such crude measures of state costs and benefits.

The square defines populists as the opposites of libertarians.

Populists are neither civil nor economic libertarians. Their motto seems to be "Someone should pass a law" and in most cases someone already has. The U.S. *Federal Register* grows by more than 1,000 pages of new or revised federal rules and regulations each week. Populists favor the military draft and some forms of censorship. They may favor laws that ban or limit certain sex acts or lifestyles. They share with liberals the support of taxes and business regulations and the forced licensing of doctors and lawyers and other professionals. They strongly oppose open immigration and ending drug prohibition.

The best-known populist was the Archie Bunker character on the 1970s TV sitcom *All in the Family*. He liked no one except a large and intrusive government that would pass and enforce his laws and wage his wars.

Billionaire Ross Perot is something of the modern patriot populist. He is a moderate or centrist populist but still a populist. His Reform Party counts as a centrist populist party. Perot does not like free trade or drugs or foreign oil or tax breaks or big salaries for executives and congressmen. He likes public spending on roads and cities and small firms and computer research. Perot also likes higher income taxes to pay off the national debt and likes higher gas taxes and oil taxes and a strengthened Internal Revenue Service to better collect these and other federal taxes.[28]

Former wrestling champ Jesse "The Body" Ventura ran on the 1998 Reform Party ticket for governor of Minnesota and won. Ventura called for more economic freedom in terms of lower taxes and less regulation and for more civil liberty. He expressed sympathy for ending state prohibition of drugs and prostitution. Thus his position was more libertarian than populist. Most Reform Party populists support Ross Perot's call for more and not less drug prohibition.

Extreme populism shades into totalitarianism just as extreme libertarianism shades into minimal-state "minarchy" or even free-market anarchy. Those extremes make up the two binary corners of the 2-D fuzzy cube and lie at opposite ends of a diagonal. They reflect the old joke in political science that the only consistent political positions are totalitarianism and anarchy. They also show how market or individualist anarchy differs from the more popular communist or collectivist anarchy. The individualist anarchist wants no government. The collectivist anarchist wants a big or even totalitarian government to enforce no government.

Fascism was just a shorter name for totalitarianism. The *Columbia Encyclopedia* defines fascism as a "totalitarian philosophy of government that glorifies state and nation and assigns to the state control over every aspect of national life."[29] Italian fascist Benito Mussolini made this clear in his book *Fascism: Doctrine and Institutions* and even called out the polarity between fascism and modern libertarianism (or classical liberalism): "[Classical] liberalism denied the State in the name of the individual. Fascism reasserts the rights of the State as expressing the real essence of the individual."[30]

Hitler and Mussolini and Franco suppressed economic liberties as much as they suppressed political liberties. They share their place in the fuzzy square with Joseph Stalin and Mao Zedong and Muammar Qaddafi and Saddam Hussein and dozens of other dictators who had enough power to truly run their countries. Modern China has since moved into the conservative quadrant as it has pursued free markets but not lessened its grip on civil liberties. This has happened with many countries in Southeast Asia from South Korea to Singapore.

Putting fascism on the far right of the left–right spectrum always had more to do with name calling than with logic. The fuzzy square shows that fascism depends as much on ideas from the extreme left as from the extreme right. The old left–right spectrum would have us believe that the left and right end in totalitarianism.

But then where would we put anarchy? The people of Iceland lived in anarchy with a private or "consensual" government until they voted to accept the king of Norway as their king in 1263.[31] The people met to hold public "things" and "all-things" that came up with laws and tried disputes. Their decentralized government had a legislative and judicial branch but no executive branch. They had the luxury of such anarchy in part because they faced no threat of foreign invasion. Ireland has also known calm periods of anarchy in its bloody history.

Some form of the fuzzy square has been around for at least a quarter century and many have come to call it a Nolan chart.[32] Both economists and political scientists have used it though none of them saw the formal tie to fuzzy logic. Nobel laureate economist Paul Samuelson used the fuzzy political square in the final chapter of his popular 1970s college text

Economics. But the openly liberal Samuelson chose to define the conservative quadrant as fascist and the populist quadrant as "serfdom."[33] This was not standard use of language or the political square.

Political scientists William Maddox and Stuart Lilie used the political square to study the structure of American politics in the 1970s with questionnaires that evoke one's place in the fuzzy square. Should the government control content on TV or radio or the Internet? Should we legalize prostitution or gambling or drugs? Should we repeal the income tax or minimum wage laws or farm subsidies? Should the United States pay for the defense of Europe or Asia?

Maddox and Lilie found that in 1980 about half of Americans fall under a "third party" ideology.[34] Populists led with 26% of voters. Liberals followed with 24% of voters and libertarians with 17%. Conservatives trailed with 17% of voters. The other 15% had no opinion or fell in the middle of the square.

That raises a quintessential fuzzy question: What lies in the middle of the square?

The political center lies in the midpoint. And it shades off in degrees on all sides. The midpoint is the fuzziest point of all in a fuzzy cube. The midpoint is unique in this sense. It lies the same distance from each binary corner. No other point in a cube has this property. The value ½ lies the same distance from 0 and 1 in a 1-D fuzzy cube or in the left-right spectrum. The same holds for a fuzzy square or cube of any dimension. The cube midpoint has no counterpart in binary logic or math. It accounts for the unripe lemon that is as much yellow as not and for the fully neutral voter.

The cube midpoint runs still deeper than this. It accounts for the "paradoxes" that have plagued binary logic since at least the time of the ancient Greeks. Does the Cretan lie if he says that all Cretans lie? If he tells the truth then he lies and so he does not tell the truth. But if he lies then he tells the truth. This forces us to conclude that the Cretan both lies and he does not lie at the same time. That is impossible in binary logic. But it is fine in fuzzy logic if he both lies and tells the truth to midpoint degree 50%.[35]

The same fuzzy logic holds for the person who sees himself on the 1-D fuzzy line as both a liberal and conservative or who sees himself in terms of the 2-D fuzzy square as not just a liberal and a conservative but also as a populist and libertarian. These "paradoxical" positions land him in the cube midpoint or political center. He also lands in the center if he sees himself as neither a liberal nor conservative on the 1-D fuzzy line or if he sees himself as not only neither a liberal nor conservative but also as neither a populist nor a libertarian.

This analysis does not stop at two dimensions.

We can add other variables to the fuzzy square and increase the dimensions of political analysis. More dimensions make the cube more realistic. They may also make analysis unwieldy or hard to test. What would make a good third variable? What view or concept would help split each quadrant? The third variable can be anything that we can measure or test with a questionnaire or with any other tool.

Abortion might make a good third variable. Are you pro-life or pro-choice? Those two extremes define their own choice spectrum from 0% to 100%. They would turn the fuzzy square into a proper 3-D fuzzy cube with 8 quadrants. Most conservatives are pro-life and so are a good many populists. Most liberals and libertarians are pro-choice. But there are still enough who are not to fill out the 8 quadrants. This holds even more so on the issue of late-term or "partial birth" abortions.

There are enough exceptions that it might make sense to treat the abortion issue as a third axis. Or we can just lump it in with civil liberties. Either approach improves on the simple 1-D left-right spectrum that has held sway for over a century in political debates and humanities courses and newspaper op-ed columns.

Fuzzy Square Dynamics: Is Government Destiny?

The fuzzy square gives a new way to think about the politics of the past and the future. Social trends trace out paths in the square. Most states decreased their economic freedoms after the Bolshevik Revolution in Russia in 1917. Eastern European states increased their economic freedoms after the Soviet Union fell in 1991. So their points in the fuzzy square moved to the economic right.

This shows that the fuzzy square gives us more than just a new way to think about politics. It gives us a framework in which to state and test social hypotheses. It lets us define social movements as dynamical systems that both feed on themselves and change when the environment changes.

Here is a simple but plausible hypothesis: The four binary corners are unstable.

Limited governments tend to grow. Even the best written constitution can do little more than slow their growth or reduce their size. The People's Republic of China produced a constitution in 1983 that contains ringing calls for freedom but that has done little to secure those freedoms. And the complete suppression of either political or economic freedoms leads to its own demise. Almost all totalitarian states have fallen of their own weight. Rulers also have a farmer's incentive to grant their subjects enough freedom so that they produce a good harvest of goods and services. What holds for the catfish farmer might hold for the ruling political class: The long-term level of freedom in a society may correlate with its maximal sustainable yield of taxes and personal control.

Here is a more speculative hypothesis: Voter self-interest tugs society toward the square's center.

The square's midpoint may act as a type of dynamic fixed-point attractor. That means all roads may lead to the midpoint. Or voter self-interest may make society spiral or bounce around from quadrant to quadrant. Social paths may converge to such a closed loop or cycle in the square. Or they may wander in a chaotic equilibrium. Only testing and data can tell.

Fuzzy political cubes of any dimension let us explore the starkest political question of all: Is government destiny?

Many people seem to assume that government is destiny. This stems at least in part from a lifetime of statelike patterns of reinforcement and social control. People learn to accept statelike authority almost at birth. Parents control their children's behavior in a type of household government that involves lawlike rules and punishments.

Soon teachers control their behavior in an even more formal system of government. Harvard psychologist B. F. Skinner observed that schools and universities can extend the state's patterns of reinforcement to students: "Schools supported by a government may be asked to apply their educational techniques in supporting the government and to avoid any education which conflicts with governmental techniques of control or threatens the sources of governmental power."[36] This state-sponsored conditioning ranges from the content of "approved" textbooks for children all the way up to the billions of dollars of funded research grants that reward professors and their deans and graduate students. Students and teachers have little incentive to question the state's authority. They have even less incentive to question its inevitability.

But the ultimate future of government rests on more than conditioned behavior. It rests on economic efficiency. States are power monopolies.

They form and survive because a power monopoly offers at least one service more efficiently than competitive market processes offer it: protection from invasion. Markets fail to provide national defense because they fail to exclude protection from free riders who do not pay for the service. That one-line argument is the bane of the anarchist. And it is the one justification for state power that has withstood all criticism. National defense is a public good and markets under-produce it.

We will have governments so long as we have atoms to protect. But will we need states to protect our bits? Chapter 15 explores this question in the context of "uploading" brain-based minds into computer chips. The cyber-worlds of the future may not need state control just as today many people feel that the Internet needs little state control and does not need a cyber-world government to control or manage it.

The question still remains in the digital age: Will overall government grow or shrink in the next hundred years?

A cynical but potent answer is that government will grow so long as we fail to control the hominid within us. It helps to recall from time to time that we share over 98% of our genes in common with chimpanzees and that constitutional liberalism is rare in the animal kingdom.

Most Americans take their constitutional liberalism for granted. They tend to assume that most people on the planet live under the rule of law and enjoy civil rights and stable property rights and the other benefits of constitutional liberalism. But little more than half of today's 193 countries are democracies. And about half of these are "illiberal" democracies where people have only the privilege of voting for the next dictator. Social analyst Fareed Zakaria has described this in blunt terms: "Constitutional liberalism has led to democracy but democracy does not seem to bring constitutional liberalism."[37]

A monopoly on power is the ideal forum for those in power to give play to their will to dominate and to the other modes of aggression that lurk in our Darwin-sculpted endocrine systems. Man has known civilization for only the last 500 or so generations. Those 10,000 years count for less than a fraction of a percent of our dark and violent evolutionary history. Aristotle got it right when he said that man is a political animal. That is a polite way of saying that we are a species that loves power. Social control and governments grow from that love. Modern governments are the greatest concentrations of power in history.

Could the early hominids have dreamed of such power?

There is more evidence for a hominid theory of government: Governments have been remarkable killing machines. One study found that on average states in the 19th century killed about 3.7% of their subjects. States in the 20th century have killed about 7.3% of the world's population. But states tend to kill fewer of their subjects if the subjects are more productive. A 1% increase in real GDP leads on average to about a 1.4% decrease in "democide."[38] Hence what holds for the catfish farmer may indeed hold for the ruler or more broadly for the system of state control. The state may farm us. The liberal-democracy version is that we may farm ourselves.

There is a statistical answer to whether government is destiny. Government simply grows. The International Monetary Fund has gathered data on the growth of governments since 1870. The average industrialized government spent about 8.3% of its country's GDP in 1870. That average "public" share had more than doubled to 20.7% in 1937 during the birth of the modern welfare state. It had more than doubled again to 42.6% in 1980 when Ronald Reagan and Margaret Thatcher came to power. The average state's share of GDP still climbed to 44.8% by 1990. The trend continued: The worldwide average climbed and reached 45.9% of GDP in 1996.[39] That was about when President Bill Clinton advanced his social hypothesis that "the era of Big Government is over." He surely meant that what had ended was a century-old social taste or presumption in favor of state solutions to social problems. The trend data suggest that the era of Big Government has only begun to mature.

This growth in state power cannot continue forever for the same reason that environmentalists like to point out that we can't keep cutting down rain forests faster than rain forests grow new trees. That does not mean the trend will reverse. The evidence to date is that governments simply do not do that. People rarely give up power or give up benefits from those in power just as the alpha-male orangutan does not decide one day to hand over his harem to weaker males.

The trend in state spending may mean that governments get more efficient and fall into some steady state. The average industrialized state may spend something like 50% or even 60% of its GDP for the next few decades or for the next 100 years. We never ran out of oil or other raw materials as some scientists in the 1960s and 1970s claimed we would by now. Growing demand led to innovation and that led to growing supply.

So now a given volume of oil costs less than a like volume of bottled water. Governments adapt more slowly than markets adapt but they do adapt. And we risen hominids are far more likely to run out of oil before we run out of government.

Government efficiency may strike many as an oxymoron. But it occurs because of the productivity gains from information technology and privatization and because of the competitive effects of liquid global capital and the rise of the Internet. Capital sloshes around the world with increasing speed and decreasing friction. Global capital tends to keep each state in its own Skinner box. The capital markets quickly reward a country's good market behavior by lowering its interest rates and quickly punish bad market behavior by raising rates and thus increasing the recession and unemployment that follows. The Internet will soon let all the citizens of the world applaud or condemn all the governments of the world and do so in real time. So states must become more efficient to adapt to the changing digital world. They can do that without getting smaller. Indeed the Internet might lead to world government.

I used a simple math model to see where a state tends to evolve in a fuzzy 2-D cube or in any fuzzy political cube of n dimensions. I tried to capture philosopher David Hume's insight from his 1776 essay "Of the First Principles of Government" that *government rests on opinion*. The idea is that government grows with pro-government sentiment and shrinks with anti-government sentiment. If the pro-government sentiment has a fuzzy degree like 70% then the anti-government sentiment is 30%.

The model says that government growth depends on two terms. The first term involves pro-government sentiment. It has a so-called logistic form. This is a simple and standard way to allow for diminishing returns and limits to growth. The second term involves anti-government sentiment and has an even simpler form. It says that the anti-government sentiment makes the government shrink in a linear way if all else is constant.

The result is that David Hume was right: The amount of government in society does rest on opinion. It rests on the balance or ratio of anti-government to pro-government sentiment.[40]

This result says nothing about how to define or measure the millions of factors that shape a society's average pro- or anti-government sentiment. The result just says where those sentiments lead in a fuzzy political square. The result is robust in the sense that more complex models should reduce to this model as a special case. It is also robust in the sense that where the state evolves does not depend on where it starts. All paths quickly lead to the same place in the square. The starting points or "initial conditions" do not matter. What matters is the balance of pro- and anti-government sentiment.

So is government destiny?

The answer depends. It always depends. This simple math model implies that government is destiny if pro-state opinion exceeds anti-state opinion. The model implies that a government will collapse if anti-state opinion exceeds pro-state opinion.

Of course the real answer depends on much more than what we can hope to pack into a math model. It depends on the swirling and high-dimensional nonlinear social froth that our ancestors will one day point to and call our "history." That depends at least in part on our genes. And in time it may depend on whether we still have genes and hormones and neural reflexes and all the other evolutionary wetware that make us both human and the very close cousin of apes.

4

The Fuzzy Tax Form

But in this world nothing can be said to be certain except death and taxes.

BENJAMIN FRANKLIN
Letter to Jean Baptiste Le Roy: 13 November 1789

To tax and to please, no more than to love and to be wise, is not given to men.

EDMUND BURKE

No constable or other bailiff of ours shall take the corn or other chattels of any one except he straight away give money for them.

Magna Carta: Chapter 28

Nor shall private property be taken for public use without just compensation.

United States Constitution: Fifth Amendment

The taxpayer does not know, and has no means of knowing, who the particular individuals are who compose "the government." To him "the government" is a myth, an abstraction, an incorporeality, with which he can make no contract, and to which he can give no consent, and make no pledge.

LYSANDER SPOONER
No Treason: The Constitution of No Authority

The hardest thing in the world to understand is the income tax.

ALBERT EINSTEIN

The income tax has made more liars out of the American people than golf has. Even when you make a tax form out on the level, you don't know when it's through if you are a crook or a martyr.

WILL ROGERS

We can applaud the state lottery as a public subsidy of intelligence, for it yields public income that is calculated to lighten the tax burden of us prudent abstainers at the expense of the benighted masses of wishful thinkers.

WILLARD VAN ORMAN QUINE
Quiddities

One takes up fundamental science out of a sense of pure excitement, out of joy at enhancing human culture, out of awe at the heritage handed down by generations of masters, and out of a need to publish first and become famous.

LEON M. LEDERMAN
"The Value of Fundamental Science"
Scientific American
November 1984

A scientific academy invested with a sovereignty, so to speak, absolute, even if it were composed of the most illustrious men, would infallibly and soon end in its own moral and intellectual corruption.

MICHAEL BAKUNIN
God and the State

Blessed are the young, for they shall inherit the national debt.

HERBERT HOOVER

AN OLD PROVERB says that he who has a partner has a master. Few partners are tougher than the government.

We work to earn an income and then have to share it with the state. We pay the tax and hope the state spends some of it on what we want. But there is not much chance of that. The state spends our tax money on what it wants and does not directly ask us to help it choose. It demands only that we pay. Those who pay have no direct say in where their tax dollars go.

A fuzzy tax form can help change that.

The search for breakthroughs in technology shows how a fuzzy tax form might work. It shows how a fuzzy tax form can help reform the way the state spends our tax money and to some degree help align such spending with our fuzzy social choices. Fuzzy tax forms might help soften the rough edges of ever more powerful scientific bureaucracies. Fuzzy tax forms might even lessen our long-term tax burden.

A fuzzy tax form is a rare case where adding more fuzz to a political system would give less power to those in power.

Picking Winners

How do you pick winners in a game if you do not know the rules of the game? How do you pick them if the game has no rules?

That is the case with research in science and engineering. There are no rules for breakthroughs. No one knows how to find a breakthrough in research on AIDS or electric cars or black holes or diamond windshields.

Some researchers think they know.

Most researchers would be glad to take your money to show you why they think they know. But we would not need the researchers if we knew how to find the breakthrough.

That is the unforeseen nature of a breakthrough. A breakthrough solves a problem in some new way that breaks through what we know. It is that rare moment when new science answers an old prayer. It depends on sweat and luck and skill and shifting market forces and what other people do and have done.

I have watched my colleagues try to pick winners for the government. I have tried to pick winners for it myself. There is something seductive about spending other people's money. The ideas come easily and the risk of failing seems low. The ideas do not come so easily and the risk does not seem so low if your job or career or firm depends on how you pick. Spending tax money is one of the perks of high-tech academia and government labs. The only perk lacking is police power. Even here the Food and Drug Administration has already set an awesome precedent for future scientific bureaucracies.

I have seen only one case where a researcher did not think it a good idea that the state should throw more money at a new "winning" field. That case gives a glimpse of the dreams and schemes and raw politics that all too often rise to the fore when one tries to further the common good by spending other people's money.

The field was neural networks. These are brain-like learning computers. A few thousand of us worked in the mid-1980s on math and computer models of how brains learn and recognize patterns. That changed by 1987 when the field exploded in the world of science and in the popular press.

The neural field had just gone public with its first large conference in June of 1987. I had organized the conference and helped make sure that the press covered the event. The momentum spilled over to a much smaller neural conference in Denver in late 1987 where a few hundred of us gathered. Most of us had something to do with the large neural study that the Defense Advanced Research Projects Agency (DARPA) had just called for. Quite a few of us met one day in a crowded hotel meeting room to argue strategy.

The issue was tax money.

How many millions of dollars should we or the DARPA study recommend that the government put up to study neural networks?

And there was a pecking order in that little technocracy. Professors were there from big and small schools. There were researchers who thought they were professors but were from government labs. And there were the far more humble staff and founders of small start-up firms. They listened more than they talked.

Each person knew they could spend millions on their own research niche. Some wanted to push equations around on paper or model sense organs or brain chunks. I fell into that group of equation pushers. Others wanted to build chips that could screen airport luggage for bombs or detect faulty aspirin bottles on an assembly line or tell a crow or a goose on a radar screen from a Soviet fighter jet or a friendly jet. The thawing Cold War still funded well over half of these independent minds. And each person could look around the room and see there were a hundred or so other researchers with like plans.

The talk soon went from what the neural research world could produce in the near term to what it should ask for right now.

Then inflation set in.

Someone first said we should ask for $10 million. Many thought such a low number was not worth DARPA's time and effort. It would make the field look small compared to the "big science" programs in physics and medicine and biology. The National Science Foundation might not take it seriously.

Then someone said the study should ask for $100 million. It was a good round number. Even the physicists would respect it. Then someone else said $100 million should be just the tip of the iceberg. We were after all talking about how we might solve the mysteries of the brain in the next few years. We might with luck even build a few insect brains in silicon or in crystal holograms.

Soon the new number was $1 billion.

Then Carver Mead of Caltech spoke. Mead was by far the most famous person in the group. He had built a silicon retina and had helped start chip companies in the past and had just helped start the new neural chip company Synaptics. Mead had taught courses with physicist Richard Feynman at Caltech and had cowritten the primal text on modern computer chips.[1] His coauthor had run one of the largest programs at DARPA. Mead gave the group prestige and leadership. And it looked as if he might also give us that billion dollars. The crowd hung on each of his words.

But he said no.

Carver Mead said the worst thing the state could do would be to throw a lot of money at the new research field. The money would both feed the hype and reward for it. Real neural breakthroughs might not come for 20 or 30 years. The money would meanwhile harm real research progress. It would lead to a type of Gresham's Law where bad researchers chased out good ones.

It would lead to the bad effects of any easy-money policy. More researchers would publish and fewer would perish. More money would go to less gifted researchers with weaker research ideas. The pace of research would depend less on talent and the forward lurches of a few great ideas. The pace would depend more on the small steps forward and backward of mediocre but well-subsidized ideas. And it would depend on the politics of who deals the cards. The piles of money might help us in the short run. But it would bloat and distort the whole field in the long run. It would risk an angry backlash among state sponsors and the broader scientific community.

No one wanted to speak when Mead finished. He had just contradicted most of the crowd. Yet each person there knew Mead had spoken the plain truth. They also knew that there were far more of them than there were Carver Meads and that the truth would not win at this session or at dozens more like it around the world. The truth might win in the end and in time it did. But that was years later and millions of dollars later and thousands of sometimes marginal neural papers later.

The world still waits for the first silicon insect brains.

Such meetings of hungry minds take place each day in all countries. Leaders mean well and so do most of the specialists they hire to help them achieve their research goals. But the good intentions tend to give way to the subtle forces of "trickle-down" bureaucracy.

The Manhattan Project of World War II stands as one of the few cases where the state scheduled a research breakthrough and got its atom bomb largely on time.[2] That project had a unique mix of talent and practical goals and extreme wartime pressure. Most research efforts are far more diffuse and use far lesser talents. They have far more fuzzy goals and constraints.

Today each new administration wants to help the country by picking winners in technology. Former CIA director John Deutch stated this baldly in the context of defense research: "[Critics] raise the specter of an industrial policy with the Pentagon picking winners and losers. But all federal research and development programs pick winners and losers."[3] The state wants to clean the environment and boost computer sales and cure disease and speed the so-called defense conversion from a contracting aerospace industry. The state wants in each case to give tax money to a few firms and researchers.

Some good may come of it. But the results may be on a par with what we would get if the state tried to pick and fund the best films or the best pop songs or the best novels. Producing films has much the same high risk as funding new start-up firms. Something like 9 out of 10 such efforts fail. We may feel that Hollywood puts out too many mindless films. But imagine what it would put out if studios and producers could vie for state funds to hedge their risk.

Think for a moment what you see when you walk through a video store or bookstore or when you scan the stock quotes in a newspaper. You do not see the losers. You see only the winners. You see only a few best-sellers out of the over 60,000 new books that appear each year. Each film or book or firm has passed through dozens or hundreds of filters. You do not see the far greater number of screenplays and book manuscripts and business plans that did not make the grade. They fell to the sword of competition somewhere in the creative process.

The failures of defense research show what happens when state money dulls the blade of that sword. A McKinsey study found that perhaps as many as 80% of the early attempts at "dual use" defense conversion failed in the marketplace.[4] These included attempts to build buses and light rail and transit control systems. They also include attempts to jump-start the chip market with VHSIC (very high speed integrated circuit) chips and other gallium arsenide chips. Successes have included police systems to detect marijuana grown indoors and new satellite and telecommunication systems.[5] Defense research has involved something like 70% of American scientists and engineers who have a master's degree or Ph.D.[6]

You can now go to Texas and see the billion-dollar holes dug in the ground to house the defunct superconducting super-collider. They cost about as much to dig as the National Science Foundation's yearly budget of less than $4 billion.

At root lies a key question: *What is the optimal level of state-funded science?*

We know the answer is not zero. The state must fund at least some unique R&D to provide for the national defense. Defense funding can take a good deal of credit for laying the foundation of the modern Internet. Of course this early Internet funding gave a huge free ride to industry and to the people and defense departments of the rest of the world.

The need to fund national defense does not mean we need the current (non-classified) defense R&D budget. It tends to be about ten times larger than the budget of the National Science Foundation. What funding agencies call national defense involves a lot of potential national offense. And why do we still have a marine corps and a navy or have a marine corps and an army? But keeping up with the digital frontier will always be expensive. Chapter 7 suggests how these costs will increase as we try to defend against new threats from information warfare and molecular technology. The potential digital battlefield now extends outward to at least the 600 or so satellites in high-Earth orbit.

The ideal answer comes from basic economic theory. The optimal level of state funding occurs just where the marginal social benefit of subsidized research equals its marginal cost. These are important concepts and they involve elegant mathematics. But we have no good way to measure and arrive at dollar amounts for intangibles like social benefits and social costs. The result is that we have given more weight to theoretical arguments and have allowed politics and special interests to deeply skew the funding process.

The standard theoretical argument is that markets need help with research. Markets will under-produce research in basic science and technology because much of this research is not commercial and because scientists cannot stop others from using it.[17] I have heard critics call this the blue-sky argument because its raw effect is that the taxpayer must pay for the blue-sky ideas and practices of scientists. Most industrialized countries today spend about 2.2% of their gross domestic product on such research and thus on the search for breakthroughs.

This argument is not as strong as many researchers and funding agents assume. And it is doubly suspect when they receive or grant such tax money. Scientists pursue research for many reasons in the scientific marketplace. They seek fame and prizes and patents and consulting and teaching jobs. Publication priority creates intellectual property and leads to a quasi-market system for producing research and breakthroughs.[18] Academic institutions themselves received 249 patents in the United States in 1974 on such intellectual property. That grew to 1,761 patents in 1994.[19] This rapid growth in university patents reflects the much larger market for "second-mover" research that drives and often pays for "prime-mover" research.[20]

Would basic research end if state funding stopped? There was relatively little state-funded science before World War II. The NSF did not begin until 1950. It began in large part because of the evangelical efforts of MIT electrical engineer Vannevar Bush and because of White House support for his controversial 1945 report called *Science: The Endless Frontier.* Yet science and engineering managed to survive and in many cases flourished before World War II and the birth of large-scale scientific bureaucracies.

Lack of state funding did not slow the prewar flow of radical breakthroughs from Thomas Edison or the young Albert Einstein. This reflects what Alfred Lotka published in 1926 and what economists now call Lotka's law of research. It states that in any scientific field only about 5% of those scientists who publish produce about half of the publications. Lotka observed this for papers in 19th-century physics but such highly skewed distributions have continued in the 20th century.[21]

So what is the optimal level of state-subsidized research?

The answer that cannot be right in most cases is the answer that researchers give in most cases: We need more research funds than we now have. This answer stems from such naked conflict of interest that there would be little need to criticize it if so many researchers and institutions did not repeat this answer and take it for granted. How seriously can we take the like request of the subsidized dairy farmer who asks for more tax-based price supports for milk or cheese? Special pleading from the educational elite should carry no more weight. And if a funding agency gets a 10% increase in funds then what are the odds that researchers will match this with a 10% increase in productivity or a 10% increase in the number of breakthroughs?

There is a further reason to be suspicious. Researchers in one country seldom if ever call for letting researchers in other countries compete with them for new research funds from their home country. Yet this increased competition would clearly benefit the society that pays the taxes. Why should not the National Science Foundation or National Institutes for Health fund high-quality (and likely lower-cost) proposals from researchers in India or China or Brazil? Nationality does not affect the breakthrough status of a new laser experiment or surgical technique or mathematical theorem.

The whole point of the argument that markets under-produce research is that we should use tax money to produce more research and breakthroughs. The point is not to increase the living standards of local citizens who have advanced degrees in science or engineering. This gets to the social justice of the matter. One study found that the average physicist with a Ph.D. has an IQ of about 140 points.[22] How much taxpayer support do such gifted individuals need? Why cut welfare for the poor and badly educated but grant it or even extend it for the hyper-trained and intellectually gifted?

There are at least three other reasons to question the status quo call for ever more taxpayer funds to pay for research and the occasional breakthrough.

The first reason is that all such funding involves a *moral hazard:* Why should corporations invest more in R&D if they can count on the taxpayer to invest in it for them? Firms have a clear incentive to shift their costs to taxpayers while they reap the benefits of subsidized research. The benefits to preferred industries are focused and large while the cost to the taxpayers are diffuse and small. This problem grows as universities work ever more closely with private firms.

The second reason is that an easy-money policy *promotes mediocrity.* This was the moral of the story about the DARPA neural network study. The extra research money tends on the margin to fund weaker researchers with weaker ideas even if at first it attracts a few "big names." The mean level of talent falls. And published research tends to nudge toward quantity over quality. Easy money can even drive out the brighter researchers in a form of Gresham's Law. And universities and government labs tend to adapt to the funding increase. They may come to expect such increases as a matter of course (and intense lobbying) and then hire and promote and build to fit the expectation. All orchardists know that if you put too much fertilizer on a fruit tree then you might at first get extra fruit or larger fruit but in the long run you get a tree with shallow roots.

The third reason is *The Big Chill:* State intimidation grows with the amount of money that a researcher or institution takes from the state. This is the classic "chilling effect" of state power on dissent. It is hard to think of a better way to silence and control researchers than to have the government give them money and promise to give them more in the future. When was the last time you heard a scientist criticize a funding agency like NSF or NIH or DARPA or even the FDA? This silence does not make the nightly news.

Scientists should hold funding agencies to higher standards than average. They should quickly and loudly denounce stalled or failed research programs. They should give program managers tough feedback on how they select research topics and on how they solicit and rank proposals and on how they monitor research progress.

But it seldom happens.

The very power that should morally spur scientists to act as extra-vigilant watchdogs serves instead to stifle criticism and to put the would-be watchdogs on de facto but comfortable short leashes. Few scientists stand up alone or in groups to the awesome power of these growing scientific bureaucracies. Yet these same scientists are often quick to lecture their students or the public on the scientific method and the need to question authority and to judge performance on the evidence for or against it. This growing and systemic intimidation does not promote independent thinking. It does not promote the kind of intellectual "creative destruction" that gives birth to new ideas and favors breakthrough research.

The Big Chill promotes intellectual cowardice and an obedient and even subservient attitude. And make no mistake about the reach of this power. The state is crass enough to threaten to withhold funding to research institutions if the institutions do not "comply" with its policies on drug prohibition and political correctness and other forms of social control.

There is a simple way to lessen all three of these problems: Cut all research funding in half. More firms would have to pick up more of their own tabs for R&D. Only the strongest research proposals would tend to get funded and only the better researchers would tend to keep their jobs. This would raise research standards and might even kill off a few of those far more than 100,000 technical journals that have grown exponentially in number for over a century. And a lot of angry researchers would stand up to the state funding agencies and curse and criticize them loudly enough to make the nightly news. The media would have one more reason to hail the "end of science."

The subsidy-reduced researchers just might give society a few more research breakthroughs than before.

We don't know what the optimum level is of state-funded research. We likely will never know it. But prudence and experience suggest that we have long since funded beyond this level. If after all the payoff from state-funded research were as high as some have claimed then citizens should be racing to invest ever more taxes and donations in the process.

So let us now put aside the hard question of how much the state should spend on research. I focused on that question to explore the real politics and economics of modern science. The media and pop culture have largely failed to lift the veil of professional science and see this all too human scramble for funds and reputation. That may be why they still tend to portray scientists as either truth-seeking saints or fumbling nerds.

Let us now work with the research budget of the moment and try to find better ways to spend it or at least part of it. So we return to the first question: How can the state pick winners?

There is a simple way to pick winners: Punish firms if they fail to find breakthroughs.

Just pass a law and demand the breakthrough on time. This has great appeal in these days of high state debt and high taxes. More states may follow California and require a minimum number of electric cars on their streets in the future. A California law required that 2% or more of the cars sold in the state in 1998 emit no pollutants. The law has since lost much of its regulatory force.[7]

A better way to pick winners is to reward breakthroughs when they happen. Condition the social reflex.[8]

This stems from what may be the closest thing we have to a law of social behavior: *Society gets what it rewards for.* Reward researchers to write more papers and we get more papers. Reward them for breakthroughs and we would get more breakthroughs.

The question is how to pay for it.

One way is a research lottery. Each time you play it some of the money you lose goes to a pool of research funds. But there are two problems with that. The first problem is that there are too many lotteries. The lottery market is now a pond well fished. The National Council on Problem Gambling found that in 1997 Americans lost over $50 billion on lotteries and other forms of legalized gambling.[23] This is staggering proof of human irrationality and how poorly most people estimate (published!) odds. It does suggest that taxpayers might pay for some research lotteries since right now they pay so much for nothing at all. But how much more disposable income can people spend on lotteries? Americans in September of 1998 had on average a slight *negative* savings rate for the first time since the Great Depression in the 1930s.

The second problem is that a lottery would not tell us or the state how to spend the money. It would not help make fuzzy social choices. But we could use lotteries to at least partially fund research once we had the choices.

A fuzzy tax form does not have these problems. It does of course have its own problems.

The Fuzzy Tax Form: Choose as You Pay

Paying taxes is at root a binary affair. You pay all or none into the pool of "general revenues." This holds at the federal and state and local levels. You pay into the common pool and someone else spends what you pay.

This looks on the face of it like the way it must be in a representative democracy. We elect officials in an open vote and trust them to spend our tax money on the common good. We can always throw the rascals out at the next election if we don't like how they spend our money.

So your vote is your only choice.

A vote may have been enough feedback in the days of Jefferson. Citizens then had no real-time TV or radio or Internet access to the affairs of state. They were far less informed than we are. And their politicians had fewer choices to make on their behalf.

A fuzzy view finds a hidden binary assumption in this tax process.

Why should 100% of your tax bill go to general revenues? The default answer is that politicians need tax money to fund spot measures and thousands of other state tasks that most of us do not see. Someone must pay to keep up the roads and courts and air force and diplomat security and all the other muscle and fat in the largest concentration of wealth and power in human history.

That is at best a good argument why *some* tax money should go to general revenues. It does not explain why *all* tax money should go there. And that is the point at issue even if states have never raised it.

This point always bothered me when I filled out my income tax forms. Why not let the taxpayer have some say in the process? Many people have no doubt thought the same thing. My fuzzy instincts led me to suggest a fuzzy tax form.

A fuzzy tax form is fuzzy at two levels. The first level challenges the binary flow of tax dollars to the common pool of general revenues. It would allow only some fraction of the money to go there. That way politicians could still fund spot measures and be sure of a steady source of funds. The second level would allow each person to say to what degree the rest of their tax money went to a list of social options. That is a small way that a fuzzy tax form can help make good on the Fifth Amendment's promise of "just compensation" for taking one's property.[9]

A fuzzy tax form might work like this. Suppose you owe $20,000 in taxes this year. But suppose only half of that amount goes to general revenues. You have no say over that $10,000 just as today you have no say over where the entire $20,000 goes.

A fuzzy tax form would let you pick where you want the other $10,000 to go. The money can go all or in part to ten or so options that range from basic research and smart roads to debt relief and homeless shelters and cleaning up the environment. You fill in the percentages. You may want 30% of the $10,000 to help pay off the national debt and 30% to build more court houses or to fix the roads. You may want the other 40% to help cure AIDS or cancer or cystic fibrosis. The percentages must sum to 100%.

You could still use a binary tax form. You would just write 0% for each option or leave them blank and let the state spend all your money as it sees fit. You do that now but not by choice.

Write-ins on a fuzzy tax form could fund all sorts of experimental ventures. And of course write-ins and "approved" categories could include spending outside U.S. borders. Environmentalists might want to include a category that would fund debt-for-nature swaps in Brazil or Madagascar or India that help pay down some of the foreign debt of these countries in exchange for turning over large tracts of rain forests or other wilderness areas to nature conservancies.

I have proposed calculus vouchers that would pay for students in K-12 schools to take calculus courses at local community colleges or universities if students pass a standardized math exam. Enough state or federal write-ins could fund this focused experiment in "school choice" and other educational measures that could boost the scientific literacy of society: A study at

Oxford and Northern Illinois Universities found in 1989 that only half of American adults (and only a third of British adults) know that the Earth orbits the Sun and does so in a year.[24]

Below is a sample fuzzy tax form:

Fuzzy Tax Form

You owe _____ in tax. Half of that amount will go to general revenues. The other half will go to the social categories of your choice.

Enter whole number percentages next to the categories of your choice. You may write in only one new category. All the percentages must sum to 100%. Else the government will normalize your choices by dividing each percentage by the sum of all the percentages. All tax moneys go to general revenues if you leave the categories blank.

AIDS Research . . . . _____
Basic Research . . . . _____
Cancer Research . . . . _____
Computers for Kids . . . . _____
Debt Relief . . . . _____
Disaster Relief . . . . _____
Environmental Clean-Up . . . . _____
Foreign Aid . . . . _____
Health Insurance . . . . _____
Homeless Food and Shelter . . . . _____
Global Warming Research . . . . _____
Infrastructure Repair . . . . _____
Law Enforcement . . . . _____
National Defense . . . . _____
Public Transportation . . . . _____
Space Research . . . . _____
United Nations . . . . _____
Welfare . . . . _____
Other:_____ . . . . _____

Total (must equal 100%): . . . . _____

Each filled-out fuzzy tax form picks out a point in a fuzzy cube. Chapter 3 used a fuzzy square or 2-D cube of political liberty and economic liberty. This fuzzy cube is a proper hypercube. Our minds can't picture cubes of more than three dimensions but they still "exist" as a type of math object. These cubes have as many dimensions as there are social options on the fuzzy tax form.[10] They contain all possible social choices.

If you put 100% of your tax-form money on one option then your choice point lies at a cube corner. Else it lies somewhere inside the cube of fuzzy social choices. Binary tax forms give no direct social choice. That means they hollow out the fuzzy choice cube and give only the zero or null corner.

A fuzzy tax form would give a direct say to those who pay. How could a politician argue against that *in public?*

Criticisms of a Fuzzy Tax Form: IQism versus Democracy

I have proposed the fuzzy tax form to many colleagues and audiences.[11] Most people like the idea. They feel it is a good mix of common sense and raw but limited democracy. Most would like to experiment with some form of it in at least one tax year.

But I have found that most intellectuals dislike or even hate the fuzzy tax form. They give many reasons but the reasons all come down to one. They think most people are just too dumb to choose wisely. The great unwashed masses on the left side of the IQ bell curve need those on the right side to choose for them. Never mind that it is their money to start with. And never mind that they get to vote on which smart person will govern them.

Call it *IQism.* And add it to the dark list of racism and sexism and ageism and general groupism.

The smart think little of the less smart.

The smart and the less smart are of course fuzzy sets. They are vague and relative. All people are smart and not smart to some degree. And we each mean something different by the term. Not-smartness or dumbness stands as one of the last put-downs that we can get away with in a politically correct world. Think of the social outrage if the film *Dumb and Dumber* had had the title *Black and Blacker* or *Gay and Gayer* or even *Old and Older.*

IQism ranges from the high-IQ group MENSA to the low-IQ group DENSA. It holds with special force in my world of professors at the university. Tenured professors tend to look down on untenured professors who tend to look down on Ph.D. students who tend to look down on master's students who tend to look down on undergraduates who tend to look down on those with only a high school degree or even less education. IQism leads here to something of a hierarchy of frustrated specialists. Each could solve the world's problems if only he or she could control the world's resources and rewrite its laws.[12]

Intelligence is a form of power. The smart find little comfort in letting the less smart have power. They find even less comfort in letting the less smart have power over them. Who wants to take orders from someone with a lower IQ? And just about everyone sees themselves as smart.

Democracy has always posed a threat to the smart elite. This is not just the real threat of the "tyranny of the majority" that can step on personal freedoms and crush dissent and persecute minorities. It is the threat that a lot of people might thumb their noses at scientific studies and expert advice and do dumb things in their own dumb way—they might live their lives too freely. The fuzzy tax form extends this threat to the very structure of government.

A typical smart argument against the fuzzy tax form goes like this: People in cities will take advantage of people in the country. I have heard this from liberal journalists and conservative creationists and former Chinese Communist bureaucrats. The people in cities want city subsidies while the people in the country tend to want farm and road subsidies. Each group wants state funds at the expense of other groups.

That is true enough. But the argument would have more weight if a fuzzy tax form allocated all tax moneys rather than just some fraction of them. Then a fuzzy tax form might shortchange some groups or underfund some issues. The argument has no force if taxpayers get to direct 0% of their money as they do today. It has full force if they direct 100% of it. It has less force if they direct only 90% or 80% or 70% of their money. Somewhere in between lies the threshold of state pain. It might be 10% or 90%. I suggest that 50% go to choice. But only experiment can tell.

The deeper response is that a fuzzy tax form is fair. Indeed it is just. It gives proportional representation in terms of tax money just as voting gives it in terms of how many representatives a state can elect to Congress.

If 70% of taxpayers live in coastal regions then it is fair that 70% of at least part of their taxes goes to coastal concerns. It would not be fair if they controlled more than 70% of the funds as they may well do in an election system where the winner takes all. And it would not be fair if they controlled less than 70% of the funds. Why should they pay more of the costs if someone else receives more of the benefits?

User pays.

Fairness is a virtue but not itself enough to justify a fuzzy tax form. Here we must be careful with our words. Fairness is a necessary but not sufficient aspect of a social policy. Fairness just means equal treatment. It does not itself mean just treatment.

Suppose China put to death all Chinese children born in a given year to help reduce its population growth. That would be more fair than killing just the girls born in that year or killing just those children born below a certain birth weight. It would not be just. A man kills ten people fairly if he kills them at random with the same probability. He kills them unfairly if he kills them with a preference. The killing in both cases is not just.

Fairness is again a virtue. But justice does not equal fairness no matter how highly we value equality.[13]

Justice is measure for measure. Justice demands that you get what you deserve and that you do not get what you do not deserve. That is why statues at courthouses depict a blindfolded lady of justice who holds a measurement scale in one hand and a sword in the other. Getting what you pay for is just in the sense that you incur both the costs and benefits of what you get. It also helps that most trade takes place in terms of objective currency amounts. But paying for something is not just to the degree that you receive only some of the benefits but pay all the costs or the degree that you receive more benefits than costs. Then the measure of gain does not balance the measure of cost.

No tax is fully just in this sense. A tax takes your money but does not match it with an equal benefit. Some people get a free ride. Others pay more for their ride than it costs or get no ride at all. A gas tax comes close to a just tax since the more you drive on the state road system the more you pay. This still holds to some degree in Europe where the gas taxes far exceed the real cost of the gas and exceed the direct costs that the user imposes on the road system and other drivers.

A fuzzy tax form is just to at least the degree that it lets the taxpayer have proportionate say in what he pays for. It is not as just as a gas tax or other user fees. The taxpayer may not use what he gets or ever see what he pays for. So a fuzzy tax form acts as a type of *virtual* user fee. The default binary tax form is simply a fee.

A fuzzy tax form is also more efficient than a binary tax form. A basic theorem of economics shows that all taxes impose a "deadweight" loss on society. A market is efficient if supply equals demand. A tax always results in some inefficiency because it makes consumers pay a price higher than the market price and because it makes producers sell fewer goods or services than they would at the market price.

A binary tax form has no direct mechanism to match the social demand for many services with their potential supply. It relies on the goodwill of elected and appointed politicians and on their ability to accurately measure social demand. A fuzzy tax form not only gives a direct statistical sample of that social demand. It channels cash to exercise the social demand for these services and thus helps stimulate the social supply of these services. It is not fully efficient but it surely goes further toward matching social supply and demand than does a binary tax form.

A second argument against a fuzzy tax form deals with special interests and lobbyists.

Special interests would do all they could to sway voters in how they filled out a fuzzy tax form. Pressure groups would fight for the press spotlight. The result would skew social choices. The most money might go to the cause whose group screamed the loudest.

That too is true. But is that as bad as it sounds? It means there would be a healthy competition for taxpayer funds. The competition would not be perfect in the formal sense of a near infinity of informed competitors. So that would lead to skewed choices in the short run as it does with all goods and services.

Such a competitive imbalance tends to correct itself. Losers learn from winners and try new tactics. Stable "market shares" of funding options would emerge. They would slowly shift in time as public tastes changed and as funded experiments played out and as media spotlights shined elsewhere.

The focus on special interests shows one reason why we may someday see some type of a fuzzy tax form. Special interests would gain from one. We are most likely to see a literal pro-choice fuzzy tax form when the expected gains to special interests exceed the expected losses from those who would see their stream of tax money lessen.

This should not hold in a perfect world but does hold in a political world of entrenched interests. And it is why the young would tend to favor a fuzzy tax form more than older voters would. The young can expect to pay more tax money than they receive in the short term if not in the long term. The older receive far more tax money than the young can expect to receive.[14] The elderly are risk averse and may not want to risk their generous share of tax income on the democratic hazards of a fuzzy tax form.

There are at least two problems with a fuzzy tax form that do not stem from IQism or from a desire to keep tax money flowing to powerful lobbies. The first one is political: Who picks the social funding options? My ten best options differ from your ten best. Who says which ones get on the form? This could be quite a dogfight. Special interests would fight hard to keep their option on and keep off competing options.

So who sets the agenda?

The state could pick the options in the worst case. That means someone other than those who pay and choose would pick them. The state does this right now when it lets you pay $1 for the presidential campaign fund. The California state tax form now lets you give $1 or more to fund drug prohibition and education. The federal or state congress could argue about the options and cast them in a bill for the president or governor to sign. That would give at least some choice.

Letting the state pick options still risks tilting the board so that the tax money flows right back to those powerful groups that already get a lion's share of it. My old home state of Kansas might end up doing no more than giving its taxpayers the choice of sending part of their state taxes to those who grow wheat or corn or soybeans.

A better scheme would put a long list of options to a popular vote. The options themselves could come from call-ins or e-mail messages or letters to a presidential commission. Voters might pick their 10 best options and might even rank them from most favored to least favored. Then we could tally the results and go with the top 10 or so. Or we could go with the top 100 and pick 10 or 20 at random each year until we had cycled through all the options.

Such voting schemes would still suffer to some degree from the tyranny of the majority. A few good causes would not get on the tax form because too small a group voted for them. This is one reason to let each taxpayer write in at least one new option. Some of the new options could move onto the option list as time passed and other options dropped off it for lack of voter interest.

The second problem deals with administration: Who would run the fuzzy tax form? How many layers of new bureaucracy would it take to set up such a scheme for taxpayer choice and to funnel the money to the chosen social causes? The new bureau or center could grow to Byzantine size and structure if left to itself.

Perhaps the money could go straight to the bureaucracies that now deal with social causes. The money could flow through the same pipes that it would have flowed through in the pool of general revenues. Only the routes and amounts would differ.

The trouble is the state is not a frictionless system. The money flow is highly viscous and much of it tends to stick to the pipes or leak away. The person or firm on social or corporate welfare receives less than half of each tax dollar sent. I do not mean to suggest that the U.S. federal government should spend its current $125 billion or so a year on corporate welfare more efficiently than it does.[25] I think the economics of wealth production requires that the state should neither tax firms nor subsidize them. Firms should pay user's fees for state services. I simply mean that state bureaucracies impose their own large tax on the tax-transfer process.

The state will always get some middleman cut no matter how tightly we streamline the process. But a fuzzy tax form need not lead to a higher cut than the state gets right now with its binary tax forms. The fuzzy tax form sets up a direct path from those who pay to those who spend. The direct path favors fiscal scrutiny far better than does a common pool of tax dollars in the hands of special interests and politicians.

I have not said which type of tax could use a fuzzy tax form. The reason is they all could. The tax could be an import tax or property tax as well as a state or federal income tax. A fuzzy tax form would add some choice to each type and still be revenue neutral.

We can argue over how much money taxes should raise and which state actions it should fund and how tightly we should limit government. But even the most radical privatizers have not shown how to fund national defense without some type of tax.[15] So some type of tax may always be with us. Government may be destiny and the world may never see Karl Marx's dream of the state withering away.

A state or federal income tax would make a good forum to first try a fuzzy tax form. Most taxpayers feel the full bite of this tax once a year. It is the tax most think of as *the* tax even though they pay dozens of other direct or indirect taxes such as taxes on airline tickets or hotel rooms or the "bracket creep" tax of price inflation that pushes taxpayers into higher marginal tax brackets. The United States raises between one-third and one-half of its funds with the federal income tax and pays out much of that each year to cover the interest on the national debt.[16] Some states like Nevada and Texas and Florida do not have state income taxes. So a fuzzy income tax form might not be as radical a change as it sounds.

A fuzzy income tax form might even lessen some of the social damage of an income tax. Taxes affect what society rewards for in a reverse or negative way: *Society gets less of what it punishes for.*

A tax on income is a tax on productivity. It taxes labor and successful investment. It slows the wealth engine. The old joke is right: An income tax is a form of capital punishment. That is why many countries like Germany and Japan and Singapore have replaced it or lessened it with a tax on consumption. Those countries have far higher rates of saving and capital accumulation than the United States has.

A fuzzy tax form could also work with a Georgist land tax.

Economist Henry George published his book *Progress and Poverty* in 1879. It built upon an idea that English evolutionist Herbert Spencer put forth in his 1851 book *Social Statics:* No one owns the land. We can lay claim to only our improvements on the land. George argued in his book for a single tax on land to fund the state and thus not to burden labor or capital: "Instead of lessening the incentive to the production of wealth, I would make it more powerful by making the reward more certain. No matter how many millions any man can get by methods which do not involve the robbery of others—they are his. Let him have them."

Henry George's book sold millions of copies over the years. It influenced thinkers from novelist Leo Tolstoy and labor leader Samuel Gompers to philosopher Bertrand Russell and statesman Winston Churchill. It led to the single-tax movement and shaped the tax structure of many cities in the United States and Canada and Australia. President Theodore Roosevelt proposed a Georgist land tax to settle Alaska.

But land rent today accounts for only about 5% of gross domestic product while most states spend well into the double-digit percentages of their country's GDP. The U.S. federal government spends about 33% of its GDP while Sweden's much larger welfare state spends about twice that. France and Germany pay about 50% of their GDPs in taxes. The average industrialized state spent about 46% of its GDP in the 1990s.

A pure land tax would have fewer people paying taxes and have more firms paying. Small and large firms tend to own or rent most of the property in crowded cities and thus in regions with the highest land values. A fuzzy tax form could allow each paying firm to act as a person and vote its choices as it pleases. Or the fuzzy tax form could apply only to the people who paid land tax. It could also give some *pro rata* say to renters.

A fuzzy tax form seems as simple as a flat tax and that may not bode well. We find very few flat taxes in the world. Each person has an incentive to skew the tax system in his favor. He may not get the chance to do so but the incentive is there and the politics reflect that. Worldwide budget deficits reflect the politics.

Balanced Budgets and Endowed States

How hard is it to balance a budget?

You can do it with a flat tax on one hand and a *flat cut* on the other. Tax each person some percent to raise income. Then cut all programs some percent across the board to reduce spending to that level of income. A 20% tax might balance a 30% cut. If not then the state could raise the tax rate above 20% or cut more than 30% or some mix of both. Thus a flat cut could avoid ganging up on a few politically weak programs to fund stronger ones. That is the simplest and fairest scheme to balance a budget. Households and companies do it each day.

A child could balance the budget with pen and paper in a few minutes. Yet it never happens.

A fuzzy tax form might suffer the same fate. It might end up as just one more revenue patch on the national debt. It might bog down in old or new bureaucracies. It might die from the press spotlight of a good funding scandal. The same risks hold for any new tax or subsidy scheme in a republic. So we might have to use a fuzzy tax form only once every two or four years. Or we might get to use one only once in a generation.

The sure way to find out is to try it at least once.

A fuzzy tax could do more than just fund research and other preferred social causes. It could in theory replace the need for any taxes at all. It could endow the state with a type of "retirement" nest egg and so end or reduce the drag of taxes on the economy.

The U.S. government spends over a trillion dollars a year. So how large a nest egg would the state need to earn a trillion dollars a year as income?

Suppose the nest egg consists of stocks and bonds (in other countries) and mutual funds that invest in stocks and bonds. An 8% tax-free real rate of return is a reasonable long-term yield since stocks alone tend to earn at least between 8% and 10% over the long haul.[26] So we would need a $12.5 trillion nest egg in current dollars.

We could grow this nest egg in 30 years if we paid in $110 billion each year to the annuity pool and increased the amount to keep pace with inflation. A fuzzy tax form might earn this amount. Or we could abolish the departments of agriculture and transportation right now and direct the $100 billion or so to the fund and direct the same amount there each year.

A fuzzy tax form can set up such a tax annuity for our heirs. It cannot give them the discipline to keep them from doubling how much they let their government spend. Maybe they will learn such thrift if Social Security and Medicare and federal pensions go bust in the next century. Or maybe their thirst for state funds will only grow.

A tax endowment also raises the old hard question of who watches whom. Do we want the state to pick its own stocks? That would create a new layer of special interests and favoritism and bureaucracy.[27] The system might avoid much of that if it invested only in broad index funds both in the United States and abroad. And of course it could not invest in its own notes or bonds. If the scheme worked then it could lead to an ironic relation between socialism and capitalism. The state as investor could end up owning much or even most of the means of production.

But just think if past generations had set up such an annuity for us.

Research Bounties: Buying Breakthroughs

Consider again the problem of picking winners in science and technology. A fuzzy tax form gives a simple but powerful way to fund breakthroughs.

It would let you buy them. Take some of those new tax dollars for research and set up bounties or contests with huge cash prizes.

The principle comes from B. F. Skinner's operant conditioning: Don't try to pick winners. Grow them. Reward for them. Reinforce their spontaneous appearance.

A large cash prize would motivate researchers far more than would the pay raise they might get if they landed a large research contract. A fuzzy tax form could fund these research bounties. Still larger cash prizes would motivate whole research teams and companies. They get paid if and only if they produce a breakthrough.

The prizes may not have the prestige of Nobel prizes or suffer from the obscure politics of the Swedish Royal Academy of Science.[28] But the prizes could make up for that lesser prestige in sheer dollar amount. Here quantity truly has a quality all its own.

Whirlpool showed how it could work with refrigerators. The firm won a $30 million prize in June 1993 for its design of the most environmentally safe refrigerator.[29] Refrigerators use about 20% of the electricity in a household. A group of 24 electric utility companies put up the $30 million prize money. Each utility put up from $150,000 to $7 million based on the number of their customers.

The contest drew about 500 responses from around the world. Each tried to meet the contest goals of a new refrigerator design that used 25% less energy than current models and that did not use chloro-fluorocarbons. Sunlight breaks down airborne CFCs into chlorine molecules that destroy the ozone in the sky. Whirlpool focused on the compressor because it consumes up to 90% of the energy in a refrigerator.

Whirlpool also used a fuzzy system to help control the compressor and defroster.

What worked for them in this case could work for us in many others. There are already something like 3,000 research prizes in North America.[30] Studies show that most people favor some form of state subsidy for basic research in medicine and other fields.[31] People know that some of their taxes go to research but they have little idea how such trickle-down bureaucracy works.

A fuzzy tax form would let taxpayers check off research bounties and get more breakthrough for their buck. It might well save tax moneys while it produced more focused breakthroughs. Research bounties would still leave untouched most research funds for basic research. But they would have the key effect of boosting the supply of research breakthroughs where there is a proven and long-standing social demand.

This would start a "paradigm shift" in research funding. It would start a methodological shift from the current funding practices that resemble the respondent conditioning of Ivan Pavlov to practices more like the operant conditioning of B. F. Skinner.

Pavlov taught dogs to salivate by coupling the bell or light with the reinforcing food. He in effect rewarded the stimulus and forced the dogs to salivate against their will and thus act "involuntarily." Skinner taught pigeons to pluck out tunes on a keyboard by dropping a food pellet when but only when the birds pecked the right key. He rewarded the right response. He would wait as long as it took for a pigeon to "voluntarily" emit that response.

Both techniques are forms of behavioral engineering. So are all schemes that transfer funds from taxpayers to researchers. We must face that bald fact. But Skinner's operant conditioning is less coercive in its application and more market-like in its incentives than is Pavlov's respondent conditioning.

A Skinnerian approach rewards only research breakthroughs. The current funding method is more Pavlovian and rewards research proposals and the ability to sell. It rewards proposal writing and political skill and other subsidy-seeking behavior that precedes breakthroughs. The crucial point is that it rewards these prior stimuli even if they do not lead to breakthrough responses.

The essence of operant conditioning is that you catch more flies with honey. We might well harvest more breakthroughs with well-funded research bounties than if state agencies try to pick winners from a pile of research proposals and thus reward the subjective promise of breakthroughs rather than reward the breakthroughs themselves. These Skinnerian experiments won't cost much if they fail to produce breakthroughs because we won't have to pay the bounties. A fuzzy tax form lets us set up such experiments with the strokes of enough pens.

So put up a few million dollars in prizes for the design of an electric car or the best way to clean up oil spills. Put up a billion dollars for a cure for lung cancer or AIDS. We spend billions more right now to treat these diseases.

If that fails then put up $10 billion for a cure for any cancer or AIDS. There comes a point in theory where we all stop what we are doing and go search for a cure or help set up or invest in the search. So if a prize of $10 billion fails to spur the cure or breakthrough we want then put up $50 billion or $100 billion. Who cares whether high school whiz kids or career scientists or profit-seeking firms find the winning breakthroughs? The terms are only cash on delivery.

And it takes just one cure.

5

The Rights of Genomes

Population, when unchecked, increases in a geometrical ratio.
Subsistence only increases in an arithmetical ratio.

THOMAS ROBERT MALTHUS
The Principle of Population (1798)

We have been God-like in our planned breeding of our domesticated
plants and animals. But we have been rabbit-like in our unplanned
breeding of ourselves.

ARNOLD TOYNBEE

The life of a fetus is of no greater value than the life of a nonhuman
animal at a similar level of rationality, self-consciousness, awareness,
capacity to feel, etc.

PETER SINGER
Practical Ethics

While in his cultivation of plants and animals the human breeder
enormously speeds up evolution and produces creatures by inbreeding,
which are of use to man, these creatures are very incompetent at living
in the wild. No corn plant, no sheep could survive in nature without
the watchful ministrations of a human caretaker.

MAX DELBRÜCK
Mind from Matter?

Great biological diversity takes long stretches of geological time and the
accumulation of large reservoirs of unique genes. A panda or a sequoia
represents a magnitude of evolution that comes along only rarely.

EDWARD O. WILSON
The Diversity of Life

Science is not God. Our deepest truths remain outside the realm of science. We must temper our euphoria over the recent breakthrough in animal cloning with sobering attention to our most cherished concepts of humanity and faith.

PRESIDENT BILL CLINTON
"Science in the 21st Century"
Science: volume 276, 2 June 1997

The human chromosomes can be thought of as strands of pearls with four colored balls. Sequencing involves taking each of these chromosomes and determining the color of the balls on each string. It will be a relatively simple matter to translate the nucleotide signals into amino acids, the amino acids into proteins, and from the proteins determine the essence of human beings.

PROFESSOR LEROY HOOD
University of Washington at Seattle

W HO OWNS YOU?
Property rights depend on the borders between mine and thine and between the private and the public. Fuzz blurs these manmade lines and creeps from there to the notions of rights and ownership.

This chapter looks at the legal nature of the self and the degree to which you own your self. The next chapter looks at the degree to which you own things beyond your self. It does little good to own a part of the world if you do not even own your own organism. Jesus said this to Satan about owning the world versus owning your soul. It still holds today for the evolving tug of war between the public and the private.

Genes before kings.

Most views of ownership start with ownership of the self. If you own anything then you own yourself. Then you may or may not own what issues from you or from the smart card or government medical file that holds a digital copy of your genetic blueprint or maybe of last week's backup of your brain's neural chatter. These views require that we know what we mean by self and thus know where the self starts and where it ends.

Where does the self start? Where does life start?

These questions land one in the fuzzy problem of abortion.

A Little Bit Pregnant

Millions of women decide each year to have an abortion.

The average number of abortions in the United States fluctuates around 1.5 million per year. The number has fallen in the 1990s as the number of U.S. abortion providers has fallen.[1] But still about one-fourth of U.S. pregnancies end in abortion. The World Health Organization claims that worldwide about 50 million women have abortions each year. About 30 million of these women have "legal" abortions and about 20 million have "unsafe" abortions. Surgical abortions kill about 70,000 mothers a year.[2]

Our global village adds about 90 million new babies to the world population each year. So it takes only three years for the world population to grow by more people than all those in the United States. The world aborts as many fetuses in five years as there are U.S. citizens.

These are large numbers and they grow larger each year. They are statistical footprints in the sense of Joseph Stalin's infamous quip that one death is a tragedy but a million deaths is a statistic. Behind the numbers lie hundreds of millions of painful choices that women and men make each year. And behind those choices to do or not to do lies the old black and white logic of the abortion debate. Pope John Paul II has been perhaps the staunchest of those who hold a binary view of a fetus's right to life. He surely is the most powerful: "The right to life is the *fundamental right*."[3]

The debate divides the 60 million Catholics in the United States just as it divides people of all faiths and creeds around the world. Abortion is against the law in Egypt and in most Muslim countries. It is technically legal in Israel and Kenya only to save the mother's life. All the major religions oppose abortion in principle and often in practice. Communist China forces some abortions as late as the sixth month of pregnancy to meet its population goal of one child per couple. Almost half of all pregnancies in China ended in abortion in the last decade.[4]

The debate comes down to where you draw the line between life and death. Is the fetus alive? Does it own itself? Yes or no?

A fuzzy view does not get caught in this Aristotelian trap. It draws a curve instead of a line. I have argued elsewhere for this view that life is a matter of degree.[5] I will now summarize that argument and then look at two objections to it.

A fuzzy view sees the fetus as growing from 0% alive at or just before conception to 100% alive at or near birth. At each month the fetus is both alive and not alive to some degree. Not-life shades smoothly into life and at death it shades back. This defines life as a smooth curve that rises from left to right as the fetus grows.

A binary view sees the fetus as passing from 0% alive to 100% alive in one big jump. The binary view requires that someone draw a hard line between life and death. Many people want the state to draw these hard lines for them and for the rest of us. That would remove the fuzz or vagueness from the concept but only lay bare its relativity. You or the state can draw a life line anywhere you please from conception to birth. The odds are good that it will not please those who draw a line somewhere else.

Extreme pro-life fans draw a hard line at conception. That line gives the fertilized egg the same legal status as the crying baby. Extreme pro-choice fans draw the line at birth. That line would abort a child days or even hours before birth. Most people would draw a line between these extremes. But all such lines are arbitrary and abrupt.

The 1973 Supreme Court decision *Roe versus Wade* drew the line at three months or the end of the first trimester.[6] India draws the line at 20 weeks. Most physicians refuse to perform abortions in the third trimester and so draw the line at about 24 weeks. U.S. physicians perform only a few thousand of these late-term or "partial-birth" abortions each year.[7] Even many pro-choice advocates have doubts about late-term abortion on demand.

Polls can estimate a fuzzy life curve. Ask 1,000 people at random at what month life starts. Plot the answers and you get a staircase that rises from left to right with one step per month.[8] The staircase approximates the life curve. Ask 10,000 people and you get a better approximation. Who would not want to see the life curve of their state or their country or their country's medical society?

The abortion debate got stuck decades ago in the binary stalemate of life lines. Life curves might move us closer to a consensus. They will not make the black-white decision we might want them to make. But they can help move the debate to the "knee in the curve" or the month where the degree of life grows most or the region where it starts to taper off. It looks as if that knee may be near the end of the second trimester. Polls can tell and it is a simple matter to take them.

Future life curves may involve less public debate than they involve private choice. The new RU-486 abortion pill will in time make each woman her own obstetrician. Women will buy the pills at the drugstore or the pharmacy or on the black market. That will make abortion more a moral decision and less a legal one. Published life curves in the instructions might give women more (or less) confidence about taking the pills since women must take the current pills soon after conception. The future will no doubt see many new abortion methods and pills.

I have heard two main objections to this fuzzy view of life curves. The first is that a life curve is not enough. You need a hard line or action threshold in the real world just as at some point you have to draw the line and slam on the brakes when the car in front of you slows too quickly.

That claim is true and a life curve gives you information to help you draw that personal line. But again it will not draw that line for you. You are more likely to trust the knee in the curve if you respect the sampling population that jointly drew the curve. A medical society's life curve may have more value to you than would a life curve drawn from a random sampling of TV viewers. The state might find reason to favor either group if the state draws a legal life line through a life curve.

The other objection is that a fuzzy life curve measures opinion and not life. Who cares if 80% of the people in a nation or a medical workshop think that life starts in the seventh month of pregnancy or that it starts there to a large degree? What matters is where life in fact starts. No one knows that point and hence we still debate.

This claim demands truth and not opinion. It is a more serious objection than the first one. And both pro-choice fans and pro-life fans can make it.

One response is that a fuzzy life curve can measure gray truth as well as gray opinion. Truth and opinion are both matters of degree. A related response is that opinion may be the best we can do when no one knows the truth. Who knows the truth about "life"? Decades of debate and research have failed to find it.

Maybe there is no truth of "life." Maybe the term "life" is just a term of last resort and stands for no thing or process in the physical universe. We can at least challenge whether there is a truth beyond what we can measure of a fetus's brain state or of a public's legal precepts and opinions. The issue is one of legal rights and not physics or chemistry.

Yet Chicago University legal scholar Richard Epstein has argued against abortion in terms of just such legal rights. He asks what abortion laws would we choose if we sat behind a "veil of ignorance" and did not know our lots in life.[9] We might end up as aborted fetuses if we chose the wrong laws. This could still happen even if a fetus did not achieve a high degree of life until just before birth.

Such thought experiments depend on fuzziness in a way they do not intend. The risk of our facing an abortion grows or falls with the relative number of abortions per births. The odds are small but not too small that we might face abortion in a world like ours that aborts 50 million fetuses a year. The odds would fall to a tenth of that in a world that aborted just 5 million fetuses a year.

You would in effect replace your fuzzy life curve with your fuzzy *risk* curve when you voted on abortion laws from beyond the veil of ignorance. The benefits of abortion would at some point outweigh its risk to you. Then you might rationally allow abortion or theft or murder or any other action that had a slight risk of harming you. That would not be a matter of morals or law but of prudence or cunning or simply playing the odds. You would still face drawing a life line or curve for those abortions you allowed.

A fuzzy life curve can also arise in law or opinion if a third party aborts the fetus against the mother's will. Suppose a fiend builds a small killer sonogram. He carries it in his coat pocket and fires it at the pregnant women he passes as he walks through a crowd along a street or in a busy shopping mall.

How do we rank the enormity of his crime? When does he murder?

We would demand more punishment for forcing the abortion of the pregnant mother in her third trimester than we would for forcing the abortion if she had just conceived the fetus the week or the day before. The crime's enormity would grow with the age of the fetus in the womb. A judge might well appeal to a standardized fuzzy life curve to weight the fiend's sentence. The judge might use the knee in the curve to help draw the line or ultra-fuzzy band between murder and not-murder or between first- and second-degree murder. Courts already count some third-party killing of a fetus as manslaughter or murder.

Genes and DNA also seem to favor the view of truth over opinion. The sperm and egg form a unique gene print at conception. The man and woman each give 23 chromosomes to their fetus. Scientists can now pick a single sperm from a semen sample and inject it into an egg. So how can a woman be just a little bit pregnant? This is the binary example of legend.

How can life be a matter of degree?

We can grant that the gene print is a binary code even if some of the genes do not belong there or if some are missing. A woman can have a partial fertilization and thus be just a little bit pregnant. The result is likely to end in miscarriage. But that is not the point of the gene map.

The point is that the gene print does not change even though the fetus it controls does change. The same holds for a blueprint that workers use to guide them as they build a house. The fetus and the house pass through stages. Fuzzy curves can describe those changes. But do they describe the gene print or the blueprint?

They may not describe the gene print but that does not mean that the gene print contains the truth of the matter. We can after all create trillions of gene prints with a computer. No one suggests that they have legal rights although someday they may. We might even turn that around and award a smart enough machine intelligence a bundle of legal rights by giving it its own human gene print. And no one suggests it is a crime to burn or delete a house blueprint that we draw with a computer even if it may be a crime to burn a house.

Or something could go wrong and the gene print might change as the fetus grows. The gene print could in theory change each second. It would still make sense to debate the legal rights of the fetus even though the first gene print no longer existed. Thinkers from Aristotle on have been careful to tell the possible from the actual.

Fuzzy life curves are tools. A life curve lets you measure any notion of life you please. The final word on them will come from how well they work in practice.

So let the polling begin.

Random Walks in Genome Space

Gene maps open the door to a fuzzy view of life or self in the larger sense of searching the human *genome space*. A complete gene map or gene sequence is a genome. Genome space consists of all possible genomes. There are vastly more genome codes in it than there are subatomic particles in our universe or even in an imaginary universe the size of ours somehow packed solid with such particles.

The Human Genome Project seeks the complete gene map of human beings. It spends about $200 million a year and expects to do so for as many as 15 years.[10] Scientists found in 1995 the first complete gene maps of two free-living creatures. That year scientists also worked out the first approximate physical map of the human genome.[11] Then common yeast gave up the 12 million base-pair gene code of its 6,000 genes in April of 1996. Thousands more life forms will soon give up their codes to science. Each creature defines just one genome point in its own genome space.

The first mapped creature was the bacterium *Haemophilus influenzae* in late 1995.[12] This bacterium has a genome sequence of about 1,830 kilobytes of data. One byte is 8 bits of data. Each bit or binary unit gives a yes or no answer to a binary question like "Is the sequence base pair an adenine-thymine pair?" A kilobyte is 1,000 bytes or 8,000 bits (it can also mean 2^{10} or 1,024 bits). This bacterium has only one type of membrane. It devotes about 30% of its 482 genes to coding proteins in the membrane.

The second mapped creature was the human gut worm *Mycoplasma genitalium*.[13] It has a genome sequence of only about 580 kilobytes of data. The more accurate statement is that the gut worm has 580,070 base pairs in its nucleotide sequence. This gut worm is the smallest known genome of any free-living organism. Its sequence could fit on a small computer diskette with many kilobytes to spare. Then came common yeast. Anyone can access these or any other sequenced genomes on the Internet.

The year 1997 began with announcement of the cloned sheep Dolly and saw more creatures give up their gene prints.[14] The first complete genome from a eukaryotic creature (having cells with genetic nuclei) came from the budding yeast *Saccharomyces cerevisiae*. Microbiologists sequenced the bacterium *Borrelia burgdorferi* that infects ticks and causes Lyme disease. They sequenced the first creature that metabolizes sulfur and the first gram-positive bacterium and the bacterium *(Helicobacter pylori)* that causes peptic ulcers. Perhaps the biggest breakthrough was the complete genome sequence of the pathogen *Escherichia coli* K-12. This bug is the laboratory workhorse of biotechnology. *E. coli* has only 4,288 genes and lives in the lower gut of many animals. Some strains can infect a human's nervous or gastric or pulmonary systems.

The year 1998 saw the cloning of mice and calves and saw still more complex creatures give up their gene prints.[15] Two venereal diseases gave up many of their secrets when microbiologists sequenced the helical syphilis bacterium and the chlamydia bacterium. Syphilis was something of the AIDS of the 19th century. Chlamydia is the most common bacterial venereal disease in the United States and a leading cause of preventable blindness in much of the world. Microbiologists also sequenced the "white plague" tuberculosis bacterium and the typhus bacterium *Rickettsia prowazekii* that infects lice and killed millions of people just after World War I and World War II. Microbiologists partially sequenced the mustard plant *Arabidopsis thaliana* and the malaria parasite *Plasmodium falciparum* that infects up to half a billion people (as many as 1 in 12) and kills about 2 million people a year. The year ended with the first sequence of an animal. The animal was the tiny dirt-worm nematode *Caenorhabditis elegans*. Almost a fifth of the worm's more than 19,000 genes match yeast genes. That suggests nature has built more complex creatures by adding new DNA onto the DNA of simpler creatures.

Microbiologists sequenced only a few percent of the human genome in the 1990s. They first sequenced an entire human genome in late June of 2000. It will take a few more years before we carry our digital genome blueprints on credit cards or smart cards.

The human genome contains about 100,000 genes and on the order of 4 billion base pairs. The base pairs consist of 4 nucleotides or nucleic acids that pair off in a fixed way. Adenine (A) pairs with thymine (T) to form the base pair (AT). Cytosine (C) pairs with guanine (G) to form the base pair (CG). Fungi have only on the order of 20 million or so base pairs. But frogs and other amphibians have much longer genomes than humans have. Some have on the order of 100 billion base pairs.

The size of a creature's genome space depends on the length of its genome or the length of the list of base pairs. The sizes of these spaces are beyond astronomical.

Consider our own genome space. Each human genome or point in human genome space is a list with on the order of 4 billion or (4×10^9) base-pair slots in it. The number of points in genome space is 4 raised to this power or $4^{4 \times 10^9}$ or on the order of $10^{2,408,239,965}$. It has almost two and a half billion 0s after the 10.

This number almost exceeds comprehension. There are only about 10^{87} subatomic particles in the whole universe. The huge number called a "googol" comes in at a modest 10^{100}. If you packed the universe with particles then the particle count would still come to no more than a trivial amount above 0% of the number of genomes in human genome space.

These numbers give us an idea of what is possible. The actual number of all genomes that walk or crawl or sprout or fly on earth is much smaller.

Harvard's Edward O. Wilson puts this total diversity of life as a base-pair count on the order of 10^{17} or 100 quadrillion.[16]

Scientists expect to map the first human genome in the first years of the 21st century. Then we will have worked out all the 4 billion or so base pairs for just one point in human genome space. And then it will be a short step to working out your own sequence.

Right now the U.S. Department of Energy stores most of our gene facts and our gene "resources" at New Mexico's National Center for Genome Resources in Santa Fe. Your gene print fits on the tiny coils of DNA in all of your body's cells except your red blood cells. Your gene print consists of less than a gigabyte of data (base-pair sequences) and would fit on a compact disk. That storage size will soon shrink as memory chips and other storage media shrink. That is why someday your gene print will fit on a cheap disposable smart card.

The cloned sheep Dolly will surely prove to be a small door to a very big room.

Future science may use your gene print to grow you a new finger or heart or hairy scalp or to warn you of health risk factors or a pending disease.[17] Insurance firms may use your gene print to price your premium or to deny you a policy. Employers may use it to screen your job application or to place you in the firm. The calm and depressed work in this group. The manic and tightly strung work in that group. Governments may use your gene print to confirm your identity or to predict whether you will vote a certain way or whether you are a crime risk or a tax risk or a security risk. The FBI and other secret state agencies use computer profiles that way today.

Now think of the genomes of your parents. Your genome is a mix of their genes. The mix is often close to 50-50 but still often far enough from a pure 50-50 mix that you favor one parent over the other. This happens because the mixing process is not perfectly random and because each self-ish gene has its own plans on how to make it into the mix. But assume you are the statistical average of your parents' genes. What does this look like in genome space?

You lie on average at the midpoint of the line between Mom and Dad.

MOM   YOU   DAD

All three of you are points in genome space. We can picture the points as black ink dots on a sheet of white paper. The paper has only two dimensions while human genome space has almost 4 billion dimensions. Strange things can happen in the structure of such large spaces. A multi-dimensional ball or box in such a space has almost all its mass on its surface. Yet the ball is a solid and filled smoothly with points. But ink dots on a white sheet still give the main idea.

Love and sex and conception end in a new dot between two old ones. You favor Mom over Dad if your dot lies closer to her dot than to his. The species unfolds through time when billions of dots move through genome space. Billions of pairs of dots meet at random and leave new midpoint dots in their wake. These billions of randomly hopping parent and child dots move through genome space much as locusts hop through a corn field. They make up the gene river and at any moment the gene pool.

Your family tree defines a short random walk through genome space.

A ball or sphere contains the whole hopping process in human genome space. The ball grows larger with time as the gene pool expands. It distorts and bulges out as humans evolve and thus as new dots form farther away from the old center of mass. Then it takes a larger ball to contain the process. Yet our species' ball takes up less mass in genome space than a marble takes up in our universe.

We would not recognize as human most creatures that would come from far regions of our genome space. Some would have eyes or brains where we have nipples or feet. Others would have only small vestigial limbs and Earth-bound sense organs. They might have flippers or gills or wings that would best help them live in a sea of water or sulfuric air or hydrogen gas. Some would have new sense organs and new bulging brain knots to best exploit signals that came from heat or plasma flows or from magnets or from rigid-body rotations.

We have just begun to search genome space. And we still search it at random.

The Rights of Genomes

We also search genome space in only a discrete way. Why hop between two dots? Why not move one dot closer to the other dot? Why not slide genome dots slightly in new directions? Why not take conscious control of our genome search?

Why not let our genes change to some degree?

I have explored this many times in fiction but did so based on fact.[18] Imagine a birthing vat in the future. You and your spouse have mixed your genes either the old way or some new way and now your child grows in the vat. The computer has studied your two genomes and knows the complete genome of the fetus. The computer knows what you two look like now and has studied photos and videos of what you looked like when you were younger. It may also know what your parents or siblings looked like.

Each day you can ask the computer to show you what it predicts your son or daughter will look like in the future. You may want to see his or her face at age 5 or 20 or 50. The older faces will be less accurate than the younger faces. A neural or fuzzy computer could even endow the face with a hybrid voice and hybrid facial expressions. Who would not want to talk to their child before he or she is born? Such talks might add new vigor to the abortion debate.

Now comes the fuzzy part.

Suppose you decide to let the child's genes drift or move away from their 50-50 mix. The child's genome starts out as the midpoint or average of its parents' genomes and then slowly moves away from their two dots. You might want a computer system to search nearby genomes that you feel improve the child's face or bone structure or pattern of fat cells.

This type of search has the formal name of *constrained* optimization. You want to maximize or minimize some criterion of the child's health or looks or even character so long as it still gets most of its genes from its parents. Mom and Dad constrain or restrict the gene search to regions of genome space near them.

How would it hurt to make a few small changes to the child's genome? But how small is small? It is a matter of degree. You could view the changes as those you or your spouse would make to your own genomes if you had the means to do so. Today you have to find a new spouse to change your child's genes. The future will likely offer more options.

A bold gene plan could cut the child's genes into thirds. Each parent would give one-third from his or her own genome. A smart gene search would find the other third. This plan would preserve the symmetry of equal shares of spousal genes. No one would feel that their spouse had cheated them out of their share of Darwinian fitness as they might feel in the case of adoption or surrogate birth or cuckoldry.

You and your spouse could go further and just give the child a fourth of your genes and let the smart search find the other half in genome space.

That gene plan would net an optimized nephew or niece or grandchild or even an optimized uncle or aunt or grandparent. You two could go still further and each give the child just one-eighth of your genes and net an optimized first cousin. Or you could have one child favor you 90% and have the next child or its twin favor your spouse 90%.

Parents could at last trade some of their raw quantity of genes for genes they feel have higher quality. Is it really better to have a 50% blood tie to a child than to have a 49% or 48% or 33% tie to a healthier or smarter child? The child might have the body of Hercules or the mind of Newton.

The rub is the child can gain more for itself if it has fewer of your genes. The child has the freedom to gain the most for itself if it has none of your genes. Your final legacy may only be to seed a new region of genome space. (Chapter 13 shows another way we can seed such abstract spaces of high dimension.)

The state may not let you vary these genes at will even if the child gains from it.

All parents would want to rid their child of genes that cause disease. But not all social groups would want to allow it or would feel that it out-weighed the risks of abuse. The debate would start with the standard shock sound bites about Nazi gene control and Frankenstein lab experiments. That would first put the debate on the nightly news and on the covers of magazines. It would diffuse from there into our schemes of morals and values and laws.

Some religious groups would try to ban what they would see as changing God's gene designs even though they accept eating hybridized wheat and corn and chicken. Think of the outrage if pregnant mothers began to treat their fetuses with new mixes of hormones and other chemicals to shape the brain or the senses of the fetus. And that would change only how the child grows and not how its genes change.

Other groups would rightly fear that some parents would take advantage of their child's genes. The jock dad might try to breed his own lumbering football superman at the expense of the child's IQ or its skeleton or its social graces. The prom queen might trade the same features for a daughter who had stunning looks and extreme curves. Poor coffee farmers or hard-rock miners or coral carvers might want to grow their own docile work force. They might have one or two smart leader children and make the rest idiot slaves. And who knows what pimps and psychopaths and biker parents might come up with.

The spousal midpoint will always be the moral and legal benchmark.

Laws today restrict midpoints only in the case of incest. Even here countries differ on how far two points can lie in genome space before they can produce a new midpoint genome. Pakistan tolerates a higher degree of incest than do most Western states. The future question will be to what degree will states allow parents to move their child from the genome midpoint to some preferred point in genome space.

This question will carry extra force since parents and science may not have the power to reverse a gene-change decision just as parents cannot reverse their decision to abort a fetus. But the abortion issue of life degrees does not arise here because the parents plan to bring the fetus to term. The fuzz lies in how far the child drifts from its "natural" midpoint set of genes.

States might at first pass laws that allow parents to move a child's genes away from its midpoint genome only if the new genes prevent a disease or physical defect. This would put the final say in the hands of medical committees. These groups would tend to evolve toward freer gene search as their newer members gave more weight to both the public and black-market demands for more gene search. And they would tend to evolve that way as new legal test cases pushed and stretched the fuzzy borders between healthful and less healthful genes and between parental duty and parental caprice.

States would have a harder time if they tried to stop you from changing your own genes. Right now that would also be harder for you to do. You can remove a fertilized egg and freeze it or put it in a petri dish and probe it or inject chemical signals into it. Those gene changes can pass into the DNA in all the child's future cells. You cannot do that so easily with your own genes just as you cannot easily shrink or expand the size of all rooms in an office building.

Future science will find a way to change your genes but right now we know only that the math says it is possible. We can define bumpy surfaces or "landscapes" over genome space. Each point or genome has a height value on the surface. The height may measure a genetic cost or benefit or some mix of both. The new math technique of "simulated annealing" shows how to use a controlled random search to search those bumps for the highest peaks or the lowest valleys.[19] Chapter 10 discusses this search technique.

The new field of DNA computing suggests that we can turn this around. Someday we might use the complexity of DNA molecules to conduct massive parallel searches through genome spaces and through other sets of alternatives. My U.S.C. colleague Leonard Adleman showed that we can program DNA in a test tube to both store and process digital information: "One gram of DNA, which when dry would occupy a volume of approximately one cubic centimeter, can store as much information as approximately one trillion CDs."[20] New DNA chips offer a more standard way to search genome space for better blueprints.[21]

But we still don't know what a single point looks like in human genome space. And we may debate for decades or for centuries which cost surfaces to define on our genome space and how to search their peaks and valleys.

Right now you own just one point in genome space. You have full or 100% rights to that point alone. That will change in time. You or your kin will one day own a fuzzy ball of genome points centered around your genome point of birth.

This may be a matter of patent law. Scientists and corporations have already won patents on parts of genomes. They will someday surely win patents on genomes that do not occur "naturally." Such patent holders could sue for patent infringement any firm or person who brought to life a creature or person with too close a genome.

You will own most of the genome points in your fuzzy ball but you will own them only partially. The fuzzy ball might look like a blue ball that has a dark blue dot as its center and that shades out with lighter and lighter blue dots as the dots approach its surface. You will tend to own those lighter-blue genome dots to a lesser degree than you own the darker-blue genome dots.

How far would this ball extend into genome space? The state may have to draw some hard legal lines to answer that.

Conflicts over property rights would force this. Someone else might take the average of your genome and your spouse's or of your genome and theirs. An old lover might do that by design or a stranger might do it by chance. Movie or music fans might want to mix their genomes with genomes close to those of movie stars or rock stars. Or they might want to have a child with 99% of their idol's genes.

So how big a ball do you own right now in genome space? How much can the fuzz of someone else's genome ball overlap with the fuzz of your ball? We know only that you own the genome points near the surface of your ball and only to some degree.

The real question is who else owns them.

The Rights of Whales

He that is nourished by the acorns he picked up under an oak, or the apples he gathered from the trees in the wood, has certainly appropriated them to himself. Nobody can deny but the nourishment is his. I ask then, When did they begin to be his? When he digested? Or when he ate? Or when he boiled? Or when he brought them home? Or when he picked them up? And 'tis plain if the first gathering made them not his, nothing else could. That labor put a distinction between them and common.

> JOHN LOCKE
> *The Second Treatise of Government*

What is the right of the huntsman to the forest of a thousand miles over which he has accidentally ranged in search of prey?

> JOHN QUINCY ADAMS

Equity does not permit property in land. It may be quite true that the labor a man expends in catching or gathering gives him a better right to the thing caught or gathered than any one other man. But the question at issue is whether by labor so expended he has made his right to the thing caught or gathered greater than the pre-existing rights of all *other men put together.*

> HERBERT SPENCER
> *Social Statics*

When the battle was over we bandaged our wounded with cloths, for this was all we had, and sealed the wounds of our horses with fat from the corpse of an Indian that we had cut up for this purpose.

> BERNAL DIAZ (1492–1580)
> *The Conquest of New Spain*

The question is commonly thought of as one in which A inflicts harm on B and what has to be decided is: How should we restrain A? But this is wrong. We are dealing with a problem of a reciprocal nature. To avoid the harm to B would inflict harm on A. The real question that has to be decided is: Should A be allowed to harm B or should B be allowed to harm A? The problem is to avoid the more serious harm.

RONALD H. COASE
"The Problem of Social Cost"
Journal of Law & Economics, vol. 3, October 1960

Article I

The States Parties to this Treaty undertake to prohibit and prevent, at any place, any environmental or geophysical modification activity as a weapon of war;

Article II

In this Treaty, the term "environmental or geophysical modification activity" includes any of the following activities:

(1) any weather modification activity . . . designed to increase or decrease precipitation, increase or suppress hail, lightning, or fog, and divert or direct storm systems;
(2) any climate modification which has as a purpose, or has as one of its principal effects, a change in the long-term atmospheric conditions over any part of the earth's surface;
(3) any earthquake modification activity which has a purpose, or has as one of its principal effects, the release of strain energy instability within the solid rock layers beneath the earth's crust;
(4) any ocean modification activity which has a purpose, or has as one of its principal effects, a change in the ocean currents or the creation of a seismic disturbance of the ocean (tidal wave).

U.S. Senate Resolution 71 on Environmental Warfare
Report Number 93-270
Passed 11 July 1973

The world is nearly all parceled out, and what there is left of it is being divided up, conquered, and colonized. To think of these stars that you see overhead at night, these vast worlds which we can never reach. I would annex the planets if I could.

CECIL RHODES
Last Will and Testament (1902)

*The material factors which ultimately limit the expansion of a techni-
cally advanced species are the supply of matter and the supply of energy.*

FREEMAN DYSON
"Search for Artificial Stellar Sources of Infrared Radiation"
Science: vol. 131, 3 June 1960

W HO OWNS THE MOON?
We all do. No one does. The United Nations does. Whoever
signed the Moon Treaty of 1979 does. Whoever gets there first owns it.

The idea is that the owner owns the moon all or none. Note that the
ownership is both legal and binary. The property rights are binary. So someone has to draw lines through them and through the moon. But who draws
the lines? And should they draw them with a fine pen or with a spray can?

We can ask these questions of all property rights. Property rights are relative or even arbitrary. The Cheyenne warrior of the 19th century draws
one line in the dirt and the Union soldier draws another. And property
rights are often vague or fuzzy to a greater degree than it may help to
admit. Hunting deer on land involves less control or ownership of the land
than does growing corn on it.

A glance at a globe shows this tug of war between our binary mind
schemes and the smooth or fuzzy world. We cut the land masses into
almost 200 binary regions or countries. Laws and taxes change when you
cross these lines. Men with guns often patrol the lines. Countries may fight
wars when the hard lines soften or change. Zoom in closer and each country draws lines through states and counties and cities and on down to many
of their parking spaces.

The globe ranks near the digital computer as an emblem of our binary
instincts.

I had an early fuzzy experience as a child when I played with a globe. I
had just advanced from learning the 50 states on a flat wooden map of the
United States. My father would hold up one of the orange wooden states
and ask me to guess its name from its shape or guess its capital from its
name. I liked the way the states fit snugly and did not overlap. It was tidy
but the carvings were irregular. I never asked who carved up the 50 states
that way. And no one ever brought it up.

That changed when I first ran my fingers over a globe. I saw that the
United States was just one country among many on a connected land mass.

And again all the countries fit snugly even if I was not always sure which countries were in Europe and which were not.[1] But now there was a problem. The blue oceans had no clear borders. The equator cut the Pacific Ocean into the North and South Pacific but the Pacific itself had no clear border.

Where did one ocean end and the next begin?

I did not get a good answer and I still have not gotten one. It made sense to speak of the Pacific or Indian Oceans as long as you stayed clear of their borders. They were vague objects. They were fuzzy subsets of the world's surface saltwater. The fuzzy patterns were both real chunks of the world and handy tools of thought.

The oceans remain a testbed of our binary and fuzzy notions of property rights. The surface of these oceans cover about 100 million square miles. Who owns the oceans? Who owns their energy and fish and gold powder? Who owns the manganese nodules on the sea floor and the oil beneath it? What are the rights of dolphins and whales? Who owns them?

This chapter looks at these issues of fuzzy property rights. A fuzzy view cannot solve the many problems that arise when the lines of mine and thine overlap to some degree. You are on your own when your neighbor plays his stereo or trumpet too loudly.

But a fuzzy view can help understand the problems and can show where or how we might try to reduce some of the fuzz. It can also help deal with the deeper problem of how to apply the black and white truths of logic and math to the gray flux of the real world. Here fuzz tends to hurt rather than help. The oceans and fishes and mankind might be better off if we carved up the oceans into 200 or so binary volumes and allowed each country to own one volume and manage it and perhaps trade rights in it.

The central issue of property rights is how to trade them. The Coase Theorem below shows where this can lead if the property rights are not too fuzzy. But the first issue is where they come from: What creates ownership?

The Pursuit of Property: Fuzzy Labor Mixings

The self gives rise to ownership of the non-self. We acquire property rights in objects by how we find them or use them or trade for them.

British philosopher John Locke put forth this doctrine of property rights about 300 years ago. He did not invent the idea just as Aristotle did not invent the idea of binary truth. The idea of such property rights goes back for hundreds of years in British common law and in the legal traditions of other societies. Locke captured the idea in words in 1689 in his *Second Treatise of Government*. A decade later he helped write the constitution for the new British colony of Carolina. All cultures have used like notions of property rights for thousands of years even if they did not have their own John Locke to sketch where those notions came from.

Locke's view starts out in a wild "state of nature" where no one owns anything. All the land stands fallow. Then along comes Robinson Crusoe washed up naked from the sea. Crusoe has no possessions. He owns only himself. That means he can fully use and fully dispose of his labor. He controls his labor.

Crusoe uses his labor to gather and build things. He gathers fallen coconuts and palm fronds from a stand of palm trees. He takes eggs from seagull nests in the rock cliffs and takes the white pearls and flesh from the oysters in the bay. He throws a thin sharp piece of driftwood to spear reef fish and shore birds. Crusoe weaves a lean-to shelter from the palm fronds and builds a bed from tree branches and from more palm fronds. He snares wild goats for meat and tans their hides to make shirts and pants and water pouches.

Locke sees these labor mixings as the origin of private property: "The labour of his body, and the work of his hands, we may say, are properly his. Whatsoever, then, he removes out of the state that nature hath provided and left it in, he hath mixed his labor with, and joined to it something that is his own, and thereby makes it his property."[2]

Crusoe acquires more and more property as he mixes his labor with more things. He mixes it with the unowned land and fish and plants and game. His reserves build up enough that he can in time trade with the friendly natives. Trade confers ownership. He can swap some of his goat skins and pearls and powdered herbs for some of their woven cloth and arrows and spices.

Crusoe ends up with two types of owned things and both come from labor mixings. The first type are the skins and the seed pods and the other things with which he has mixed his labor. The second type are the arrows and sinew ropes and the other things for which he has traded some of his owned objects for someone else's. Gifts count as no-cost trades for one party. So Crusoe either makes something or buys it. Else he does not own it.

These two Lockean processes can continue until all traders own all things. Then people and states trade property rights with one another. This includes the labor we sell for a wage or a tip or a consulting fee. The law tries to make sure that the players keep their deals honest and live by their contracts.

Fuzziness disturbs this Lockean paradise. Robinson Crusoe may fight with the natives over who owns the palm trees or the goats or the fish and crabs in the bay. Crusoe may believe he owns these things. He has mixed his labor with them to some degree. And these things live on or pass over land that he has come to own or protect. The natives may feel the same about their labor mixings and the land they frequent.

The problem lies in the fuzz of the mixing process.

Assume you own yourself and your genome 100%. This makes the ideas simpler even if it often does not hold in practice. Slaves own themselves 0% or to only small degree. Many religious people think that God owns them and that He in effect only leases them to themselves to some degree. Most governments act as if they own their subjects to some degree. They demand part of the fruits of their labors and sometimes demand jury labor or military draft labor from them. States also demand that their subjects not take certain drugs or trade certain services or take their own lives.

Mixing your labor with nature is not a binary process. Does Crusoe own a palm tree if he sleeps beneath it or in it or if he prunes and waters it? Does he own it if he chops it down? Does he own the wild goats and rabbits and doves and the other *ferae naturae* that he shoots or snares or keeps in a pen or that come to drink the spring water he sets out for them?

How much human labor has to mix with how much of the non-human object to convey ownership?

These are clearly matters of degree. We can draw hard lines through the fuzz and get working answers in many cases. We agree that you own the wild flowers you pick on the mountain but not that you therefore own the mountain. You do not win title to the ocean if you swim through it just as you do not win title to the sky if you fly through it.[3] We do not ram or bomb the foreign fishing trawler so long as it stays with its scoop nets in the international waters at least 200 nautical miles from our shores.

Land rights may not carve up as cleanly. Most wars have involved land claims or outright land grabs. Iraq invaded Kuwait over an old land claim as well as a new thirst for more oil. Peru and Ecuador still fight skirmishes over their borders near the headwaters of the Cenepa River and have fought them since they fought each other in outright war in 1941.[4] China has warred at one time with each of its neighbors. Sikhs and Pakistanis and

Indians fight over borders that predate those the British drew for them in the days of the British Empire. The United Nations drew a line through Palestine in 1947 and gave the Israeli chunk to the Jews based on a land claim almost 3,000 years old.

There were few large land squabbles over 10,000 years ago. There were fights within small bands and blood feuds between them. Wars came only when large groups formed and claimed their Lockean property rights to the land and their labor mixings.

Nomads do not have the same need to defend the land they use just as the modern desert Bedouins do not today. You do not own land just because you walk across it. But your group or tribe or sect may feel it owns or has crossing rights to at least the part they walk across on a regular basis. These claims grow more acute and the property rights grow less clear when the land contains a rare good such as a desert oasis or a diamond field or an oil reserve or a Kauai ocean view.

Land rights can further blur through the doctrine of "multiple use." A rancher might graze his cattle on the same land where loggers cut some of the trees or where oil firms drill for oil or natural gas.

Mining can involve the most complex use of land in multiple use. Most conflicts deal with the surface of the land and what grows or moves on it. Mining deals with the fuzzy subsets of land beneath the surface in the earth's crust. How far down do you own if you own the mineral rights to a piece of land? Can you dig or drill straight down to the mantle? Can you tap the common pool of oil or gas that lies under your land and under your neighbor's?

The U.S. government had reached a working peace with the Sioux Indians in the early 1870s over the Black Hills of South Dakota. That changed when prospectors found placer gold in its streams in 1874. Thousands of men and women rushed to the Indian land for what lay beneath it and not for its prime deer and elk hunting. The Indians lost most of their land and all the gold. George Hearst's Homestake Mine in South Dakota remains to this day the largest producing gold mine in the Western Hemisphere.

The Indians in South America have not fared better. Miners and loggers have treated much of the Amazon Valley as a free good. They made deals with eager if not greedy Indian chiefs for the Indians' claims to local placer gold and mahogany forests. Then the miners dumped mercury and cyanide into the rivers as they leached gold from placer diggings and crushed ore.

Many of the Indians have become miners and loggers themselves and now mix their labor with the Amazon's wild and unowned "state of nature."[5]

Fuzzy property rights confuse ownership. This makes it hard to trade or bargain or define a contract. These actions have their own fuzzy costs and the property fuzz only compounds it. The result can be social waste and even "market failure." Then governments often step in to draw their own lines. This gives a working solution but not an optimal one.

The Coase Theorem of modern economics shows how property rights can lead to optimal resource allocations. Fuzz plays no part in it.

The Coase Theorem: A Shopper's Paradise

Ronald H. Coase was a professor of law and economics at the University of Chicago when he published "The Problem of Social Cost" in October of 1960.[6] He was still a professor there when that paper's so-called Coase Theorem won him the 1991 Nobel Prize in economics. The Royal Swedish Academy gave him the prize for the "discovery and clarification of the significance of transaction costs and property rights for the institutional structure and functioning of the economy."

Most economists win Nobel prizes for mathematical theorems or for new math tools like game theory or price expectations. Coase is the only Nobel laureate in economics who does not work with math symbols.[7] His 1960 paper was purely verbal and his theorem has a verbal form. No one seems to have captured it in cold math symbols and that may be just as well. Theorems are about pure structure. They describe the formal relations among things or among sets of things. Theorems can use any symbol scheme or even word scheme to convey that structure.

You can state the Coase Theorem in many ways and economists often do.[8] All ways show how the market process of deals and bargains and auctions leads to a social optimum or equilibrium called Pareto optimality or Pareto efficiency.[9] (Italian economist Vilfredo Pareto lived from 1848 to 1923 and helped found welfare economics.) People or any bargaining agents have no joint incentive to change a Pareto optimum once they fall into it. They cannot even bribe one another to change it.

A social state is efficient or Pareto optimal if and only if we can make no one in the society better off unless we make someone else worse off. There are no more ways for the agents to achieve unanimous gain. You are not in a Pareto optimum if you or the state can improve your lot in such a way that it does not reduce the lot of anyone else. Then the losers can persuade or bribe the winners to swap goods or other gains with them.

The idea of the Coase Theorem is that each person keeps making such selfish changes until the whole group reaches a stable Pareto optimum. The Coase Theorem states that agents or traders or people will reach a Pareto optimum if two conditions hold. The old fiction of "perfect competition" is not one of the conditions.

The Coase Theorem has a one-line statement: *If property rights are well defined and if the transaction costs are zero then the market outcome is Pareto optimal.*

The first condition demands that someone draws hard lines through all things and services. Property rights are well defined if all things have full owners. That means the property rights are binary. Then the lines of mine and thine are precise to the last micron. This tends to hold for land parcels and store items and other goods that come with a bill of sale.

It does not hold in a tragedy of the commons. People own the streets or parks or beaches in common. But what does that mean? It means that no one owns them themselves. Each person can use them but no one has exclusive use to them. No one has an incentive to treat their street or local park as well as they treat their yard or car or home.

The tragedy is that they have an incentive to take the benefits of the commons and leave the cost to others. So the streets can crowd with parked or moving cars. The streets can also have potholes that cars drive around and that no one tries to fix. The parks and beaches can overfill with crowds and litter on weekends and holidays. They may lie mostly vacant the rest of the time.

The second condition demands that the cost of doing business is zero. There is no friction in the swapping process. Transaction costs are the costs of making a deal. They range from the fees you pay a tobacco shop to ship and handle a box of Cuban cigars to the new taxes and paperwork and legal constraints that a small firm faces when it hires a new worker.

Transaction costs come in many forms. Most of us have to pay a brokerage fee if we want to buy or sell a house or a share of stock or a derivatives contract. You may have to drive many miles to trade dollars for goods at a shopping mall or antique store or car dealership. The state levies import tariffs and taxes on many of these goods. You have to pay airfare and other costs before you can trade at a vacation site in Hawaii or Bonaire or New

Zealand. You also have to pay taxes on the air ticket and the hotel room and the rental car.

The costs are greater if you barter or swap goods rather than pay with money. You may have to sit through more than one swap meet until you find someone who will trade you the right set of kitchen knives for your set of baseball cards.

The Coase Theorem rests on these two ideal binary conditions. It demands that people or agents are ideal traders who can trade all goods and services at no cost.

The result is an ideal outcome. The free market works perfectly and has no market failures. No scheme of computers or governments or space aliens could beat the market outcome. Those schemes could help only one of the traders at the expense of at least one of the other traders. The Coase Theorem leads to a type of Adam Smith state of nirvana. The traders may not be perfectly happy but they are maximally efficient.

I have stated the Coase Theorem as a formal theorem to focus on its binary structure. Ronald Coase did not present his theorem this way. And most students of law or economics or business do not see it in this abstract form. They hear of the result the way Coase first showed how it applied. They hear of the "reciprocal nature" of social costs when someone pollutes the air you breathe or the water you drink.

Suppose you live in a community downstream from a glue factory. You pump your water from the stream and have built your home close to its shore for its scenic beauty. Then the glue factory gets a new owner and starts to dump glue toxins and boiled horse parts into the stream. The toxins and horse parts trespass on your rights. Or do they?

This is a classic case of market failure. The standard wisdom is that the state must step in and shut down the glue factory or impose on it a steep tax or pollution quota. Economist Arthur Cecil Pigou enshrined this view in his famous 1932 book *The Economics of Welfare.*

Coase did not agree with Pigou. Coase saw that the social costs are reciprocal and apply to all parties. The glue maker imposes a cost on you and the community. But you or the state would also impose a social cost if the glue maker shut down or produced less glue. You may be better off if the glue maker moved or reduced its output. Society may not be better off.

The Coase Theorem suggests a broader view. The question is not just how much should the state make the glue maker pay. The question is also how much should you pay the glue maker to stop polluting the stream. The members of your group might each agree to pay a fee to the glue maker to make up for his lost income. A larger group may have to tax its members to prevent holdouts and free-riders since all would gain from any deal it made with the glue maker.

The Coase Theorem has a simple proof that shows how the theorem can apply to the glue maker and other cases. Suppose property rights are well defined and all things have full owners. Suppose it costs nothing to trade or bargain.

Now suppose the social group is not in a Pareto optimum. Some people can improve their lots through actions or trades that do not harm the lots of others. Suppose John and Mary are two such people. John would be better off if he swapped some of his blue sapphires for some of Mary's gold coins. Mary feels the same about the trade. They both know the terms of the trade and would incur no cost in making it. So they trade. This happens over and over again in the social group. People improve their lots with deals and swaps and bribes and kickbacks and side payments until they can gain no more through trade.

Then the group has reached a social equilibrium.

Some persons may still have wants that others can satisfy but the wants are no longer reciprocal. Suppose John wants all of Mary's gold coins. He would trade all the blue sapphires he has left for all the gold coins she has left. But Mary has slaked her thirst for blue sapphires and wants no part of the trade. We could use force to make John better off. We could take some of Mary's gold coins and give them to John. But that would make her worse off.

This holds across the group. The agents have bargained themselves out and the swap meet comes to a halt. We could not arrange the goods to make anyone better off without making someone worse off. So the group has reached Pareto optimality. And that proves the Coase Theorem.

The case of the glue maker depends on the relative costs to the group and to the glue maker and to the public that her product serves. The stream complicates the issue because no one owns it. The group could simply sue the glue maker for pollution trespass if the group owned the stream. And the glue maker could pollute it as she pleased if she fully owned it. This suggests that many market failures might turn to successes if the lines of private property were less fuzzy.

Economists have often pointed to property or contract fuzz to explain why the Coase Theorem may not apply to a market failure. The textbook case is the apple farmer and the bees. I first heard this fable as I worked on my degree in economics even though some economists had already refuted it.[10] The apple farmer needs bees to pollinate his apple blossoms. So he gets a free ride from the hives of nearby beekeepers. Beekeepers also get a free ride when their bees carry the nectar from nearby fields and orchards. The market fails to achieve an efficient outcome.

A look at rural phone books shows that the Coase Theorem often still holds here. Beekeepers contract out their hives to farmers who grow apples or almonds or clover. The contracts state where and when the beekeepers will move their hives in the orchards or clover fields. The contracts are not perfect and so they are somewhat fuzzy. Nearby farmers can gain from stray hive bees as well as from the wild bees that attend most orchards and fields. Almond growers in California need so many bees that they have to rent hives from out of state.

The world contains many exceptions to the Coase Theorem. It contains in a fuzzy sense only exceptions. Each person we see or hear or smell imposes social costs and benefits on us. We may delight at the sight of a woman's face and hair or at the sound of her laugh. Or we may wince at her choice of earrings or nail polish or perfume. Our neighbor's car pollutes our air to some slight degree just as each tree in our neighbor's yard cleans our air to some degree (unless it burns or rots and gives up its carbon dioxide).

These costs and benefits are too slight to sue over. They are too slight to draw up and enforce contracts that will trade them. The transactions costs are too high. The state may impose laws of nuisance liability to deal with extreme cases as when someone yells "Shark!" at a public beach or "Fire!" in a theater. But none of us takes out an insurance policy to cover such damage. And the costs and benefits also suffer from fuzzy and poorly defined property rights.

Yet these tiny costs and benefits shape the quality of our lives. They affect our tallies of well being when we sum our likes and dislikes and our pleasures and our pains. The Coase Theorem may hold when we shop for a new car or a job or when we trade coins for potato chips or a can of diet cola in a vending machine. It may hold only slightly or not at all when we chat with friends at a party or drive past storefronts and billboard signs or listen to the neighbor's lawnmower.

We live in a world of fuzzy rights and fuzzy costs and benefits. We need a Coase Theorem to help trade those fuzzy rights or to guide us in how to defuzzify some of them. We need a Coase Theorem that better matches fact.

A Fuzzy Coase Theorem

The Coase Theorem is a binary theorem.

It faces all the problems that each equation in science and engineering faces when it tries to match logic to fact. We are more likely to detect these problems with the Coase Theorem because it applies to us and ours and not to the distant atoms of physics or the ghostly 0 and 1 bits of electrical engineering. The Coase Theorem constrains what the science-fiction writer can say about future societies and what the political ideologue can say about our own. Or at least it should.

But the theorem rightly has its critics.

Some economists have claimed that the Coase Theorem proves too much or nothing at all.[11] It just says perfect bargainers will perfectly bargain. No one knows all things about all people and their trades. The ideal assumptions never hold in practice. But that is the lot of any theorem. And you get no more from the conclusion of a theorem than you pack into its premises.

Other critics push this further. They feel that all math theorems are empty tautologies of the form "A is A." So the theorems say nothing about the real world.[12] The logic never matches fact.

These claims reach far beyond the Coase Theorem. They challenge the de facto scientific method of using math to model the real world. I have called this the *mismatch problem:* We use black and white math and logic to model a fuzzy and sloppy and gray world. Scientists know that the best equations of physics are only approximations cast in the shorthand of math symbols. Their accuracy is always a matter of degree.

But our binary symbol schemes admit no degrees.

Fuzzy logic shows a way to soften the edges of this mismatch problem. I will present it first in pure logic and then show how it can apply to the Coase Theorem and its related problems of rights and bargains. This matching logic holds just as well for a statement of physics or spectral analysis or materials science.

The simplest logic scheme is the *modus ponens* rule of inference: If the apple is ripe then I will eat the apple. This apple is ripe. So I will eat the apple. *Modus ponens* has the same form for all symbols. If *P* then *Q*. *P* is true. Then *Q* is true. It moves us down one binary step of the logic ladder. The truth of the premise flows to the truth of the conclusion in a valid inference. The Coase Theorem has this if-then form and so do all theorems. A theorem says a result is true if a premise is true.[13] The premise itself may consist of many statements.

But what if the premise is only partly true? All apples are ripe and not ripe to some degree. What if the apple is only 60% ripe or 70% ripe? Then how true is the conclusion?

Fuzzy logic shows how the truth of the conclusion falls as the truth of the premise falls. The relation "If *P* then *Q*" need not hold 100%. And the lone assertion *"P* is true" need not hold 100%.

Suppose this statement is true 90%: "If the apple is ripe then I will eat the whole apple." Suppose you pick an apple and judge that it is at least 70% ripe. So the statement "The apple is ripe" is true at least 70%. Its truth value or truth score is at least 70%. The statement may be true to a greater degree but you are not sure. Then how true is the statement "I will eat the whole apple"?

A theorem of fuzzy logic says it is true at least 60%.[14] You can conclude that the truth lies somewhere between 60% and 100%. The conclusion has less meaning or content as this truth gap grows wider. It would have no content if we could conclude only that the theorem was true at least 0% because all statements have a truth value somewhere between 0% and 100%. It would have the most content if the theorem were true at least 100% or nearly 100%. The truth gap grows as the truth score of the premise falls.

Fuzzy logic also shows what to conclude when the dual rule of inference *modus tollens* holds only to some degree. This rule denies or negates the then-part or consequent of the if-then statement. If the apple is ripe then I will eat the whole apple. I do not eat the whole apple. So the apple was not ripe. This too has a more general form. If *P* then *Q*. *Q* is not true. So *P* is not true.

Suppose again that the statement "If the apple is ripe then I will eat the whole apple" is true 90%. But now suppose I eat only 60% or less of the apple. So "I will eat the whole apple" is true at most 60%. Perhaps I eat mostly the ripe parts of the apple. Then how true is the statement "The apple is ripe"?

A second theorem of fuzzy logic says it is true at most 70%.[15] The conclusion "I will eat the whole apple" would be true only at most 30% if the statement "I will eat the whole apple" were true only at most 20%.

The world had to wait over 2,000 years for these two theorems of multivalued logic. There was no formal progress in the field of approximate reasoning from Aristotle to the first part of the 20th century. That was because scientists made almost no formal progress in black-white reasoning for most of that time.

English mathematician Augustus De Morgan did not publish his book *Formal Logic* until 1847. Then English mathematician George Boole published *An Investigation of the Laws of Thought* in 1854. These two books formed the basis of modern symbolic logic and cast it in strict black and white terms.

The term Boolean logic now means the same thing as the terms binary or bivalent or 2-valued logic. The focus on just the two truth values *true* and *false* or 1 and 0 led others in time to explore the math of three or more truth values and thus to explore the first forms of fuzzy logic.

The closest thing to fuzzy logic was the gambling math of chance or probability. That did not arrive until after Columbus landed in the New World. The Italian gambler Gerolamo Cardono wrote his *Book on Games of Chance* sometime in the early 1500s. The book was the first math text on probability and did not appear in print until 1663.[16] Yet even the modern math of probability does not address partial matches of logic and fact. It just puts odds on whether black-white events occur as when we say that "There is a 30% chance tonight that the pipes will burst."

Randomness differs from vagueness.

Whether the pipes burst differs from how much or to what degree they burst. The random view sees the burst event as an all-or-none affair and gambles on it. The vague or fuzzy view see all pipes as burst pipes to some degree and as unburst pipes to some degree. The two degrees just have to sum to 100%. You can hold a slightly burst pipe in your hand and for all practical purposes be certain that the pipe has burst. Then the "randomness" has vanished but the vague object remains in your hand.

Language can confuse this difference. We might say that the odds are 80% that the apple is ripe and mean that the red apple is 80% ripe and that it is not ripe 20%. But the statement's logic does not say this. It demands that the apple is either all ripe or all unripe or all not-ripe. It says that the odds are 80% that the apple is 100% ripe and that the odds are 20% that the apple is 0% ripe.[17]

There is nothing random about a partially ripe apple that sits in a barrel. And there is nothing vague about whether an all-ripe apple sits there.

The upshot of these two theorems of fuzzy logic is what we have all assumed. The result or then-part of a theorem or math model holds to some degree if the if-part or premise of the theorem holds to some degree. Some of the binary structure can apply to our gray world and gray word schemes. We cannot compute the exact bounds in most cases. But it is good to know that they are there when we trust the physics of a roller coaster or the biochemistry of laser eye surgery.

We can now state a fuzzy Coase Theorem: *If most property rights are well defined and if transaction costs are small then the market outcome is approximately Pareto optimal.*

We can view this statement as a formal relation. The fuzzy terms *most* and *small* stand for partial truth values and the if-then claim itself has a fuzzy weight. Or we can view the statement as the binary Coase Theorem but with fuzzy slots that let us plug in how well a real set of premises matches the theorem's premises. This view lets us keep the 100% certainty of the if-then claim of the binary version.[18]

Partial matches at the if-part lead to partial results in the then-part. So the market outcome may not be perfect. It will tend to degrade as fewer property rights are well defined and as transaction costs rise. People can still bargain themselves into near-Pareto optimality so long as the premises are not too fuzzy. This shows how the Coase Theorem tends to break down and suggests ways to improve it.

This fuzzy Coase Theorem ranges between two extremes. The binary extreme gives the laissez-faire paradise of the Coase Theorem that may never hold in practice. The other extreme is the fuzzy communal case. People here own little or nothing if at all. Then the theorem need not lead to anything and likely will not lead to an efficient Pareto outcome. The same empty result would hold if transaction costs were so high that no one could afford to trade what they owned for what they wanted.

Fuzzy Pareto optima lie between these extremes. These include the social states where you breathe your neighbor's grill smoke or pipe smoke or have to listen to him play the drums or listen to his dog bark. Your property rights in the air are too fuzzy and too slight to prevent this and it may not be worth the bother of trying to change that. It also includes social states where you do not trade with someone in a distant city or country because the transaction costs are too high.

Transaction costs tend to fall with the onset of new technology.

Radio and TV helped firms compete head to head to sell products. They did not give perfect price competition. People still had to wade through the hype of commercials. Home shopping networks and the Internet give something closer to pure price competition. You can search all airlines for the flight you want or run your own math routines to watch the stock or bond you want and then enter your order to buy or sell.

The Internet also lets people share price data on the same goods in distant countries. It lets the New York banker hire someone in India or Africa or Chile to write software or process data or manage local accounts. This brings more traders into the market and extends the reach and richness of a fuzzy Coase Theorem. That can help squeeze some of the fuzz out of the theorem as agents evolve new working patterns of rights. Software contracts tend to rely more on judgment than the letter of the law when they state who will perform what or who owns which lines of code.

The real advance of a fuzzy Coase Theorem is that it shows how fuzz corrupts the binary theorem. And that may suggest ways to remove some of the fuzz.

Fuzz acts as a type of noise in the market process. A little bit of fuzz in property rights can help smooth edges between mine and thine. But too much fuzz in property rights impedes the exchange of rights. Digital computers work best if we can round off a noisy signal to a fully present pulse of energy or a fully absent pulse of energy. That can ignore some signals but it can also streamline the computing process. The fuzzy Coase Theorem suggests that we should explore a like digital scheme in the legal world of rights.

The more fuzz we can reduce the closer we can move to a Pareto optimal outcome. We do not have this luxury with a theorem that describes a physical process. We cannot change nature to make a theorem more accurate.

But we can draw lines through some of the fuzz of property rights to make the fuzzy Coase Theorem more accurate. The gains may be worth the costs.

Defuzzify to Optimize

Social outcomes might improve if we defuzzify some property rights.

Public auctions are one way to do this. The Federal Communications Commission began the biggest auction in history in December of 1994.[19] It put on sale a small chunk of the electromagnetic frequency spectrum in the megahertz range for use in wireless personal communication systems that control pagers and telephones and faxes. Large communications firms hired game theorists to help them cast their bids. The state reaped billions of dollars as a new set of binary property rights fell into private hands.

Society did not fare so well in the case of mining in the United States. President Ulysses S. Grant signed the General Mining Act in 1872 that still lets firms and persons stake a mining claim for only a few dollars an acre on federal land. Some mining claims go for as little as $2.50 an acre for a placer or gravel mine and $5 an acre for a lode or hard-rock mine.

The federal government owns about a third of the land in the United States. Mining claims account for about one-fourth of 1 percent of the total U.S. surface area. Grant meant to spur the homesteading of the West. He saw mining as the "highest and best" use of much of the land despite its many costs to the land and the water and even to the air.[20] But Grant spurred much more than the Coase Theorem would have others bargain for.

Firms have mined about half a trillion dollars' worth of minerals on federal land since 1872. They have not paid federal royalties on an ounce of it. They would have had to pay a state royalty for the same mine on land that any of the 50 states owned. And miners have to pay a 12.5% royalty for coal or oil or gas they mine on federal land. Even the ancient Greek city-state of Athens charged a 4% royalty on the silver mines it privatized around 500 B.C.[21] The Spanish crown demanded a full 20% royalty for the silver and gold mines it privatized in the New World.

Fuzzy property rights in the West lie behind many such cases of "cowboy socialism."[22] Again multiple-use rights compound the fuzz. Loggers have bought the rights to cut trees at a fraction of what they would have had to pay in a private market. Many logging firms have gone further and used federal tax money to pay for logging roads and timber waterways to help them gain more Lockean labor mixings. Ranchers still pay less than $2 a head per month to graze cattle on the federal commons. It costs many times that to graze a steer on private land or to feed a dog in your backyard.

A fuzzy Coase Theorem would better apply in these cases if the state opened such land gifts to public auction for private rights in them. The auctions could follow the example of the auction for scarce frequency spectrum. And the auctions need not bar foreign traders from bidding.

We can also defuzzify some of the fuzzy commons in the air. The United States moved toward this when in 1992 it first allowed some polluters to trade in "smog credits."[23] The law passed as an amendment to the Clean Air Act of 1990. The law uses market competition to reduce the corrosive sulfur dioxide in the air that burns eyes and noses and lungs. The law gives major polluters a budget of smog credits. Each credit lets its bearer emit a ton of sulfur dioxide. The state makes a firm pay steep fines for each ton of sulfur dioxide that it emits beyond its credit limit.

The binary rights let firms compete for air quality. Cleaner firms can sell their extra smog credits to firms that pollute more. The cleaner firm strives to further reduce its smog so it can sell more of the credits. The less-clean firm tries to reduce its smog so it has to buy fewer of them. The success with trading in sulfur dioxide credits has led some researchers to call for a like market system for the far more ambitious goal of cutting global greenhouse gas emissions.[24]

Smog trading stands in sharp contrast to most state attempts to reduce smog. Mexico City passed its "Today Don't Drive" law in 1989 to keep certain private cars off the street on certain days. Studies showed that the law did not reduce the city's famous smog and may have made it worse. Smart agents bargained their way around the law. Most people bought a new car to use on their off days or they took a taxi. The new cars and the taxis used as much gas as the banned cars would have used. The number of cars on the street grew from 2 million in 1989 to almost 3 million in 1995.[25]

Countries may someday use smog trading to deal with greenhouse gases like carbon dioxide. They might also use like trading in dumping credits to govern how firms or even states themselves dump waste and toxins into the oceans. The Coase Theorem does not care who gets what property rights and it does not demand the binary ideal of perfect competition. It just cares that someone or some state gets them and with as little fuzz as possible. An owner has the best chance of matching costs to benefits.

This suggests that some form of defuzzification might work on a more massive scale.

The Rights of Whales

The law of the sea says that all ships may use the high seas. It has always been cheaper to move goods over water than over land. That is why wealth and populations concentrate along coasts. The law of the sea reflects this. Ships can sail and dump and fish at will. They own what they catch because they have mixed their labor with the unowned common sea and its bounty. So the ships overfish and pollute at will.

Can we defuzzify this tragedy of the commons? Will the whales go the way of woolly mammoths? DNA evidence suggests that these big-brained singing mammals may have reached a cultural level of social learning that has affected their genetic diversity.[26]

Consider first Arthur C. Clarke's whaling scheme in his 1957 science-fiction novel *The Deep Range*.[27] Clarke had a socialist world government put gated ultrasonic fences in the ocean to corral the whales in deep 3-D ranges. The world state ranched the whales as a source of diet protein. Clarke used a world state to ignore the hard problem of who owns the seas. But his idea of ultrasonic fences is a clever way of showing who could rent or trade them. The fences draw hard legal lines through the sea fuzz.

Aqua-farms are an early form of such fenced-off volumes of sea. Many small and large firms now raise shrimp or clams or abalone or salmon or other fish on small closed tracts of the sea coast. Thailand and Ecuador lead the world in shrimp aqua-farming and sometimes in aqua-polluting.[28] Aqua-farming is the fastest growing form of agriculture. It accounts for something like a fourth of the fish that the world consumes. More aqua-farms will appear as fishing pressure grows and as wild fish stocks shrink.

Countries have also viewed their close coastal waters as their own and allowed their citizens to fish them. The fuzzy question is how close is close enough to claim rights to these fisheries.

Most countries laid claim to the first 12 miles of ocean off their coasts. Increased fishing pressure led the United States to claim in 1977 a fishery border that lies 200 nautical miles off its coast. The maritime Law of the Sea adopted the 200-mile border in 1983 as an "exclusive economic zone" for all oceanfront countries. This grew out of the United Nations Convention on the Law of the Sea of 10 December 1982 and has more moral force than legal force. Even peaceful Canada has had armed skirmishes with Spanish and other fishing fleets over the fuzz in its 200-mile sea border. The nations of the earth protect less than 1% of the marine environment. Many environmentalists think we should set aside as much as 20% of the marine environment to deal with pollution and to help manage fishing.[29]

So right now many countries have the legal right to set up their own fenced-off deep ranges within the 200-mile borders. They may need a few breakthroughs in ultrasonic gates before they can auction off some of their deep range. Or states can go ahead and draw hard lines through their new 200-mile chunks of sea and then trade fishing credits or dumping credits.

Again the Coase Theorem does not care who owns the sea chunks or the fishing credits.

A country could even divide the sea chunks into tiny units and give one or two to each citizen to trade. Most citizens would sell their fishing rights to fishing firms or to fishing mutual funds. Something like this happened in the new governments of Russia and Poland and the Czech Republic when they broke up the old state-run firms held in common. The state gave each citizen a property voucher in a state-run firm. The citizen could keep the voucher or sell it.

Vouchers would help only where a country owns part of the sea. They would not help for the greater part of the sea that still lies there for the taking and the polluting. The United Nations might vote to take control of it and then give out sea-volume vouchers to member nations. A fuzzy tax form in some member nations could help pay for these vouchers or help pay for the naval force that enforced them.

A fuzzy tax form could also help wealthy countries fund debt-for-nature swaps with poorer countries.[30] Today less than 5% of the earth's land surface enjoys legal protection and still less of its coastal waters do. Germans or Americans or Britons could direct some of their tax bill to international stewards such as Conservation International or the World Wildlife Fund. The stewards would buy foreign debt as they do now at a deep discount in exchange for some degree of property right in eco-hot spots. These regions could include coastal sea volumes as well as large tracts of forest or jungle or desert.

States could also defuzzify some of their common lands. The United States could do that with some of what remains of the Wild West. The federal government owns about half of the land west of the Rocky Mountains. Russia could do this with some of the great stretches of Siberia to help control the worldwide rush to exploit its oil and forests and precious metals and diamonds.

States could slice large land tracts into smaller land tracts and give each citizen his own tract or give him just the mineral or grazing or logging rights on it. One fuzzy rule might be to give logging rights to some forest tracts but allow the owner to cut only every third tree or every tenth tree. Strip or layered logging would offer like fuzzy schemes. The binary property rights would let each person make decisions based on all the opportunities the land offered.[31] The evidence is that the self-interest of long-term investments leads to more efficient and sustainable use of natural resources. Mexico loses about 1% of its forests each year while more careful forest management in the United States has led to a 30% increase in timber volume in the last 50 years.[32]

The binary slices would trade accuracy for simplicity. They would trade the fuzz of who owns what for a simple scheme that lets people swap and bargain their way to a more efficient world. They would give better living through defuzzification.

You might even get to own your own whale.

The Japanese have made the first move toward tracking lone whales across the ocean and thus have cracked the fuzzy door to a property right in them. The Japanese plan to launch a whale satellite. It will track blue whales that have sensor-filled harpoons stuck in them.[33] Future sensors and satellite eyes will keep better and better track of these large mammals. Ecologists have also used genetic tags or markers to help identify and track humpback whales.[34]

The United Nations might divide these "free-range" whales among the 200 or so nations of the earth. Then nations could trade limited whale rights as they can now trade fishing rights in their coastal waters. New Zealand has pioneered the use of such Individual Transferable Quotas for fish harvests. The Icelandic taste for whale meat would give owners a further reason to watch and grow their herds.

The distant future may also let us boost the IQ of some of our whales with chip implants or smart genes. This is the science-fiction dream of "uplifting" a species to higher intelligence.[35] Would you eat a cow or a chicken if it had the IQ of a chimpanzee or a 10-year-old child? Where do you draw the rights line through the spectrum of intelligence? How high of an IQ is high enough for a creature to secure its own Lockean rights?

The rights of whales may come from our fuzzy rights in them.

The Sky Is Not the Limit: The Greening of the Galaxy

A fuzzy Coase Theorem need not stop at the land or the seas.

John Locke called the ocean "that great and still remaining common of mankind." That also holds for the sky and the satellite belt and the moon and the whole solar system. Ownership seems to fall with distance from the earth's surface just as it seems to fall with distance from our skin. The United States and Russia and other countries have had no qualms about dumping thousands of pieces of space junk into space and high earth orbit.

The sky has so far been the limit of our fuzzy property rights. Airplanes and air forces gave rise to the countries' claims to the "air space" above their land. This became law of sorts after World War I when 33 countries signed the International Convention on Air Navigation in 1919.

Countries still have not agreed where the legal air ends and space begins.

The United Nations General Assembly passed a set of space laws in 1963 in a unanimous vote. The laws do not allow a country to plant a flag on a planet or asteroid or other space body and claim it for its own. Then the United States and other countries signed in 1967 the Outer Space Treaty and agreed not to use outer space for any military purpose. The United States largely ignored that agreement in its pursuit of a Star Wars or Strategic Defense Initiative in the 1980s. The Outer Space Treaty forbids Earth-based governments to lay claim to objects in space. It does not forbid private ownership but it is not clear who would enforce such property rights.

The Senate also passed a token bill in 1973 that banned any U.S. effort to change the weather or seas or ground as a weapon of war. The bill has no legal force since the House did not pass a like bill and then-President Richard Nixon did not sign it. Perhaps members of the Senate had watched repeat showings of the 1965 spy-spoof film *Our Man Flint*. There actor James Coburn saves the world from three mad scientists who have learned how to control the weather and who then blackmail the world's governments.

Changing the weather or seas does not pose a threat today but someday it could. A crazed Chinese general might set off a massive nuclear earthquake on the Chinese mainland if he thought it would lead to a tidal wave that would wipe out the coast of Taiwan or Japan. Future cloud seeding or sun-mirror schemes might allow one country to even slightly decrease the average rainfall of its neighbors. That might be enough to tilt a trade balance or to allow the country to extort payment to stop the weather change. The Coase Theorem would apply.

Some scientists have called for no-fault insurance to protect losers from winners in any large-scale scheme to change the weather.[36] And you can now buy hurricane futures on the Chicago Board of Trade to hedge against a coastal disaster. Some firms offer act-of-God bonds to insure against such large-scale disasters.[37]

Today there is little threat of large-scale weather modification. We have yet to do better than seed a few clouds with pellets of dry ice or silver iodide. The pellets produce raindrops that tend to evaporate before they hit the ground.

The United States and former Soviet Union have spent hundreds of millions of dollars trying to make rain and clear fog and suppress lightning and dampen tornadoes and hurricanes. The science was weak and the tests were on too small a scale to produce real change in the weather. The tests also faced class-action lawsuits if they worked and sometimes even if they failed. Honduras accused the United States of stealing its rain in 1973. It claimed that the United States had used cloud seeding to lessen a Florida hurricane.[38]

Chaos theory offers little hope for controlled weather change. Chaos occurs when small changes to a system's input leads to large changes in its output. Small changes in a cloud or fog column or sea surface can lead to large and unforeseen changes in the weather. The hurricane may grow and head in the wrong direction. The light rain might turn to a flooding torrent or the tornado may split into two. Chaos theory would favor using weather change more for war than for peace. But even here no one could be sure what they would get. They would always face the uncertainty that the pool player faces before he breaks the racked balls.

The old drive to set up colonies will further push the borders of many fuzzy property rights as it has throughout history.

The sea remains the best place to start once you have given up on the surface land and ice caps. The private home ownerships on land today may someday find themselves afloat in the Caribbean.[39] Futurists have put forth many schemes for setting up new colonies or tiny states at sea. The Law of the Sea requires that they lie at least 200 miles offshore unless a coastal state grants access to such a colony.

"Free Oceania" is the latest scheme to build a city at sea and a tax haven for all. Oceania would float far out in the Caribbean Sea where few hurricanes pass. Its sponsors have shown off their design as a toy model in Las Vegas and based it on the design of real floating hotels in the waters of Australia. The sponsors have yet to raise the first billion dollars they need to launch the project. Such a scheme would have to deal with corrosion and harsh weather and extreme changes in pressure.[40]

A floating city would also have to deal with the threat of military invasion from any of the coastal countries in the Americas. A floating city would need its own security system to protect its citizens and property. The United States or Mexico might see such a floating city as a security threat or a health risk or a haven for drugs and crime. These or other countries might one day launch a preemptive strike if the floating city had its own stockpile of smart weapons.

The so-called Millennial Project has far more ambitious plans for sea colonies. It sees them as the first of eight steps in colonizing both the solar system and the stars in our Milky Galaxy in the next 1,000 years. The sea colonies could produce renewable energy from cold and warm ocean water and could grow tons of protein from blue-green algae.[41]

Other steps would build large eco-spheres in space and sculpt domed colonies in moon craters. Then the plan calls for making good on one of the oldest dreams of NASA and science fiction. It would heat up Mars to melt its polar ice caps and in time terra-form the cold Red Planet into a warm blue-green Eden. Humans and robots would have a new world where they could forge their Lockean rights and bargain their way into new Pareto equilibria.

The last steps of the Millennial Project call for putting a lampshade over our Sun and over as many of the hundreds of billions of stars in the Milky Way galaxy with which we can mix our labor. Physicist Freeman Dyson long ago proposed that an advanced species would take most of the mass of its solar system and recast it in a thin sphere or cloud around its star. Such a "Dyson sphere" would make the best use of the star's energy and shift its light toward the red end of the spectrum.

Dyson proposed that we search for alien empires in the sky by searching for such red-shifted or even green stars. Man's property claims would expand outward in degrees from our solar system in a type of greedy sphere. Solar systems would give way to robot-made Dyson spheres and could lead to the "greening of the galaxy."[42]

No telescope has ever found such a green star or star system. That may be the best evidence that there are no advanced alien cultures in this corner of the universe.

And that means that no one else may own it. The bulk of the universe may lie fallow for us and our heirs. It may be our great Lockean "remaining common." Imagine a whole galaxy or the whole universe cut into binary or fuzzy pieces of mine and thine. The Coase Theorem would move it from one efficient outcome to the next. The empty vacuum of space would house one massive shopping mall.

But first things first. We have trillions of gray areas here on Earth and over our backyards that we would have to defuzzify to fully invoke the Coase Theorem.

We still do not know who owns an e-mail message.

Smart Wars

War is nothing but a duel on an extensive scale. War is a mere continuation of politics by other means.

> KARL VON CLAUSEWITZ
> *On War*

All warfare is based on deception.

> SUN TZU
> *The Art of War*

If I can deceive my own friends I can make certain of deceiving the enemy.

> GENERAL STONEWALL JACKSON

One had only to do the unexpected thing quickly. This was the secret of successful war.

> JOHN STEINBECK
> *Cup of Gold*

Great generals strike where they are least expected against opposition that is weak and disorganized. The successful general chooses the line or course of least expectation and he exploits the line of least resistance.

> BEVIN ALEXANDER
> *How Great Generals Win*

The strength of an army, like the power in mechanics, is estimated by multiplying the mass by the rapidity. A rapid march augments the morale of an army and increases all the chances of victory.

> NAPOLEON BONAPARTE

The movement of humanity, arising as it does from an infinite number of human wills, proceeds continuously. Only by taking an infinitesimally small unit of observation—the differentia of history, the homogenous tendencies of men—and attaining to the art of integrating them (taking the sum of these infinitesimals), can we hope to arrive at the laws of history. It was the sum of men's individual wills that produced the French Revolution and Napoleon, and only the sum of those wills that first tolerated and then destroyed them.

> LEO TOLSTOY
> *War and Peace*

Given all the independent variables that come into play, grand strategy can never be exact or foreordained. It relies rather upon the constant and intelligent reassessment of the polity's ends and means. It relies upon wisdom and judgment.

> PAUL KENNEDY
> *Grand Strategies in War and Peace*

American military superiority on the conventional battlefield pushes its adversaries toward unconventional alternatives.

> ASHTON CARTER, JOHN DEUTCH, and PHILIP ZELIKOW
> "Catastrophic Terrorism"
> *Foreign Affairs:* November 1998

Information warfare is about power. He who controls the information controls the money.

> WINN SCHWARTAU
> *Information Warfare: Chaos on the Electronic Superhighway*

In war events of importance result from trivial causes.

> JULIUS CAESAR
> *De Bello Gallico*

YOU CANNOT DODGE a bullet or shoot one down. A shield or building can block it. Or the bullet can run out of energy before it hits you. But you cannot block it yourself or outrun it.

It costs more to defend against a bullet than it costs to attack someone with one.

Cheap bullets and revolvers helped destabilize the Wild West of legend if not the West of fact. God made men and Colonel Colt made them equal. The Colonel also made quite a few of them trigger-happy. A drunk with a gun could end your gene line with one hand and end it at a distance. The best you could have done was hope he missed or try to end his gene line first.

The future of war may involve a like loss of stability. Smart cruise missiles and their heirs will shape future warfare as the revolver helped shape the Wild West. They will not cost much more in relative terms than did the first revolvers. And it will be as hard to shoot down a swarm of them as it is to use a rifle to shoot down a swarm of gnats. Cruise missiles will be the first weapon of choice in a smart war.

Fuzzy logic may help bring this about.

Fuzz lies at the core of the concept of war. And fuzzy techniques will find their way into how we model battles and wars just as they will find their way into the nose cones and machine IQs of cruise missiles. Technology drives the structure of war now more than ever: faster and cheaper and smaller and smarter.

Technology will someday drive the structure of war right past the edge of chaos: How can you shoot down a bullet with a bullet?

The Fuzz of War

What is war?

Where is the line between war and not-war? Does a war or battle start with an attack or with a threat to attack? Does it end with a treaty or with the death of the last warrior?

All the concepts are fuzzy. War is fighting on a large scale. A battle is a war stage or war chunk. It is a fight on a smaller or shorter scale. How large is large and how small is small? Do the borders of conflict lie in land or in air or somewhere in a wide area computer network?

Prussian General Karl von Clausewitz defined a battle as a "conflict waged with all our forces for the attainment of a decisive victory." He wrote in the early 1800s and meant crowds of red soldiers marching against crowds of blue soldiers. Soldiers had just begun to use clocks and still used bad maps and crude signal schemes of mirrors and messengers.[1] Men and machines need not wear uniforms to fight modern battles. And states can wage conflicts even if they do not declare war on one another.

The fog of war extends to the fuzz of war.

Suppose Iran or Italy uses the phone lines to crash the New York Stock Exchange or to start a run on U.S. banks. Is that an act of war?

What if hackers in Mexico City wipe out the air-traffic control system at Los Angeles Airport? What if they wipe out the control systems in other airports and then hundreds of planes crash? How many airports would have to close or planes have to crash before the United States threatened Mexico with force to stop or catch or turn over the hackers?

Modern warfare has begun to shift more from metal and chemistry to information. It has shifted more from *atoms* to *bits* in the language of MIT's Nicholas Negroponte.[2] And that means no country is safe from the information warfare of any other.

Wars used to occur in large part along coastal regions but even that no longer holds.[3] A terrorist can use the Internet or satellite system or wireless networks to reach out and touch someone on the other side of the planet. A small country can hack into or deceive or subvert the intelligence of a larger country. The attack force does not have to depend on roads or stores of food and water. The attack leader need not be in eyesight of the battle.

The point is that the cost of attacking will fall. A country can fight with bits far more cheaply than it can fight with atoms.

This shift in cost has also helped make the nuclear bomb obsolete as a strategic tool.[4] A country can still threaten to level a city with a nuclear bomb but it does not need one to take out a thick bunker or command center. A smart enough missile can use TNT to take out a command center if it can place its ordnance at the right doorstep.

It all depends on the size of the kill circle.

Each missile has a CEP or circular error probable or kill circle. A 95% CEP is a circle or ellipse such that it is 95% likely that the missile will hit within it. Military planners often call CEPs "cookie cutter lethality functions." They plan strikes by covering target regions with CEPs that overlap. Most people who live or work in large cities like London or Los Angeles or Tokyo can be sure that a nuclear planner has drawn a CEP that includes them.

CEPs have shrunk in size over the years as chips have shrunk in their circuit design. The kill circles have shrunk from hundreds of feet in diameter to just a few feet.[5] Old nuclear missiles had CEPs that measured in the hundreds of yards or even larger. The missiles had to explode in great bursts of energy to make 95% sure that they destroyed their strategic target.[6]

Shrinking CEPs have done as much to do away with nuclear weapons as have the treaties between the United States and the former Soviet Union. They have also helped make weapons smarter and cheaper and in greater demand among poorer countries.

This involves a modern irony. War or conflict becomes more fuzzy as weapons become smarter and more precise. Poor or rich countries may rely less on diplomats to resolve disputes than on small pinpoint strikes or information sabotage or stealthy tit-for-tat volleys. Most countries do not go to war over a small border skirmish. Smart pinpoint strikes will push this envelope to new limits.

The shift from atoms to bits has also changed how we plan wars and model them on computers in war colleges and in aerospace firms. The pace of battle has quickened from minutes to seconds to even milliseconds or nanoseconds. And the complexity has grown to exponential levels. Warfare grows more nonlinear with each year and with each advance in computer chips and laser sensors and turbojet design.

A commander can know at best a few fuzzy rules of thumb that describe how a battle may unfold. War colleges teach vague principles of how to balance troop mobility and supply lines and ground and air forces. The rules were fuzzy in the days of Hannibal Barca and Alexander and are at least as fuzzy today if they still apply at all.

The age-old "art of war" remains an art despite the trillions of dollars that states have spent on computers and scientists and gadgets to make it more of a science. The massive war-game computer networks have no more chance of predicting battle outcomes than like networks have of predicting stock prices in a global stock market.

But they still try.

The Math of War: From Tolstoy to Chaos

The term "mathematical warfare" may suggest numbers smashing into one another or two accountants racing to add up figures. It refers in fact to the logical branch of military science.

Mathematical warfare is the mathematical analysis of all types of military conflict. Many journals publish papers on it. The Military Operations

Research Society meets each year to present the latest papers on the mathematical structure of war. Some of these abstract results have helped shape the structure of the modern military industrial complex.

The idea is to invest in math as a new type of war bond. Spend enough money on enough analysts and on enough computers to simulate enough battles and on enough prototype weapons. Then you might find a clearing in the fog of war. You might get to glimpse the first few seconds of the next battle.

The goal of mathematical warfare is to capture the complete structure of battle in a set of equations. Then leaders or graduate students could predict the outcome of a battle in advance. That also means the enemy leader could predict the same outcome with the same equations. And that would lead to a type of mathematical arms race. Each side would have to come up with a new equation that builds on the equation of the other side and then bests it.

Scientists and thinkers have searched for the math of war for hundreds of years. Newton showed why the cannonball or arrow or bullet traces out a parabola on its way back to earth or to an enemy soldier. Leonardo da Vinci solved cubic equations in like cases with methods we still do not know. Other men have looked at mobile armies or air and water fleets and drawn an analogy with physics or even fluid mechanics.

Few courses in literature teach that Count Leo Tolstoy put a military force equation in his magnum opus *War and Peace*.[7] I doubt if any courses mention that Tolstoy was an early champion and forerunner of mathematical warfare. Most of the forewords to *War and Peace* that I have seen claim that the book is about marriage. Yet Tolstoy goes to great length in the book to say what it is about. The book is a long verbal exercise in the integral calculus.[8]

Tolstoy's society moves forward in both war and peace through the *sum of wills*.

The Russian officer Tolstoy knew much of the math and science of his day and had the time to reflect on them during and after the battles of Sevastopol in the Crimean War. He had the genius to see the calculus at work in great social waves. And he had the artistic gift to sketch those waves in hundreds of men and women and their intertwined hopes and schemes and loves and fears.

War and Peace is a novel of show and tell. Tell them your great new idea or theory and then show them how it works in fiction. It belongs to the long line of advocacy novels that runs from self-reliant Daniel Defoe's *Robinson Crusoe* in 1719 to behaviorist psychologist B. F. Skinner's *Walden Two* in 1961.

Tolstoy found a new tool in the "new mathematics." He thought he had used it to show that Napoleon was more myth than military genius. Tolstoy viewed the French leader and all persons as machines or automata that move with the forces in the social wave. He refused to believe in the "great man" view of history that one man can shape events.[9]

Tolstoy was wrong.

He was right that history moves through the combined effects of billions of human wills and outside forces. He was wrong to think that that movement reduces to a simple *sum* of wills. The integral or sum of integral calculus is too simple a tool.

Tolstoy made the common mistake of the first-time math modeler. He assumed a linear model to describe a nonlinear world. Just add up all the "infinitesimal" wills in each breast and you get the macro-will of society or history. The whole is just the sum of its parts.

This does not hold in a nonlinear system. The whole is more than the sum of its parts. The whole can also be less than the sum of its parts. And some of the smallest parts can shape the whole some of the time. A great man or a common man can push the wrong button or throw the wrong stone or ask the right question and set in motion causal chains that change the world and the history books.

Chaos theorists sometimes call this the "butterfly effect" in a nonlinear system. Small changes in input can produce large changes in output. How an orange monarch butterfly flaps its wings in Mexico may in time affect how much rain falls in France. Tolstoy should have said that history proceeds as a nonlinear combination of wills. But then he could not dismiss out of hand the nonlinear effects of "great" men like Napoleon or Ghengis Khan or the reigning Pope.

Tolstoy also tried to back up his linear case with the claim of social determinism. Napoleon and kings had no choice in what they did. Their freedom fell as they gained more power over the masses until they became mere "slaves of history."

But determinism just means all events have causes.

Ants and dogs and military leaders are free to choose as they please. But they are not free to please as they please. Wants and drives change with myriad forces inside and outside the skin. That does not take Tolstoy where he wants to go. Causal determinism holds with equal force for a nonlinear system.

Napoleon may have gotten the idea to invade Russia and seize Moscow from a joke or dare at a dinner party or from a report on the military strength of his neighbors or from a random chemical reaction in his brain. The source of the choice does not matter. Only its effect does.

One fuzzy branch of Napoleon's decision tree said *Go to Moscow* and the other branch said *Don't go to Moscow.* Napoleon drew a hard line through the fuzz and for some reason went with the Moscow branch. Russian and world history would have differed if he had gone with the other branch or if he had then dropped dead of a stroke or heart attack.

Modern mathematical warfare has slowly come to see Tolstoy's linear mistake.

Its first advance came during World War I and did not become well known until after World War II. The advance itself was a nonlinear result but a mild one. It said troop strength did not grow in a linear way with the number of soldiers in a troop. It grew in a square or quadratic way.

Common sense said that a troop gained more strength from 20 more men than from just 10 more men. But how much more? Common sense ends here. A linear model would say that the new increment of troop strength doubled when it went from 10 new men to 20 new men. The math said that the increment grew from 100 to 400. It quadrupled rather than just doubled.

This is the so-called Lanchester's square law: Troop strength grows with the *square* of the number of new recruits.

British scholar Frederick W. Lanchester published his famous equations in his 1916 book *Aircraft in Warfare: The Dawn of the Fourth Arm*. The equations used simple concepts from calculus to describe the mass structure of modern warfare. The equations then imply the square law of recruit strength as a theorem. Tolstoy would have loved the equations. He could have written quite a war novel with them.

Lanchester also showed that the mass hackings and clubbings of "ancient war" lead to a linear force law.[10] These were simple bouts of attrition where greater numbers gave only a proportionate advantage.

Modern war leads to the square law. Lanchester seems to have had in mind the trench warfare of World War I. He assumed each soldier in the red army could fire on each soldier in the blue army and vice versa. This defines a complex war of attrition.

A snowball fight at a grade school gives a simple example. Each red kid can throw a snowball at each blue kid. Each red kid fears in turn that each of the blue kids will throw a snowball at her. Two new kids who join one side will give that side a fourfold increase in snow power. This effect starts to break down when the two sides grow too large for each snow thrower on one side to hit all those on the other.

Lanchester conflict is also sometimes "stochastically determined" for large groups of soldiers.[11] The actions of lone soldiers are random but large groups of soldiers have stable statistics. This formal result holds only in the ideal case where some of the stronger assumptions of random sampling theory holds. Still it has helped fund hundreds if not thousands of complex battle simulations. It is also the main idea behind Isaac Asimov's famous *Foundation Trilogy* in science fiction where future mathematicians use "psycho-history" to predict the long-term statistical structure of society. Some form of this swirls in the mind of the long-term stock investor.

Lanchester's square law has held to varying degree for trench warfare and tank and air fights and even for some of the massed battles of the American Civil War.[12] But the Battle of Iwo Jima remains the best fit between a math model and the grim kill scores of combat.

U.S. troops invaded the Pacific island of Iwo Jima on 19 February 1945. The Japanese fought them for a month almost to the last man of their 21,500 or so soldiers. Both sides attacked the other in roughly Lanchester fashion and in an isolated island setting. The battle was among the fiercest in the war.

Iwo Jima was as close as math warfare has ever come to having a lab experiment.

The math curves plot out very closely to the real mortality curves.[13] The U.S. troops received two small waves of new recruits in the first days of the battle. The Japanese received none. You can plot the number of U.S. survivors over the 36 days of the battle and see the curve jump up twice as the new recruits arrive. Then the curve slides down as enemy attrition takes its toll. The predicted curve does the same.

These two curves spurred a great deal of research in the 1950s and 1960s. Many researchers looked for similar battles where the Iwo Jima results might apply at least to some degree but no results matched those of Iwo Jima. The Iwo Jima model remains the chief predictive triumph of mathematical warfare even though it was an after-the-fact retrodiction.

The Lanchester model has not fared so well off the shores of Iwo Jima.

The Lanchester model breaks down when one side or both sides cannot fire at all the soldiers on the other side. That happens in almost all battles since they have long and complex fronts. It also happens when one side digs in and defends rather than counterattacks.

Then there is the problem of guerrilla warfare.

You can add more math to the Lanchester equations to model guerrilla warfare.[14] But the equations are in general too complex for the human mind to solve. A computer must crunch through them and still tends to spin off on tangents at each crunch.

Here each red guerrilla can fire from the jungle at each blue soldier of the conventional army. But the blue soldiers can fire only at the jungle region itself. This often held to some degree in Vietnam and held to a lesser degree in guerrilla bouts in Central America. Still more complex math can model bouts of red guerrillas versus blue guerrillas.

It all breaks down in the math: Nonlinearity grows with accuracy.

War is one of the most nonlinear processes we know. You can assume away much of its structure with simple math that we or computers can work with. The game of chess does that. Or you can use more complex models to paint more accurate pictures of the real world. A few thousand math models since Lanchester's have tried to do that. Both extremes of the trade-off have flaws and yet can show enough promise to sustain their own lines of research.

This trade-off holds in all fields of science. But mathematical warfare starts with a bigger subject than most. It starts with the shifting swarm process of war itself. Each researcher must face afresh the Von Clausewitz question: What is war? He can try to model the whole elephant or just focus on the ear or trunk or tail. The whole process is too big and complex to model well. But it is too nonlinear to model in parts.

The best you can do with Lanchester models and their kin is to capture some of the shadow of battle. The battle itself depends on too many millions or trillions of variables and a like number of parameters. The true math surface that defines it would have too many bumps and kinks in it.

So what is a math warrior to do?

Many analysts answer this with a slogan: When all else fails use a computer. If that fails then use a bigger computer. Don't worry about proving theorems. Don't let human math skill constrain the structure of the process.

Make it as real as you can. Add all the variables and time lags and nonlinear profiles you can. The computer simulation will sort them out.

War models can now include tens or even hundreds of thousands of variables and policies and command thresholds. These variables capture more structure and tend to give more weight to defense analyses.

They also give rise to chaos.[15]

Chaos means small changes in red or blue forces can change whether red or blue wins the battle. Suppose you rig a game so that red has the edge over blue. You might let red have more soldiers than blue and let them all kill with the same or nearly the same skill. You run the game and red wins.

Now you replay the game but with a change. You add a few more troops to the red side and leave the blue side as it was. Who wins this time? You would think that red would win. How could it lose if it won before and now it has even more troops?

But red may well lose to blue. Increase red's troop strength some more and then it may win again. Increase it some more and it may lose. The battle outcome can jump back and forth this way as you increase the red troop strength and keep the blue troop strength fixed.

Math warriors call this a non-monotonic effect in a combat model.

I learned of this nonlinear effect in Lanchester models when I lectured at the RAND think tank in Santa Monica. I had worked with large war models more than a decade before in aerospace and had written about their nonlinear effects but only in fiction. I had also seen chaos seep into most branches of applied math in the same decade.

I made a joke at dinner to my RAND colleagues about the danger of such large-scale models. They did not take offense at the remark or try to rebut it. They instead told me of the debate raging in the math warfare world over non-monotonic effects.[16]

The young Turks had found these and other chaotic effects while the older analysts tried to explain them away. The chaos or the so-called structural variance has always been there but analysts may have dismissed it as "noise" (random disturbance) or missed it outright. That happened with chaos in fields from the physics of gravity and airflow to the design of electrical and neural circuits.

All this suggests Tolstoy was more wrong than anyone thought.

The Lanchester combat models are among the simplest math models and the best tested. If they can spin off into chaos then so too can more complex models with more players and more policies and decision variables.

These complex models have captured at most only the rough structure of older battles. They have not predicted a real battle in advance. They have at best retrodicted the mass structure of a few simple battles in the past.

Computer simulations are no better than the math models they run.

They are often worse. The computer adds its own round-off errors and forces the battle to unfold to the beat of its own clock cycle. These effects can compound the nonlinear effects buried in the unknown structure of large-scale math models. The result can give a modern twist to the terms intractable and inscrutable. And defense spending on it stands only to grow in the 21st century.

Math itself has failed to explain warfare or even fully describe its effects. Indeed its main lesson seems to be that warfare is too chaotic to model well. But math has still helped shape warfare when we apply it to the tools of war like radar and target recognition and a bullet's ever-changing ballistic coefficient.

A small but key part of that math is fuzzy logic and its kindred modes of machine intelligence. They help boost machine IQ. They add to the smart part of what we call smart weapons.

And smart weapons pave the nonlinear way to smart wars.

Smart Wars: Cheaper to Attack Than Defend

A smart war is our smart weapons versus theirs.

The cruise missile is the archetypal smart weapon. It is a flying bomb that sees and hears and one day will think.

The Pentagon ranks the cruise missile as the number one "proliferation threat."[17] Cruise missiles fly fast and so close to the ground that most radar systems cannot tell them from the hills and trees and other "clutter." Modern Tomahawk cruise missiles can fly about 550 miles per hour and travel up to 1,000 miles. Their low flight makes them hard for ground forces to see when they fly over. A low-flying red hawk passes right over your head. The same red hawk seems to float in the blue sky when it flies much higher up. Trying to shoot one down is much like trying to hit a bullet with a bullet.

The Pentagon has two main worries about cruise missiles. The first is that they are hard to shoot down and getting harder: You can't shoot down a bullet with a bullet. The second is that they cost less each year and get smarter each year. So all nations want them and most are buying them. Quite a few nations make them. So the threat grows. It "proliferates."

The growth in cruise missiles stems from the falling cost of what the military calls COTS or commercial off-the-shelf items.

The fuzzy *Anarchist's Cookbook* has a new entry. You can build your own "dumb" cruise missile right now for just a few thousand dollars. Start with a small airplane like a Cessna and a good TV or video camera. Add to it a handheld GPS navigator from a fishing boat and a few tons of TNT and you have a poor man's cruise missile.[18]

That suggests the Pentagon misspent billions of dollars on its Star Wars or Strategic Defense Initiative and its outdated Cold War threat of hundreds of ballistic missiles in the heavens. A swarm of low-flying cruise missiles could beat any trillion-dollar Star Wars shield. The Reagan–Bush Star Wars program tried to build a roof on the house of security but forgot the sides. Star Wars also helped kill funding for the Air Defense Initiative or Air Wars program in the late 1980s that would have at least studied the cruise missile threat.

The cost of cruise missiles keeps falling because computer chips keep getting denser and cheaper and faster.

This stems from the so-called Moore's Law of chip density that we will look at in detail in chapter 15. Moore's Law is a trend that has held in some form since the advent of the microprocessor in the early 1970s. Gordon Moore helped found Intel in 1968 and helped build the first chip. He first called out what we now call Moore's Law when in 1964 he observed how fast one could put transistors on a semiconductor.[19]

Moore's Law says that the number of logic circuits on a chip doubles every two years or so. This chip doubling has grown in recent years. Chip density now takes only about 18 months to double. A like effect holds for chip costs and packaging. Moore's Law means that chip densities have grown at an exponential rate. Moore's Law will slow someday but some form of it will continue well into the 21st century.[20]

Chips and boards fit in the nose cone of a cruise missile. Moore's Law lets one double the computing power every couple years even though the volume of the nose cone remains fixed. The machine IQ doubles roughly every two years. No defense program makes this happen. It is a gift from the world chip market.

This chip power affects many parts of the cruise missile from how the missile turns to how it budgets its fuel. The biggest effect is in how it finds its way to the target. This involves both navigation and target recognition. And you can pack the math and software of both navigation and target recognition into the missile's chip set.[21]

An older cruise missile followed profile maps or TERCOM (terrain contour matching) maps until it got close to its target. The chips were so weak that the cruise missile could use only a few TERCOM maps in flights up to 700 miles. Its flight connected the contour dots between the few TERCOM maps. Weaponeers drew the maps from data they got from the Defense Mapping Agency.

The small chip brain used TERCOM maps and onboard gyroscopes to get in the ballpark of the target. Then it used a DSMAC (digital scene matching area correlator) that flashed a light or other energy source at the terrain and matched the reflection to small preset maps in its chip brain. The matching scheme and the missile can crash in the face of blowing snow or sand that covers the mapped terrain. The Tomahawk cruise missiles of the 1991 Gulf War could not fly over the desert sands of Iraq to reach their targets in Baghdad. They needed landmarks to match against. So they had to fly instead over the mountains of Iran.

Newer cruise missiles can use satellite signals to guide them across foreign lands and bring them close to their targets. The missile picks up signals from at least 4 of the 24 Navstar Global Positioning Satellites (GPS) that circle the earth.[22] Those 4 GPS signals can pinpoint the missile's position to within feet or sometimes even inches. The GPS signals correct the missile's inertial guidance (gyroscope) system and the math in its Kalman filter. The Kalman filter predicts the missile's next location and then compares the prediction with its actual position once it has flown by.

GPS signals can guide a Tomahawk cruise missile across the Iraqi desert or over the hills of Bosnia or Kosovo. It cannot tell a cruise missile whether the blob it senses is a tree or a tank. That takes an act of pattern recognition and a big boost in machine IQ.

Enter fuzzy logic and neural networks.

The dumb way to recognize patterns is to store a lot of pictures and match new scenes against them. This requires that you can measure the image "distance" between a picture and a new scene. It turns out there are infinitely many metrics that measure this distance but researchers use only a few of them. Most of these schemes require that the chip compare each tiny "pixel" or picture element of each picture with the pixels of the new scene. The DSMAC technique does this for just one stored picture. Even then the number crunching can exceed what the onboard chips can do in real time. Other schemes try to match just the abstract features or edges in pictures.

The dumb way to recognize patterns may be a good way if the chip brain can store enough pictures and match them fast enough to the new scenes. But there are a vast number of ways to sense a tank in a hill forest. You can take the picture from thousands of angles. The tank can hide behind endless patterns of trees and rocks and brush. And tanks can come in many sizes and many colors. There is no way to code all these tank patterns in advance. And there are no chips on the digital horizon that could match so many pictures in real time to the novel scenes that pour into the sensors.

A smart missile must abstract the fuzzy pattern of *tank* and perhaps a few hundred other target patterns. It must generalize from its training experience. It must learn.

This leads to the field of ATR or automatic target recognition. The newest ATR schemes use some form of "brain-like" neural network to learn target patterns.[23] The system trains on hundreds or thousands of radar or satellite images.

A simple training scheme would show the neural net hundreds of photos that contain a tank somewhere in the image and hundreds more that did not. The net would learn to answer yes when the photo contained a tank and to answer no when it did not. Then the net would tend to give the right answer when it saw a new image that did not differ too much from the training images.

The Defense Advanced Research Projects Agency (DARPA) has begun to use neural and fuzzy systems for cruise missile ATR but only at the research phase.[24] A smart cruise missile may need to store or train on thousands of images to learn the abstract pattern of the target from any angle. It has hours or weeks to train off-line before the shooting starts. Dumber matching schemes need to store as many as 50,000 images to achieve the same skill.

A cruise missile's machine IQ involves more than just good navigation and pattern recognition. It needs to make the flying bomb look "stealthy" to enemy radar. The missile's skin needs to absorb or scatter radar signals. The missile also needs to reason. It needs to rank new target lists in battle or plan new bombing routes and bombing strategies. Newer cruise missiles have both stealth and two-way comm systems that let them talk to the launch base and to one another.

The machine IQ goes up and the cost goes down.

I worked on cruise missiles at General Dynamics in the early 1980s as I finished my Ph.D. Cruise missiles still cost millions of dollars a piece then and were largely dumb flying torpedoes. General Dynamics was the world's largest defense contractor and had the prime contract to build Tomahawk cruise missiles at its Convair Division in San Diego. Convair even called its employee park Missile Park and had small cruise missiles for kids to sit on in the swing sets. Hughes Electronics bought Convair in the aerospace restructuring of the early 1990s. Raytheon has since bought Hughes.

The weaponeers at Convair used to joke that they were "merchants of death." War protesters had once called them that on a picket sign somewhere and the weaponeers were quite proud of it. Their hands never touched a Tomahawk cruise missile. But their drawings and software and raw guesses helped plan the missiles' missions and flight paths and fiery endings.

The real cost of a cruise missile has fallen about an order of magnitude (a factor of 10) since those days of the death merchants in the 1980s. Cruise missiles now cost about $100,000 if they use the body of a drone or "unmanned aerial vehicle." China and other countries have been quick to pursue these low-end cruise missiles.[25]

The United States sold 65 Tomahawk cruise missiles to England for $320 million or about $5 million each. That was the first time the United States sold Tomahawk cruise missiles to a foreign power. Many Los Angeles residents were surprised to learn in November of 1998 that the British navy would test-fire unarmed versions of some of these missiles and do so 80 miles off the coast of Newport Beach.

It appears that the Tomahawks traveled from the HMS *Splendid* across the Pacific Ocean and on north of Los Angeles and over the Mojave Desert where they landed by parachute at China Lake Air Warfare Center. News of this test-firing did not seem to cause much concern among the citizens of Los Angeles. It did lead a news crew to interrupt my lecture on probability so that we could tape a brief sound bite about the nature of cruise missiles and the wisdom of testing even unarmed ones near populated centers. The U.S. Navy claims it has test-fired unarmed Tomahawks at China Lake each year for over a decade.

The cruise missile is going the way of the pocket calculator and for the same reasons.

The cost of a cruise missile can only fall as Moore's Law and GPS guidance and research in smart materials take their toll. The unit cost should fall by an order of magnitude some time early in the 21st century. That would put the low-end cost of a super-smart cruise missile at about $10,000 or far less than the cost of a new car. Its size may also shrink along with its cost. Smaller and cheaper and smarter.

Somewhere along the way we cross a fuzzy threshold: *For the first time in military history it will be cheaper to attack than to defend*. The age of the smart war will have begun.

It has always been the other way around. It has cost far more to attack than to defend.[26] And the world has been a more stable place than it might else have been.

Ancient armies would often wither from hunger and dysentery as they marched across strange lands to attack an entrenched foe. The Greeks had to launch a thousand ships to lay siege to Troy. The Trojans lost but they likely put far less wealth and manpower into their defense than the Greeks put into their attack.

Modern warfare offers more striking evidence of the principle. Afghan rebels used cheap handheld Stinger missiles to shoot down multimillion-dollar Soviet helicopters. The Gulf War allies used the far cheaper Patriot missile to shoot down the far more expensive Scud missile of Iraq. Bazookas and other handheld rockets have destroyed thousands of tanks in small and large conflicts around the world.

Colonel Colt's revolver may have been the first step toward the smart wars of the future. No one but a few Aikido masters have ever dodged a bullet. The Nazis came next when they showered London with well over 2,000 V-1 proto-cruise missiles. U.S. nuclear weapons and ballistic missiles had the same status during the 1950s. The Soviet Union had no real air defense against such nuclear attack until the early 1960s.[27]

Cruise missiles burst on the TV scene in 1968 when Egyptians used the Soviet Styx cruise missile to destroy the Israeli ship *Elath*. India also used a Styx cruise missile in 1971 to sink a Pakistani destroyer. The world watched in 1982 when the Argentineans used a French Exocet cruise missile to sink the British ship *Sheffield* in the Falkland Islands war.

The world also watched as the United States fired Tomahawk cruise missiles at Iraq in 1991 and again in 1993 and 1996. The United States fired some more at the Serbs in Bosnia in 1995 and in Serbia in 1999.

The U.S. Navy fired 80 or so Tomahawk cruise missiles in August of 1998 to attack terrorist training camps in Afghanistan and to destroy a pharmaceutical plant in the Sudan that the Clinton administration claimed produced the nerve gas VX (itself the result of a 1958 U.S. patent that one can now find on the Internet).[28] At least two of the 80 cruise missiles were duds and entered Pakistani airspace before they crashed in Pakistan. The United States launched the summer 1998 cruise missiles in retaliation for the terrorist bombings of its embassies in Africa. This was the first time a country used cruise missiles against targets in countries with which it or its allies were not at war. *Time* dubbed the missile attacks "Tomahawk diplomacy." The United States ended 1998 by firing hundreds of Tomahawks against Iraq in "Operation Desert Fox."

The world demand for cruise missiles has only grown. The supply has begun to keep pace with demand since a country can buy so many smart missile parts at the local computer store. These market forces promise a future world far less stable and more violent than the one we have known.

The Mideast may be the first region to suffer from too many cheap cruise missiles. Suppose Iran and its Islamic allies launch a few thousand cruise missiles against Israel. Israeli air defense could knock out only a fraction of those swarms since again it is hard to hit a bullet with a bullet. And some of the missiles might carry chemical or biological warheads. None need be nuclear. Enough cruise missiles could swamp Israel's cities and military compounds. Israel might well launch its arsenals in response. The odds of this outcome rise as the cost of cruise missiles falls.

The United Nations might try to broker a limit on smart cruise missiles. Such missile control would likely work no better than gun control. You can again find too many missile parts at the local shopping mall.

The United Nations has yet to broker a meaningful ban on the far dumber yet deadly land mine. There are on the order of 110 million of them buried in farm fields and forests and deserts around the world.[29] It is far cheaper to attack with these atom scramblers than to defend against them. That will only get worse as we pack them with more bits to let them tell friend from foe.

A new land mine costs only about $3 to make. But to find and dig one out costs on average more than 100 times that. That explains why each year we plant something like 5 million new mines but clear only about 100,000 or so mines. Researchers have proposed neural and fuzzy systems and chemical sniffing systems to detect nonmetallic mines.[30] Smart weapons promise a like proliferation.

The likely stable equilibrium is for all countries to have large stockpiles of smart weapons. This outcome is stable in the sense of game theory.

Countries will not have an incentive to disarm once they adopt the strategy of buying a portfolio of smart weapons. The cost of smart weapons will be too low to resist buying. Countries will want to have a smart edge on their neighbors and want to make sure their neighbors do not have such an edge over them.

The old arms race will give way to a smart-arms race. Military power will shift more from atoms to bits and the cost of attacking will still fall.

The Digital Battlefield

You can watch a digital battlefield on your TV screen. And so can your enemy.

Signals bounce from men and machines in the field to aircraft and high-altitude drones in the sky to satellites in space and then back down to dishes and cables on the ground. The United States will soon have the war eyes to watch in real time as war unfolds in all of Korea or Egypt or Yugoslavia.

The U.S. Army hopes to be fully digital by the year 2010. The most far-out plans call for satellites in space to beam hologram decoys of tanks and troops onto the ground or into enemy computers. Other plans hope that lasers can shoot down missiles and drones.

The United States made its first big move toward a digital battlefield way back in the Vietnam War with the air force's Operation Igloo White.[31] The United States paid about a billion dollars a year to drop plant-shaped sensors in the jungle. The sensors detected sights and sounds and smells and sent them back to computers in Thailand's Infiltration Surveillance Center. Even the crude computers of the early 1970s could use this data to watch daytime and nighttime troop movements and assign bombers to attack them.

Many smart wars will play out on digital battlefields. We will get to watch some of them while their smart eyes watch us. Privacy stands only to lose from advances in the digital battlefield. City Hall and cybernauts can use the same smart techniques that let sensors and drones and satellites look through walls and listen to coded chatter.

But there will be many battles that no one sees.

Invisible battles may become the norm in the face of the omniscient stare of the digital battlefield. Small countries and terrorists will not be able to afford to fight larger countries on a digital battlefield for long stretches of time. They would lack the sensor suites and the sustained firepower and would likely lose control of their supply lines. Smart weapons may lead them to start a fight but that does not mean they could withstand an all-out smart war.

Information warfare offers a cheaper strategy.[32] It completes the shift from atoms to bits in modern warfare.

Small countries and terrorists can focus on info-sabotage to taunt or cripple their enemies. They might plant logic bombs in computers and wide-area networks. They might scramble or corrupt bank accounts and phone lines and satellites and the vast databases of defense and industry. They might crash a train or shut down a city's power grid or even destroy a chemical or nuclear plant.

Information warfare has an attack-defend cost ratio far less than that of smart wars based on smart cruise missiles. Bits are cheaper than bullets and may prove even harder to block. It costs almost nothing to attack with information and no one knows how to defend against it. The attack can come from too many fronts and in too many ways.

States may fall into a stable game-theory equilibrium where each can info-attack one another and so they refrain. That may help keep order in the United Nations. It will not slow down the radicals in the world's thousands of protest groups or the more daring teenagers in all countries.

These groups and kids will never own a smart cruise missile. But they will have access to the cheap info-weapons of the month. That will be a lot of power in a lot of hands. The extreme result could be a Hobbesian information war of all against all.

And it will not stop there. The fuzzy border between war and not war will melt in other ways as science brings the world more power to share and hoard.

The next step in smart warfare may be a shift back to atoms.

Research in chips and chemistry has already begun to probe the frontier of nanotechnology.[33] Nanotech seeks to build things an atom at a time. It might build small smart machines that eat oil molecules in sea spills or kill flu viruses in the bloodstream or weave glass molecules into diamond-like textures.

Nanotech can also destroy things an atom at a time. Nanotech can often take apart molecules more easily than it can combine them.

Modern super-acids are an early form of nanoengineering. They are millions of times more corrosive than sulfuric acid. The same "smart" acids that can tear apart the oil molecules in a spill can do the same for tanks and shields and flesh. Some nano-bombs could in theory use some of the energy from the matter they destroy to drive further nano-destruction. This could set off a rapid chain reaction and leave little but gray goo in its spherical wake.

The attack-defend cost ratio would hit a new low in a smart nano-war. A good nano-bomb might cost little more than the cost of a pocket calculator and the cost of a few pounds of chemicals or plastics. The state will have little chance of controlling such cheap nano-bombs.

The cost of a nano-defense might be infinite at first for all countries and then only for poorer countries. A nano-defense might release its own nano-agents to attack the nano-destroyer as when firefighters build a fire wall to slow or contain a runaway forest fire. The nano-result may leave behind far less structure than does a forest fire.

In the end smart wars change only the cost structure of conflict. Conflict itself is just a shortcut to power. The essence of war stays the same as it has since the first land fights of prehistory. Its tools can just as well be an Internet crash or a smart cruise missile as an arrow or flint knife. Tolstoy saw that it involved the human will.

The essence of war is the will to kill.

Politicians can bury that will in words and generals can wear it on the medals on their chests. The rest of us can let it loose in a horn honk or a football game or a national election. Our hominid will to kill does not go away with time. It is conserved. The will to kill now leaves chaotic footprints in battle simulations and leads to real chaos in real battles.

Von Clausewitz was right. War is a duel on a large scale. War is what results when two people or two groups or two machines have the will to kill each other. Attack costs will fall and machine IQs will rise for decades if not for centuries. There is no reason to expect the root of the conflicts to change in the information age.

The will to kill will shift smoothly from our atoms to our bits.

Part Two

Fuzzy Science

8

Fuzzy Science

As the sciences have developed further the notion has gained ground that most, perhaps all, of our laws are only approximations.

WILLIAM JAMES
Pragmatism

The student should constantly keep in mind that making statistical inferences from the data does not strictly follow a mathematical approach. Models are subjective and the resulting inference depends greatly on the model selected. Two statisticians could very well select different models for exactly the same situation and make inferences with exactly the same data. Most statisticians would use some type of model diagnostics to see if the models seem to be reasonable ones but we must recognize that there can be differences among statisticians' inferences.

ROBERT V. HOGG and ALLEN T. CRAIG
Introduction to Mathematical Statistics

It seems that our autocatalytic social evolution has locked us onto a particular course which the early hominids still within us may not welcome. To maintain the species indefinitely we are compelled to drive toward total knowledge, right down to the levels of the neuron and gene. When we have progressed enough to explain ourselves in these mechanistic terms, and the social sciences come to full flower, the result might be hard to accept.

EDWARD O. WILSON
Sociobiology: The New Synthesis

S CIENCE EXPLAINS THINGS.
Science cuts the world into big or small pieces and shows how the
pieces affect one another. One world piece affects or causes others. These
pieces affect others in complex and interwoven chains of cause and effect.
Science explains an effect when it cites a cause and gives evidence for the
cause-effect link. Inflation goes up because the money supply grows too
fast. Earthquakes strike because land plates move and release energy. A
lightning flash gives off a thunderclap because it superheats an air column
and because pressure tends to vary with temperature.

But where do the explanations come from? They come from equations
or from their verbal ancestors. Experience may partially confirm or refute
these equations. But where do the equations themselves come from? They
may come from more general equations but in the end those equations
come from scientists. This leads to a key and perhaps impolite question:
Where do scientists get the equations?

They *guess* at them.

Brains guess at them.

Some brains guess better than others. Isaac Newton guessed well at the
equation for gravity. Albert Einstein guessed better at it. String theorists try
to outguess Einstein to find better equations that combine gravity and light
and the quantum world. A brain trained in advanced math tends to guess
better at equations than do brains trained with less math. Still our humble
and narrowly evolved little brains have just begun to guess at the equations
in God's hidden blueprint of nature.

No amount of training changes the fact that science is a guessing game.
Calling the guesses "educated" may give the layman more confidence in
the enterprise. It does not change their limited and subjective brain-based
nature.

There is no shame in admitting how much of scientific progress depends
on guesswork. Scientists would guess better and thus produce better science
if they studied the psychology of guessing and trained their brains with
guessing exercises as vigorously as they once trained them with deductive
exercises in math. Guesses may lack method and may result in far more
misses than hits. But they inject creativity into science and technology.

So why not let computers guess too?

The three chapters in this section explore that question. They look at
new results in fuzzy and neural systems and in the many fields of science
and engineering that support them. This section does not give a formal

review of these fields but rather gives a current taste of them. The tastes and ideas pave the way for the last and most speculative section of the book.

Chapter 9 presents the main problems and some of the main results of modern fuzzy systems. Chapter 10 extends these results to the case of neural or adaptive fuzzy systems. These smart systems learn from experience and face still more problems as a result. But often they guess at the shapes of things better than we do.

Chapter 11 takes a fuzzy view of the new view of physics that the world is nothing but binary information. This extreme black-white world view results from centuries of guessing at black-white math to describe the gray world around us. It also suggests that we might someday use our black-white science to create our own gray worlds.

The core idea is that science might do its job better if it used better brains to guess at its explanations. Machines can mimic and someday create these brains faster than we can evolve them or condition them. We can at most hope to guide these machine brains and keep them pointed at the harder problems and not at us. The next and final section of the book looks at why we might want someday to throw in the towel and join them.

Meanwhile the best guesses still win in science. The sad fact is that humans are not getting any better at making these guesses. The sadder fact is that most of these guesses are at lines.

Patch the Bumps

LOGIC, noun
The art of thinking and reasoning in strict accordance with the limita-
tions and incapacities of the human misunderstanding.

> AMBROSE BIERCE
> *The Devil's Dictionary*

The fact that mathematics as a whole is taken to be synonymous with
precision has caused many scientists and philosophers to show a con-
siderable concern about its lack of application to real world problems.

> EBRAHIM H. MAMDANI
> "Application of Fuzzy Logic to Approximate Reasoning Using
> Linguistic Synthesis"
> *IEEE Transactions on Computers*
> Volume 26, number 12, December 1977

Names are important, and looking back over 50 years it does seem
that the use of the names control engineering, automatic control,
and systems engineering *have not achieved for our subject the recog-*
nition that might have been expected. Names such as cybernetics
and robotics *command a greater degree of public recognition and*
apparent understanding.

> STUART BENNETT
> "A Brief History of Automatic Control"
> *IEEE Control Systems:* June 1996

Science never pursues the illusory aim of making its answers final or
even probable. Its advance is rather toward an infinite yet attainable
aim: that of ever discovering new, deeper, and more general problems.

> KARL POPPER
> *The Logic of Scientific Discovery*

The science of the future will be built by brains that cannot have had more than 10^{80} bits used in their preparations. And they themselves will advance only by something short of 10^{80}. This is our information universe: What lies beyond is unknowable.

W. R. ASHBY
"Some Consequences of Bremermann's Limit for Information-Processing Systems"
Cybernetic Problems in Bionics

THREE WORDS SUM up 30 years of research in fuzzy systems: *Patch the bumps.*

Fuzzy systems model systems or processes in science or medicine or finance or in many other fields. They model systems with rules like "If the image is somewhat out of focus then turn the lens slightly to the left" or "If the stock's price-to-earnings ratio is very low and if the firm's return on equity is high then the stock's price is medium low." An expert sees the rules as just common sense. But the world of math sees them as something else. The rules define fuzzy blobs or patches.

Most people think of math as numbers or equations. That is only the algebraic view. The same equations have a geometric view: They define graphs or surfaces. Equations also define systems. They describe how systems turn inputs into outputs. The simplest systems are linear and give rise to a flat surface. More complex systems are nonlinear. They give rise to bumpy surfaces of hills and valleys and sometimes pointy mountains or pits. Each fuzzy rule describes only part of a fuzzy system. So each rule gives rise to only part of a surface. It gives rise to a fuzzy patch.

The camera or stock system itself defines an imaginary bumpy surface in some abstract space of camera actions or stock prices. The bumpy surface perfectly describes the system. You can put the rule patches anywhere on the bumpy surface. Each expert has her own scheme or hunches where to put the patches. Many of these hunches are rules and thus are patches. Automatic learning schemes put the patches in their own places on the surface and then move them about as learning unfolds.

The best you can do is put the patches on the bumps. First cover the peaks and valleys of the surface and then fill in with the patches left over. Patch the bumps. All else is suboptimal. This chapter unpacks this formal concept of a patch covering.

I worked with fuzzy systems for many years before I saw and proved this simple fact that the best rule patches cover bumps in a system surface. But seeing it differs from using it in practice. The trouble is that in most cases you have no idea what the system surface looks like. So you have no idea where the bumps are. We have ways to search for these bumps and future research will refine them and find new ones.

There is a reason why this search for bumps will go on for decades if not centuries. We now know that fuzzy systems slam into a wall known as rule explosion or the "curse of dimensionality."[1] All math systems face this curse in some form. Fuzzy systems just face it in more vivid form than most. The number of rules grows in an exponential way as you add more variables to the system to make the system more realistic. The best we can do is to patch the bumps and often we cannot do even that.

The curse of dimensionality is the ultimate bound on all human and computer progress in science and mathematics.

We can always add more variables to a system so that it will exceed our ability to analyze or control the system. Our best math models in physics may describe all things but we can solve the math only for the simplest cases. Each advance helps push back the curse a little further. No advance can get rid of it.

This limits the root task of science: function approximation.

Science as Function Approximation

Scientists guess at how things cause other things. Then they test these hypotheses about the causes and effects. They might guess whether sunspots cause droughts. Then they look at data on sunspots and droughts to test the guess.

A hypothesis is a guess about a class or about how classes affect one another.

The simplest guesses are whether an object belongs to a class or how much it differs from other objects in that class. Do solar flares resemble sunspots? Do most sunspots last for two weeks? The class or set of sunspots is vague or fuzzy. All cool regions on the sun's photosphere belong to the set of sunspots to some degree and do not belong to it some degree.

Other guesses involve how much objects in one class affect or cause objects in another. Do the sun's rotation and magnetic field cause sunspots? Do sunspot cycles cause drought cycles on Earth?

A simple guess says that A causes B. Sunspots tend to appear in 11-year cycles and that may match some drought cycles on earth. So one guess is that sunspots cause droughts. A weaker guess is that sunspots cause droughts to some degree. More complex guesses bring in variables like convection currents or solar densities or sound speeds or any of the many processes involved in helioseismology.[2] All these guesses deal with fuzzy sets and how they relate to one another.

A still simpler guess is one we might make in the kitchen or when we look at a weather map. We might guess that high pressure causes high temperature. More advanced guesses capture the guess in math and cast the effect B as a "function" of the cause A (and perhaps write it in math as $B = f(A)$). This means someone puts forth an equation. The ideal gas law is a classic and powerful example. It says that gas temperature is proportional to the volume of the gas times its pressure ($cT = PV$ for some constant c that depends on the structure of the gas). An equation translates talk of fuzzy patterns into math talk about the same patterns.

Fuzzy rules can model these fuzzy patterns of cause and effect. If the pressure is high and the volume is almost constant then the temperature is high. If the pressure is low and the volume is almost constant then the temperature is low. All gas pressure is high and not high to some degree. The same holds for high temperature and for all other fuzzy types of temperature.

This suggests that fuzzy rules might themselves define equations even though we may not know what the equations look like. This turns out to hold in general and it defines the essence of fuzzy systems. It lets us model systems without guessing at equations.

The FAT or fuzzy approximation theorem says that fuzzy rules can always replace equations. The FAT theorem does not say that it makes sense to use rules. The rules for complex systems would make no more sense to us than would the mile-long equations they replace. The FAT theorem just says we can always replace the equations. This means that a finite number of rule patches can always cover a system's surface.[3] Systems of fuzzy rules give a universal way to compute.

Fuzzy systems let one do science without math. They help graft common sense onto math. That can be a great aid when you do not know or care how to talk and guess in math.

This does not mean that fuzzy systems are not themselves math systems. They are. Fuzzy systems have a simple math structure that lets theorists prove theorems about them and that lets programmers capture them in simple software and chip designs. Many of us have refined the math of fuzzy systems to reduce it to this simple level.[4]

It does mean that you can program a fuzzy system with words.

An endocrinologist may build a fuzzy pancreas with rules like "If the glucose level is high and if the change in glucose is small then inject a small dose of insulin." How high is high and how small is small? Those are matters of degree. Human judgment can tune the fuzzy sets *high level* and *small change* and *small dose* that define these concepts. Or neural systems can study clinical data to shape these fuzzy patterns.

The endocrinologist does not have to guess at a math model of how glucose interacts with hormones like insulin or epinephrine or growth hormone. Fuzzy systems are model-free or "black box" system approximators at this level. Their fuzzy rules build a bridge from inputs to outputs and fill the black box. Three simple rules might have the following form:

Rule 1:   If the glucose level is low and the glucose change is small and negative then the change in injected insulin is small and negative.

Rule 2:   If the glucose level is medium and the change in glucose is nearly constant then the change in injected insulin is nearly zero.

Rule 3:   If the glucose level is high and the change in glucose is small and positive then the change in injected insulin is small and positive.

Real rules might add other variables to the if-part or add other control actions to the then-part.

Most techniques in science have no room for such rules. They force you to guess at an equation and then tune it to fit the data at hand. This is just what some endocrinologists have done to detect early forms of diabetes.[5]

The big picture is that science is a branch of the mathematical field of function approximation.

The ancient Greeks were the first to guess at the structure of the world with mathematics. Pythagoras saw that math described the structure of music and triangles. He went on to claim that the world was nothing but numbers. Much of Plato's theory of ideals stems from his attempt to cast ideas like goodness and blueness as pure math constructs like circles and cubes. Isaac Newton began modern science in many ways with his guess at the functional form of gravity. Albert Einstein derived his famous equation of mass and energy and light ($E = mc^2$) as an approximation from other math guesses. Then he later put forth his own math guess at the functional form of gravity as we discuss in chapter 11.

The rub is that math is deductive but science is inductive.[6] Math facts follow from a guess in math. In science it is the other way around. A guess follows from the facts. Or at least a guess follows after looking at or thinking about some facts.

A scientist can guess at all the equations he wants. But in the end no one cares if the equations do not somehow imply something that we can test or refute with data. A mathematician need show only that her premises imply her conclusions. A scientist has to use his conclusions to argue for his premises. Philosopher Karl Popper observed that the best logic can do is refute your premises if the tests do not go as you predict.[7] Tests cannot prove a hypothesis.

I can say that if I eat honey then my blood sugar will rise. But I cannot conclude from a measurement of high blood sugar that I ate honey. Many other foodstuffs or factors may have caused the rise in blood sugar.

Aristotle called such a conclusion the fallacy of affirming the consequent. You cannot conclude anything about P if you know that P implies Q and if you observe only that Q is true. But you can conclude that P is false if you know that P implies Q and if you observe that Q is not true. Of course you may never know that P implies Q. Chapter 6 discussed how these logic laws can hold to some degree when P and Q and the implication between them hold only to some degree. And the truth of P or Q is always a matter of degree.

Affirming the consequent is just what we face when we test a statement of fact or a math model. The best we can do is throw out the bad guesses and place our bets on the remaining ones. If all goes well then we get a better and better approximation of the "real" math model. This math model may only describe the world. Or it may make up part of the blueprint of the universe.

Fuzzy systems cannot improve on the logic of this method. Each input leads to a fuzzy system output. The system emits the output as a prediction given the input as a cause. You test the output as you would test any other prediction. You compare it to real data or some other standard and see how well it fits.

The advance in fuzzy systems lies in the ease with which we can program the black box and in the power of the black box to model or approximate systems. Fuzzy systems offer the same sort of advance that statistical programs first offered. You can just feed rules or data into their black box and stand back and let the computer do the rest. Where would economists be without their simple trend lines? But statistical models still need an economist or someone to guess at the basic model. And they still tend to guess with lines and other simple math curves.

Fuzzy systems also suffer from the two main problems of statistical black boxes.

The first problem is one of confidence. How can you trust a black box to land an airplane or invest your savings or shave your hip bone? How can you guarantee how it will perform?

Smart black boxes offer no more guarantees than human experts offer. You trust the air pilot or fund manager or physician to perform well. You might sue them if they fail at their task. But you cannot be sure that they will not suffer a lapse of judgment or loss of nerve or even a stroke.

Fuzzy systems offer no guarantees because they are nonlinear systems. There are very few known theorems that describe how nonlinear systems behave. We teach students of science and engineering an almost straight diet of linear systems. So many of us come to expect a math guarantee for our tools. But linear systems exist only in textbooks. There are no pure linear processes in nature because no system has an output that always stays proportional to its input.

The best you can do is test a lot of cases on a computer. NASA does this to test the nonlinear math programs that guide its space shuttles. Car makers do this to test how well a new hood or windshield design deflects air or resists friction. Chip makers do this to see how well a new coding scheme compresses music signals or video images.

This reflects once more the old trade-off between the complexity of the world and the complexity of our math schemes that describe it. We can prove all the theorems we need about our linear models. But they do not fit a nonlinear world. A linear system has a surface that looks like a flat piece of typing paper. A nonlinear system looks like a bumpy or crumpled sheet of typing paper. Math guarantees quickly run out as the paper gets more crumpled.

Nonlinear models trade tractability for accuracy. They trade ease of math for accuracy of model. Such models may wrench our brains but they lie on our path to progress and our understanding of nature.

Fuzzy systems suffer from a second problem that faces statistical methods. They get vastly more complex as you add more variables. This is the curse of dimensionality. In fuzzy systems it leads to exponential rule explosion.

The Curse of Dimensionality: Rule Explosion

Sooner or later all math schemes suffer from the curse of dimensionality. Their complexity blows up if we add more terms or variables.[8]

We can solve the Schroedinger wave equation for the hydrogen atom. It takes some assumptions but we can do it. The hydrogen atom is the simplest atom and the most abundant in the universe. It has just one proton in its nucleus and one electron in orbit around it. Next up is helium with two protons in its nucleus and two electrons in orbit. Brains have yet to figure out a complete solution of the Schroedinger wave equation for helium or any other more complex elements. The textbooks stop at the hydrogen atom or an atom with only one electron that orbits about its nucleus.[9]

The Schroedinger wave equation is the hallmark of quantum mechanics. It describes how matter waves change in time based on how matter concentrates in space. But again it runs into its own curse of dimensionality. Supercomputers must churn for days to solve the equation for even the simplest atomic systems.

Physicists call this a many-body problem.

That is shorthand for the curse of dimensionality. Newton ran into it when he said that all clumps of matter attract one another through the force of gravity. The earth attracts the apple and to a much lesser degree the apple attracts the earth. But the earth attracts trillions of other objects and they in turn all attract one another. The effects may be small and we assume they cancel out but the math says they are there.

Even the famed rocket scientists cannot work out the math of how to send a man to the Moon. The Earth and moon make up a two-body system of matter and gravity. But the Sun is so massive that a rocket has to factor in its gravity. That gives a three-body problem that computers must approximate over and over as the rocket heads toward the Moon. A more accurate model would include the gravity of Jupiter and the other planets. This body explosion would give a math scheme so complex that no computers could hope to solve it.

The curse of dimensionality means that most math schemes do not "scale up." The math scheme becomes more than twice as complex when you double the number of inputs. The complexity tends to grow in an exponential way while the inputs grow in a linear way.

Decision trees were among the first math schemes to suffer the curse.

Suppose you want to get from one store to the next in a city with a lot of streets and intersections. You leave the parking lot and come to an intersection where you can go left or right or straight ahead. That is the first branch point and it has three paths. Each one of these paths leads to a new intersection where again you can go left or right or straight ahead. That second level of branch points now gives 9 paths. The third level gives 27 paths. The tenth level gives 3^{10} or almost 60,000 paths and so on.

The game of chess has on the order of 10^{120} paths or possible chess games. That dwarfs the total number of all subatomic particles in the universe. That particle count runs on the order of a "mere" 10^{87} pieces of matter. No standard (non-quantum) computer will ever search through all the chess games in the game space just as no computer will ever search through all of human genome space. The so-called Bremermann limit goes further and says that no matter-based computer can work with more than 10^{93} bits of information even if it used all the matter of the earth to compute for billions of years.[10]

There are vastly more possibilities than there are bits to describe them. We return to this limiting fact in chapter 11 when we look at the bit count of the entire cosmos.

Artificial intelligence (AI) schemes have searched part of the chess-game space for half a century. AI capped this effort on 10 February 1996 when a computer program first beat a world grandmaster in a chess game under tournament rules. The machine won the game but not the match. IBM's super-computer chess program Deep Blue had beaten world chess champion Garry Kasparov. Fast parallel chips let Deep Blue look at millions of board positions each second. Then Deep Blue won the next match on 11 May 1997.[11]

Deep Blue's search rate will never make but a dent in searching the whole chess space. Nor would a rate millions of times faster. But such dents lie far beyond what our brains can achieve. That will keep the chess title in machine hands.

AI tree searches have made progress in machine intelligence within the current bounds that rule explosion sets. Some software schemes make credit decisions for loan firms[12] or turn uranium hexafluoride gas into pellets of uranium dioxide powder[13] or help astronomers sort through billions of sky objects.[14] New computing power pushes back the wall of rule explosion in search trees and allows for a growing set of AI applications.[15] At least one expert system has over one million rules in it.

Bayesian networks are still richer AI tree structures that put probability weights on links that model and predict some patterns of causes and effects.[16] These smarter trees and other graph structures can help find patterns or missing data in databases. But they add the math complexity of the probability calculus on top of their tree-branch complexity.

Fuzzy systems do not suffer from the same kind of rule explosion as do AI search trees. Decision trees and AI search trees involve long chains of binary if-then links or rules. They are "deep" in this sense of long inference chains through many branch points. Most fuzzy systems are shallow trees. They are only one-layer deep.

Fuzzy systems are wide.

All the rules fire in parallel. Each blood sample fires the if-parts of all the rules that describe low or high or normal glucose levels or changes in glucose levels. These partial firings scale the then-part actions. A sum adds up all the scaled actions and takes their weighted average. This leads to a final output increase or decrease in insulin.

A large fuzzy system looks like a broom with thousands or millions of broom straws.

This broom structure works well for small sets of rules. Each system output comes from a weighted blend of many pieces of knowledge. The AI tree is "brittle" because it walks down just one path and ignores the knowledge in all the other paths. The fuzzy system walks down all of the paths to some degree. But the paths are only one step long.

Yet these small fuzzy systems have produced a rich and surprising range of applications. They control microwave ovens and brew sake and inject plastic into molds and retrieve documents. Other fuzzy systems help analyze your golf swing or inspect fabric colors or help manage a dam or help cool a nuclear power plant.

These fuzzy systems use their machine IQ to perform old tasks better or more cheaply or to perform new tasks. Fuzzy systems have found their way into hundreds of real systems around the world.[17] The applications have spread far from their initial base in Japan. Brazil now uses fuzzy systems in many of its efforts to drill for oil and process it.[18]

Fuzzy systems have a simple math structure that aids the spread of applications. This math has itself evolved over the years. It is now simple enough that most small-scale applications do not require a special fuzzy chip. Engineers can just reprogram the software in the chips that now reside in cars or consumer products or industrial systems. This math has the acronym of SAM for *standard additive model*.[19] Almost all applied fuzzy systems use some form of the SAM model. And some SAM models in turn resemble the models used in many neural networks or learning systems.

All these fuzzy systems have a simple flow structure or topology. They are feedforward. They convert an input into an output and that is all. They answer a question only if you ask it. Some sensor suites may feed them inputs and so ask them questions many times per second. But they give only one answer for each question.

The trouble comes when you add more variables or types of causes to the fuzzy system. Then it must work in high dimensions. The broom grows wider in a geometric way and each broom straw adds less structure to the whole system.

A fuzzy system's patch structure works against it in high dimensions. A rule patch ties knowledge to geometry.[20] But in most cases a rule patch cannot compress or contain too much knowledge. The patch structure shows us just where the rule covers or partially models the system surface. It also shows us that the rule ignores the vast part of the rest of the surface.

Consider the problem of backing up a truck. A fuzzy system can do that with 30 or fewer rules. Some schemes can get the rule count down to 10 or even 5. The rule count jumps over 100 when the fuzzy system must back up a truck and trailer. It can jump over 500 when it backs up a truck and two trailers. Below we discuss a new type of fuzzy system that can back up a truck and five trailers.

One way to cope with rule explosion is to divide and conquer. Sometimes it makes sense to build three or four fuzzy systems and combine them to control a process rather than build one huge fuzzy system.

My students and I did this when we tried to model smart cars that might drive in single-lane platoons on the highways of the future.[21] Platoons of fast smart cars can in theory increase highway capacity by as much as a factor of 4 or 5. But a platoon might consist of ten cars that travel at 70 miles per hour over bumps and around curves.

The platoon system is more complex than ten bowling balls connected with springs that travel at 70 miles per hour over bumps and around curves. No one knows what the nonlinear dynamics look like. And cars can join a platoon or split off and form a new one.

The legal aspect alone gives pause to car makers. Whom do you sue if the platoon crashes? Do you sue the platoon leader or the software vendor? Do you sue the car maker or the state agency that controls the smart highway? The American answer tends to be that you sue everyone and let the courts work out the details. This has helped keep many types of smart-car research stuck at the research level.

We broke the system down into small fuzzy systems. One system controlled the throttle. A second controlled the gap between cars. A third controlled the brakes and so on. In some cases we took the best math models we could find and absorbed their structure into a fuzzy rule system. Then we tuned those fuzzy rules with simulations and neural learning and a few real car tests on Interstate 15 in Southern California. More complex fuzzy systems can have a hierarchical structure where master fuzzy systems control slave fuzzy systems or slave math models.

Divide and conquer does not avoid rule explosion. It just breaks the big rule explosion into smaller rule explosions that we may be better able to manage.

This reflects the worst problem of rule explosion. It limits the causal scope of fuzzy systems.

A fuzzy system is more realistic when it includes more causes or inputs. The fuzzy pancreas could add terms that measure adrenaline or other hormones like insulin or thyroxin or growth hormone. These four terms would give a more accurate model but a less tractable one. Models of loan risk often use over 20 variables that range from payment and word records to age and income. Models of business and information management can have over 100 variables in them.

More terms also make it harder to test a prediction. A model with enough terms can explain any outcome. You can always claim that you did not hold enough terms constant or that some of the terms combine in unknown ways.

Rule explosion tends to be worse for many problems outside of simple control or prediction. Problems in communications may use many terms and process them on much faster time scales than control systems require. This holds for all forms of video compression and motion estimation. It also holds for virtual reality where the computer must update thousands or millions of features each second in a virtual world that may let a user walk through a zoo or mountain park or battlefield or swim with dolphins and sharks.[22]

Rule explosion reminds us that rules are a scarce good. We cannot grow them or learn them at will or at zero cost. We have to spend our rule budget as carefully as we spend our savings. This leads to the search for the best rule budget and the best rules in it.

Optimal Rules: Patch the Bumps

Learning moves rule patches as it tunes them.

I watched hundreds of these moving patches in simulations before I saw the pattern. Learning moves the patches on a system's surface. But where does it try to move them? Learning should try to move them to the best place. But where is that?

The patches I watched all moved toward the same thing: bumps. They moved toward the turning points or "extrema" of the surface. They patched the bumps.

One simple case used ten rule patches that had the shape of eggs. My students and I started them out in a line where they overlapped. Then we fed the patches samples from a curve with five bumps in it. Quickly five of the patches moved to the five bumps. Two more patches moved to the end points of the curve. The other patches moved to fill in three of the four spaces between the bumps. Further learning or training only made the patches jitter slowly around their bump positions. The patches never moved away from the bumps once they reached them.

The next figure shows how fuzzy rule patches can patch the bumps of a simple curve:

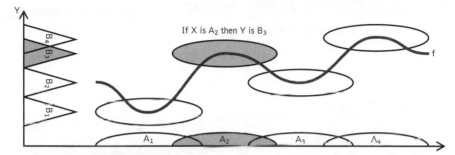

Fuzzy function approximation. A fuzzy system approximates a system by covering its graph with fuzzy rule patches. The fuzzy approximation theorem guarantees that a finite number of fuzzy rule patches can cover any system surface to any level of accuracy. Lone optimal rules cover the extrema or turning points of the system's surface. Learning changes the shape of rule patches and moves them toward optimal locations.

Here four rules give rise to four rule patches. The patches tend to overlap in practice. The overlap allows the fuzzy system to approximate more complex functions. A lone rule patch just gives a flat line. Indeed that fact allows one to prove that lone optimal rules patch the bumps of the system surface.[23]

The deeper idea is that you want each new rule to patch the next largest bump in the error surface.

The error surface is the gap or residual between the fuzzy system's surface and the surface of the physical system. You can picture the fuzzy surface as a bumpy red carpet and the system surface as a bumpy blue carpet. Simple bump patching puts the fuzzy rules at the bumps on the blue surface. This is a special case of patching the bumps on the new purple error surface.

Suppose you have more rule patches than bumps. This happens with simple systems that have only two or three inputs. Where do you put the extra rules? The commonsense idea is the right math idea. You put the rule patch where it most reduces the error or gap between the fuzzy system and the system that the fuzzy system tries to approximate. This place covers a bump in the surface if there are fewer rule patches than bumps.

Patching the bumps shows where learning should lead.

The next chapter looks at how neural or other schemes can tune a fuzzy system by changing its rules. Each change moves a rule patch or changes its shape. Bad learning schemes do not move lone rule patches toward the nearest surface bump. Better learning schemes move them there and perhaps faster and with less data than do other schemes.

But it remains easier to tell than to do.

The Search Goes On

There is one big problem with patching the bumps: You seldom know where the bumps are.

The goal is to approximate a system surface that you do not know. Bumps are part of the surface. If you know nothing about the surface then you know nothing about the bumps. So what can you do?

You have to guess at the bumps.

If you have to guess at something about the surface then the bumps are likely the best thing to guess at. That still gives a system based on guessing. But it is not as bad as guessing at the entire surface. And it tends to be far safer than guessing at the entire surface with a super-sensitive nonlinear equation. You can always guess at an equation that tries only to model the bumps. That can be as hard and the result can be as brittle as if you guessed at the whole surface.

Experts often guess at bumps when they share their knowledge or their rules of thumb. Experts tend to describe the turning points of the system or at least the turning points of *their* system.

The main job of an expert is to help you out of tough spots. A tough spot tends to occur where there is a bump or kink in the system surface. You do not need a race car driver to steer for you on the straight part of the road. You need her help to steer around the road curves and around the cars and the cows that jump in front of you.

Training data also tends to show where some of the bumps lie. Suppose you plot out some samples of how the bushels of wheat grown on a farm change with the amount of some new fertilizer the farmer puts in the ground. The samples are just dots on a page. But there will be a highest dot and a lowest dot. These dots give the first guess at the lowest trough and the highest peak. More dots can fill in between these dots. Other sample dots can define smaller troughs and peaks. The bump guesses tend to improve as more wheat and thus as more sample dots pour in over time.

Patching the bumps does not solve the problem of rule explosion. Nothing can. The minds or machines of the day can only push the envelope a little further back from the day before. They can search and grow knowledge trees a few layers deeper and wider than could those that went before them. They cannot remove the exponential nature of the rule explosion.[24] That lies in the structure of using knowledge or rules to model nonlinear systems. Some schemes have tried to break the rules into clusters or hierarchies but that just rearranges the exponential complexity.

Another way to lessen the rule curse is to improve the structure of the rules. Patching the bumps shows where to move the rules. Many learning schemes try to achieve this. But the shape of a rule affects how well the fuzzy system works even if the rule covers a bump of the system surface.

Most of this effort comes down to the search for new shapes of the if-part sets that describe fuzzy concepts like *cool* air or *high* gold price. Older fuzzy systems used simple sets like triangles or trapezoids. Others used bell curves. Triangles may be the simplest for an engineer to work with but they have an effect that few early fuzzy engineers noticed. Triangles produce a locally linear system. They produce a "piecewise" linear system that in effect tries to model a smooth bumpy surface with chunks of plywood.

I have worked with students to search for the best shape or at least a good shape for fuzzy if part sets. We have tried hundreds of set shapes and compared them with one another. The best sets to date have a complex form that often defies common sense. This search is a little like a fuzzy version of Thomas Edison's search for the best filament for a lightbulb. The results from this search are always tentative. Some form of this search will no doubt continue for decades.[25]

The search for bumps remains the long-term strategy of how to model a system or its pattern of cause and effect. Knowing that can keep us and our machine heirs from wasting time and effort searching the wrong tree branches or twiddling the wrong values in an equation.

And still the search for bumps is a simple case that may not hold in practice. For real systems are not so well behaved. They tend to change with time.

The Promise of Feedback

Rule explosion gets worse when a system changes with time. That can happen when a robot hand tries to catch a fish or when one fighter pilot tries to shoot down another.

Then the system surface does not just contain bumps. It wiggles in time and so the bumps change.

One scheme is to keep adding more rules to keep patching the new bumps. This leads to a new level of rule explosion. A simple three-variable system can require more than 30,000 rules to keep track of a chaotic system. Other schemes can use learning to track the moving bumps but this too can add a deep layer of complexity.

One way out is feedback: Let the rules feed back to themselves. Let the system take its own output as its input.

This means that we change the fuzzy system's topology or its flow structure. We convert the open system that flows from left to right into a closed system that flows in a circle. A feedback system may need only a few rules to model a changing system. The 30,000 rules of the feedforward fuzzy system may give way to only a dozen or so rules of the feedback system.

But feedback has its own problem: stability.

A feedback system may get caught in vicious circles that make it spin off to infinity or to system overload. That is instability. We have all heard the screech that comes when a microphone picks up and amplifies its own output. We take for granted the stability of our air conditioners that cool when the room air is warm and that cool less when it is cold. Engineers have worked hard to ensure the stability of the hydraulic servo-mechanisms that steer and stabilize a large ocean ship or that stabilize an orbiting satellite or a fighter jet in flight.[26]

Most feedback fuzzy systems are not stable. They quickly bounce out of control if they take their last output value as their new input value. The standard fuzzy models have no known stable versions. That has forced us to change their structure and even most of these new versions enjoy stability only in very special cases. Other feedback fuzzy systems are rich tangles of rules that connect to one another and offer little hope that we can control them or even understand them.[27]

The simplest feedback fuzzy systems extend a property of standard feedforward fuzzy systems. They weight and add up local "experts" or systems. Standard systems compute an output as a weighted sum of some stored values. The stored values describe the then-part side of the rules as in the fuzzy set for *low* in the rule "If the gold price is high then the demand for gold is low." Each input changes these weights but the stored values do not change.

Feedback fuzzy systems cast each output as a weighted sum of simple systems. These systems can in theory have any nonlinear shape but then there is little that we can say about them. In practice these systems are linear while the overall feedback fuzzy system is itself nonlinear. Then we can prove that the whole system is stable and quickly converges to an equilibrium if the then–part systems have a certain structure. Stability does not depend on the shape or structure of the if-part sets as in the fuzzy set of *high* gold prices. We still are a long way from finding optimal rules for these systems.

Kazuo Tanaka of Kanazawa University in Japan first worked out this feedback fuzzy scheme. Tanaka later applied these stable feedback systems to the now standard test case of backing up a truck-and-trailer system to a loading dock in a parking lot. The truck-and-trailer system can start at any place in the parking lot and that includes starting from a jackknife position. Then Tanaka applied this basic feedback scheme to the very delicate task of backing up a truck with five trailers.[28]

But stable feedback fuzzy systems are quite rare in math space. That is because most dynamical systems are not stable. That in turn is because most systems are not linear. And again most systems have vast numbers of variables and not just the few that our minds can work with.

This presents the fuzzy theorist with a working dilemma.

He faces rule explosion if he works with a feedforward system. And he faces instability and outright system inscrutability if he works with a feedback system. He pays for a model of the real world either in rules or in lack of knowledge of how the model works. Or he may pay in some mix of both rules and knowledge.

The only comfort is that misery loves company: Some form of this dilemma confronts all scientists and engineers. The best math to describe our world will seldom fit in our brains.

We can only hope that some of it will fit in our chips.

10

Optimal Brain Damage

Imagination is nothing but decaying sense. When we would express the decay, and signify that the sense is fading, old, and past, it is called memory.

THOMAS HOBBES
Leviathan

The plasticity of the living matter of our nervous systems is the reason why we do a thing with difficulty the first time, but soon do it more and more easily, and finally, with sufficient practice, do it semi-mechanically, or with hardly any consciousness at all.

WILLIAM JAMES
The Laws of Habit
Talks to Teachers on Psychology

When I use any such terms as "state," "money," "health," or "society" I assume that my listeners understand more or less the same thing I do. But the phrase "more or less" makes my point. Each word means something slightly different to each person even among those who share the same cultural background.

CARL JUNG
Man and His Symbols

New knowledge leads us to recognize that the theory of evolution is more than a hypothesis.

POPE JOHN PAUL II
Address to the Pontifical Academy of Science: October 1996

We define intrinsic value as the discounted value of the cash that can be taken out of a business during its remaining life. Anyone calculating intrinsic value necessarily comes up with a highly subjective figure that will change both as estimates of future cash flows are revised and as interest rates move. Despite its fuzziness, however, intrinsic value is all-important and is the only logical way to evaluate the relative attractiveness of investments and businesses.

WARREN E. BUFFETT
1994 Annual Report of Berkshire Hathaway Incorporated
(Reprinted with permission of the author)

Problems involving vast numbers of possibilities will not be solved by sheer data processing quantity. We must look for quality, for refinements, for tricks, for every ingenuity that we can think of. Computers faster than those of today will be of great help. We will need them

HANS BREMERMANN
"Optimization Through Evolution and Recombination"
Self-Organizing Systems

G OD HAS NOTHING TO LEARN. The math blueprint of the universe is fixed and only facts exist. There are no secrets from omniscience.

The rest of us have to guess at this math blueprint and search for the facts. We have to learn and we use our brains and machines to do so. Experience shapes and changes these structures. Facts and fiction and noise tune the synapses in our brains and tune the parts or software in our machines.

Learning also tunes and shapes fuzzy systems.

New data leads to new fuzzy patterns. The pattern of a high stock price slowly grows from samples of stock prices and cash flows. The pattern of low earnings may grow and change with the same data. The new patterns change the rules that relate them. The changing rules change in turn the fuzzy systems that use the rules to map inputs to outputs. New data changes fuzzy systems just as it changes our brains and the billions or trillions of patterns that swim through our brains.

Other schemes can change fuzzy concepts and rules. Random search and mutation can change or prune rules in fuzzy systems just as they change DNA coils in gene pools. This is the domain of genetic algorithms.

These schemes shake a fuzzy system at random until they find one slightly better and then they shake that system.

Neuro-fuzzy and geno-fuzzy systems have many applications in engineering and science and they have many problems. These systems have even found their way into the mix of numbers and hope called financial engineering.

Now neuro-fuzzy systems throw darts at stock charts.

Neural Networks as Pattern Computers

Neural networks learn patterns and then watch for like patterns.

A neural Pap smear system learns the abstract pattern of a cancerous Pap smear and then matches each new patient sample to that abstract pattern.[1] A neural fraud hunter learns the pattern of how thieves charge on stolen credit cards and then searches through each card's charge history looking for a match. A neural process controller learns the best mix of the dozen or so chemicals that pour into the silver-halide soup that makes up a camera's film or the chocolate soup that makes up a candy bar.

Neural nets learn these patterns from examples.

An expert tells the neural net when it makes a good or a bad candy bar. This data feeds back through the net and changes its structure. The net is now a new net. It mixes a new chocolate soup and in time a new candy bar. The expert tastes the new candy bar and tells the neural net how well it did. In time the neural net learns the fuzzy pattern of a good candy bar. It learns the fuzzy pattern as a type of intuition or habit.

Each training sample slightly changes the fuzzy pattern just as each sensation slightly changes our super-sensitive brains. The Harvard psychologist and philosopher William James wrote about this remarkable learning property of the brain in terms of moral patterns in his 1890 text *The Principles of Psychology* that helped found both modern psychology and modern neural network theory:

> Every smallest stroke of virtue or of vice leaves its never so little scar. The drunken Rip Van Winkle, in Jefferson's play, excuses himself for every fresh dereliction by saying, "I won't count this time!" Well, he may not count it, and a kind Heaven may not count it; but it is being counted none the less. Down among his nerve cells and

fibers the molecules are counting it, registering it and storing it up to be used against him when the next temptation comes. Nothing we ever do is, in strict scientific literalness, wiped out.

This is the neural basis of the conceptual anarchy.

Our fuzzy or vague concepts like *red* or *fat* or *small* are relative because our own neural networks are so plastic. Our nerve nets learn or abstract their patterns from the ever-changing flux of experience—even when that pat tern flux comes from a day-care center or rock concert or a TV set. There are many types of neural-net models but all share this pattern plasticity.

A neural net itself contains hundreds of neurons or fuzzy on-off switches that sum up the neural juice that flows into them. They turn on if enough juice flows into them or they turn on to a higher degree. A neural net may have as many as ten layers or fields of neurons. The neurons in a layer act like the flashing lightbulbs in a sign on the Las Vegas strip. The neural juice flows from sign to sign and changes the flashing patterns as it flows. The more advanced fuzzy neurons have a continuous or fuzzy flow from off to on and occur most often in modern neural models.

The neural juice flows through wire-like links or synapses. The human brain has on the order of 100 billion neurons. Each neuron connects on average to about 10,000 or more other neurons. The synapses form the connections. They grow with use and shrink and decay from lack of use or just from old age. A brain stores patterns across its great tangled webs of synapses. Patterns of sight or smell or sound or motion pile on top of one another in a synaptic web.

No one synapse stores a pattern.

Neural nets work in some sense like large harps in the brain. Learning tunes the synaptic (axonal) strings. Patterns do not reside in lone strings but in chords that play out across thousands or millions of strings. Changing the string settings changes the patterns. Thought emerges as global patterns of musical resonance. Mind emerges from these patterns that play out in parallel across the thousands of webs of strings. No one plucks the strings or tries to conduct them. The neural nets self-organize themselves. And they compute. They turn input patterns into output patterns.

Artificial neural networks are pattern computers.

Neuro-computers reduce the webs of neurons and synapses to their simplest math form.[2] Each web or net turns inputs into outputs. You feed the net a list of measurements and it gives back a list of control actions or a positive or negative ID. The net might map measurements of the chocolate soup into a new mix of ingredients. Or it might match an image or retina scan of your face against the data stored on a bank's smart card.

A loan officer might feed a trained net the data on hundreds or thousands of new applicants who want to buy a home or refinance a home. The net maps each input to the degree to which the loan applicant will repay the loan and perhaps to a risk or confidence measure of its own answer. Neural networks help search for good clients or plan a production or marketing schedule. The worldwide neural market in 1997 was on the order of $1 billion. Most large firms and countries support neural research and applications.[3]

A neural net does not need the rules that a fuzzy system uses. It uses layers of neurons to map inputs to outputs. No one neuron has much effect on the network. It is a spoke in a very big wheel.

A neuron does not define a system patch as a fuzzy rule does. It just plays its small part in passing inputs to outputs. This leads to a key advantage over a fuzzy system. A neural net need not suffer from rule explosion or neuron explosion.

Neural nets can scale up.

You can add more input neurons or causes to a neural net and not need to add ever more neurons in the middle or "hidden" layers. You may need to train the neural net a lot longer. So the computation may grow or even explode. But the architecture need not explode.

Some small nets have passed many inputs through as few as one or two hidden neurons in the middle layer. Then the one or two neurons can give back the input as the output. So if the on-off pattern of 30 or so neurons spells out the letter "A" then two neurons can often map this neural juice to 30 or so output neurons that will emit the "A" pattern. This is a so-called identity map or replicator network.[4]

Neural nets pay a high price for this mapping feat. They trade neuron explosion for system inscrutability.

There is no general way to know what the neural net knows. Open up the neural net's black box and you find only tangled synaptic spaghetti or "connectionist glop." You find the same thing if you open up a real brain. The neurons or brain cells make up a thin layer on the convoluted surface. The bulk of white matter inside consists of wet wires or the axons and synapses and dendrites that connect the neurons.

Most real neural nets are also feedback networks. So they take as input both their own output as well as sensory signals and neural signals from other neural networks. A few artificial neural networks have a feedback structure to help model signals or patterns like speech that change in time or to help buffer the flux of new neural patterns in light of the old stored patterns. But feedback nets always risk swirling off into instability and tend to involve an extra burden of computation. They also have to achieve a very delicate balance between the neurons that change while the synapses change.[5]

Artificial neural nets share a key feature with real neural nets. They forget patterns just as the best and worst brains do.

Our brains forget patterns as we grew older for at least two reasons. The first reason is that our brains begin to slowly decay after age 30 or so. Each year we lose more neurons and synapses and lose more of the glial or food cells between them. And unrepaired cell damage only mounts with time. That is just what old age is: compound bio-decay.

The other reason is pattern crosstalk or crowding out. This too ceaselessly changes the conceptual anarchy. New faces crowd out old ones in memory. Each new face we learn does not push out one old face we learned as a child in grade school. Rather it pushes out all the faces we learned but only to slight degree. This is distributed decay. Each stored pattern slowly degrades in our private conceptual anarchy.

The more we learn the more we forget.

That helps explain why an old house or school or city park never looks the same when you see it again years later. You have seen thousands of school buildings and school-like buildings since then and you tend to recall their average features. Or rather you recall the weighted average of the features you have most recently observed. You also have learned millions of other patterns across the same webs of synapses. Something has to give and what tends to give is recall accuracy.

Thomas Hobbes got it right in 1651: Memory is just decaying sense.

This is hardly an optimal engineering design. Only blind evolution could have housed our precious minds in such sloppy meat machines.

Optimal Brain Damage

Neural nets do not suffer from bio-decay. Their math is eternal and their software can last for eons.

Instead they suffer from pattern crosstalk.

They learn the first few patterns with no crosstalk and perfect recall. But the crosstalk grows and the recall blurs as the patterns grow in number and as the patterns begin to resemble one another.

A neural network learns best when the patterns are most "orthogonal" or perpendicular to one another in a geometric sense. This seldom happens in practice. Most patterns are of a piece. There are only so many basic types of faces or houses or film plots. That means a given pool of data may contain only a few orthogonal or independent axes or types. Most patterns combine features from these few basic types just as most colors combine red and blue and green.

Neural engineers have tried dozens of schemes to reduce pattern crosstalk. Some work better in some cases than do others but none work well all the time. One common way to deal with a new pattern is to relearn all the old patterns. That way the new pattern does not crowd out too great a portion of the old patterns when learning tunes the synaptic web. Instead all the patterns tend to equally crowd out one another.

This relearning can require hours or days of computer crunch time. And there is no way to be sure that the net will learn the new and larger pattern set as well as it had learned the older and smaller set. The neural net's storage and recall will in time degrade as the number of patterns gets near the crosstalk limit.

So neural networks can model real brains too well.

They can quickly learn abstract fuzzy patterns like cool air or sticky resin or enemy tank. They need only sense or "experience" samples of these patterns. No one defines these patterns with equations or programs them into the net with rules. The net learns by doing or watching.

Neural networks can also quickly match new patterns of air or resins or tanks to the stored old patterns. They act as mathematical associative memories. They match a new pattern of cool air to a stored pattern of cool air based on the new pattern's content and not on the old or stored pattern's "address" in the synaptic web. The stored pattern has no address. It lies in each synapse to some degree just as an image lies in all the pieces of a hologram.

But again neural networks forget as they learn. Their synaptic webs are so inscrutable that we have no good way of knowing what they have forgotten when they learn something new. The heart surgeon runs the same risk when she sits through a new heart workshop or learns to wield a new scalpel or reads the morning paper.

Neural engineers have been slow to accept the problems of connectionist glop and crosstalk and forgetfulness. Markets and other scientists have been more pragmatic. Few neural engineers today propose neural nets as a lone black-box tool where one size fits all. Fewer still try to just throw neurons at a problem. The neural nets grow too large and computers have to churn too long to tune them. That is why many neural engineers search for schemes to prune the massive synaptic webs.

One scheme has the fitting name "optimal brain damage." It ranks each synapse by how well the synapse helps the net perform on average and then throws out the low performers.[6] The scheme lets the neural net learn its patterns and thus settle down into an equilibrium. Then it prunes as many of the synapses as it can without changing the equilibrium too much. Pruning remains an active area of both neural and fuzzy research.

Optimal brain damage is optimal only in the sense that it prunes the net in a state of equilibrium. This means that the neural state ball has rolled down to the bottom of a local well on the error sheet in some synapse space of staggering dimension. The scheme finds a way to learn a set of fuzzy patterns with fewer synapses and perhaps improve how well the net models or recognizes these types of patterns.

But the lost synapses may be the ones that the net needs to learn a new set of patterns. That is the problem with schemes that prune the synapses in a neural net or that prune the rules in a fuzzy system. They trade the system's potential to learn more patterns in the future for a more compact way to learn the patterns of the present.

Still neural nets have emerged as powerful tools to use in hybrid systems such as neural fuzzy systems or other "smart" hybrids. These hybrids can also include the new search techniques called genetic algorithms.

Genetic Algorithms: Evolution as Random Hill Climbing

You can cast many problems in science as optimization problems.

These problems search for the best way to combine variables. Find the lowest cost mix of hops and malt and sugar and water to make a lager beer of a certain quality. Find the curve that best fits a scatter of points. Find the largest number of balls of a fixed size that fit in a box or in a tube of a fixed size. Find the cow genes that give the most boned meat in three years for a fixed diet of corn and grass. Find the neural net or fuzzy system that best maps inputs to outputs or that bests lays out the logic circuits on a chip diagram.

Few of us ever find the best of anything. Most searches end when we find an outcome that beats what has gone before and when we simply tire of searching or run out of budget. Each new chip crams more logic circuits in the same space. Each chip beats the chip before it in circuit count or cost or both. No product or technique achieves optimality. Much of progress comes down to this local search for better ways to combine variables.

These optimization problems have a bumpy geometry. The goal is to get to the bottom of the deepest cost well. Find the mix with the lowest cost. Or find the mix with the highest profit if you search for the highest hilltop.

This is much harder than it sounds.

The first problem is that no one really knows what the surface looks like. This is the same problem a fuzzy system faces when it tries to model a system. It tries to patch the bumps of the system's surface but in most cases no one knows where those bumps lie.

The same holds for most cost surfaces. No one knows what they look like or how to write down their equations. Economists sometimes use the rule of thumb that a firm's total cost curve is a cubic function of the number of goods produced. No one believes that this holds for real firms but it makes for simple homework problems in a textbook.

The second problem is local minima. The search tends to stop at the bottom of the closest well. This may be no more than a dimple on the entire cost surface. Most search schemes are greedy. They try to move in the direction where the surface falls or rises the most much as a ball traces out the gradient as it rolls down a hillside.

There is no general way to know whether the well bottom you found is the deepest well. We call this deepest well the global minimum. You would have to know what the rest of the surface looked like to know that. And a cost surface can have thousands or billions of wells in it. Many automatic search schemes get stuck in shallow cost wells.

The third problem is that most search schemes face both the first and second problems at the same time. You blindly search an unknown or even random cost surface. And you do your best to slide downhill as you search. You may think you are going downhill but instead you burrow through the cost surface or barely move as the surface shakes around you. A search scheme called the LMS (least mean-squared) algorithm shows how to search simple bowl-shaped surfaces in many cases.[7] This helps cancel the echo in long-distance phone calls and helps an antenna lock on to a radio signal. LMS schemes work well for cost surfaces with only one well but elsewhere they tend to get stuck in shallow wells.

Random hill climbing is one way to jump out of shallow wells and into deeper ones. This search scheme is greedy only on average. It does not always try to move downhill like a marble rolling down a sink. The scheme picks its next search site at random. If the new site has less cost than the old site then it moves there. But if the new site costs more than the old site than it moves there only with some probability. So sometimes the search scheme climbs uphill. This helps it break out of shallow wells and search out deeper ones.

Engineers sometimes anneal or "cool" the random hill climbing to control the search. The field of *simulated annealing* deals with many ways to control the randomness much as metallurgists melt and then cool metals into low-energy or crystalline states. The annealing schedule lowers the "temperature" of the search. It slowly changes the form of the search from pure random search to pure greedy search.

Simulated annealing acts much as real annealing. The first high temperature "melts" the cost surface. All its bumps melt into a type of flat blob. Then the cost surface regains its structure as it cools and the temperature falls. If all goes well then the first new structure that appears is the deepest cost well.[8] This has helped find the layout of silicon chips and deblur brain images and solve a wide range of other combinatorial problems.

The trouble with simulated annealing is that it can take a very long time to cool the search. Cool too fast and the wrong well first appears and the search ends in a shallow well. The same problem occurs if we view the search as bouncing a small ball or marble on the search sheet. It bounces in great hops at first. This helps the search ball bounce out of shallower wells and bounce into deeper ones. Cooling the search makes the ball bounce with smaller hops. If all goes well then the ball gets trapped in the deepest or nearly the deepest cost well and cannot bounce out of it.

The new field of genetic algorithms (GAs) applies simulated annealing to whole groups or populations of bouncing search balls.[9] Each ball acts like a genome or a list of one creature's genes. The ball moves or bounces when some of its genes change. This gives a rough model of random mutation. The GA surface defines the Darwinian fitness of each genome point rather than its cost. A population of evolving binary creatures or bit strings of 1s and 0s searches this surface in parallel and at random.

Two creatures mate when they produce a new creature or genome point made up of their genes. Chapter 5 showed one way to form a new genome from two old ones. Draw a line between the two and pick the midpoint of the line as the new genome. GAs use many schemes for mating or mixing genome points. Most schemes mate only those points that have a high fitness (or low cost). Then the schemes replace points with low fitness with the new offspring points. Random hill climbing guides the parallel GA search as GA pioneer John Holland explains:

> One conventional technique for exploring such a landscape is hill climbing: start at some random point, and if a slight modification improves the quality of your solution, continue in that direction; otherwise, go in the opposite direction. Complex problems, however, make landscapes with many high points. As the number of dimensions of the problem space increases, the countryside may contain tunnels, bridges, and even more convoluted topological features. Find the right hill or even determining which way is up becomes increasingly difficult. In addition, such search spaces are usually enormous. . . .
>
> Genetic algorithms cast a net over this landscape. The multitude of [binary genome] strings in an evolving population samples it in many regions simultaneously. Notably, the rate at which the genetic algorithm samples different regions corresponds directly to the regions' average "elevation"—that is, the probability of finding a good solution in that vicinity. . . . First, each string in the population is evaluated to determine the performance of the strategy that it encodes. Second, the higher-ranking strings mate. . . . The offspring do not replace the parent strings. Instead they replace low-fitness strings, which are discarded at each generation so that the total population remains the same size.[10]

So GA schemes breed winners and replace losers with the offspring.

This process acts like natural selection or the "survival of the fittest." If the surface defines cost then only the cheapest solutions survive. The random nature of the search weakens this somewhat. Only the cheapest solutions survive on average.

GA schemes work well on large problems in engineering or manufacturing where a system combines many variables and no one really knows how to best combine them. Brains and math models can guess at a few mixes of variables and these can then seed the GA search of the cost surface. General Electric has used GAs to improve its computer-aided design system. This has led to improved designs for the Boeing 777 jet engine and improved designs for steam turbines and hydroelectric generators.[11] Texas Instruments has used a GA system to improve the design of a silicon chip and U.S. West has used a GA system to help it design networks of fiber-optic cable. Other GA schemes help predict the fluctuation in a country's currency or help produce faces for a witness to recognize and rank after seeing a crime or help brew beer.[12]

Genetic systems can mix and match with other systems to form hybrids. A cost surface or "fitness landscape" can define the cost of how well a fuzzy system matches a known or unknown system. So it can model the error of function approximation. This leads to a geno-fuzzy system.

Neuro-Fuzzy and Geno-Fuzzy Hybrids

A hybrid fuzzy system picks its rules with something other than the human brain. A neural net picks and tunes the rules in a neural-fuzzy or neuro-fuzzy system. A genetic algorithm picks and tunes them in a geno-fuzzy system. A statistical or other system can also pick and tune them. Each hybrid improves the fuzzy system in some way.

Each hybrid also incurs new costs.

The fuzzy system always faces rule explosion as it adds new input or output variables. And it fires all its rules to some degree each time it maps an input to an output. This parallel process comes with its own cost and limits to growth.

A neural-fuzzy system adds the neural costs to the fuzzy costs. The neural network may need to run through hundreds of thousands of training cycles to find a good set of rules or to tune them. Each training loop fires all the fuzzy rules to some degree. The computer may have to crunch for hours or even days to find the rules. That is why most neural-fuzzy systems learn off-line. Few systems let the rules change while you use them but they will become more common in the future as the cost of computing falls.

There are two broad types of neural-fuzzy systems. The type of hybrid depends on the type of neural learning. The net can learn with or without a teacher or error message. Learning without a teacher is unsupervised learning. Learning with a teacher is supervised learning. Supervised learning is slower and costs more but is more powerful.

Unsupervised learning is a form of blind clustering. It groups data into clusters by grouping like with like. The data clusters define rule patches in a neural-fuzzy system. Supervised learning uses each new piece of data to move the rule patches in some direction. If all goes well then it moves them toward the bumps or turning points in the system surface.

The Frankenstein monster shows how both types of learning work.

Suppose the monster first wakes up in an international airport like JFK in New York or LAX in Los Angeles. The monster hears many voices in many languages. He hears people speak in English and Spanish and German and Hindi and in dozens of other languages. No one tells the monster which speech sample is from which language. He has to learn without supervision.

The monster soon starts to cluster like with like. He clusters English with English and Spanish with Spanish and Hindi with Hindi. The English cluster may at first contain non-English speech samples. But in time the clusters tend to improve in accuracy as he hears more speech samples. He may even break the English cluster into an American subcluster and a British subcluster.

The monster forms these clusters on his own. He somehow matches new features with old features in the fuzzy speech patterns. And he somehow senses the centers or average points of the clusters.

Now suppose Dr. Victor Frankenstein shows up and supervises the learning process. Victor knows the language of each speech sample. So he can tell the monster whether the monster has made an error or has matched a speech sample to the right language class. Victor can reward good matches and punish or not reward bad matches. The monster may come to learn fine distinctions between language classes. He in effect draws a jagged border between language samples in the pattern space in his mind's eye.

This training scheme often works best if the monster has first worked out his own clusters. Unsupervised learning lets the data tell its own story.

Supervised learning tries to fit the data into the teacher's story. Supervised learning can put a fine polish on the unsupervised clusters. But it may get stuck in its own local minimum if it applies blindly and treats all speech samples as equal.[13]

The same holds with neural networks and neural-fuzzy systems. Unsupervised clustering finds the first set of clusters and thus fuzzy rule patches. Then supervised learning can tune the rules by tuning the clusters. But this assumes that one has a teacher on hand. That does not hold in general when you try to approximate a nonlinear system or process like a school of swimming minnows or a single-lane platoon of smart cars. No one knows what the system surface looks like and so no one knows whether the fuzzy system errs when it maps inputs to outputs.

More complex systems use separate neural and fuzzy systems to train and shape each other. The neural-fuzzy systems I have described are fuzzy black boxes that use neural learning schemes to tune the fuzzy sets or rules. The same learning schemes can tune a neural black box made up not of sets and rules but of neurons and synapses. The more complex schemes combine the two black boxes in novel ways.

The most popular neural-fuzzy systems use supervised learning to train off-line. Unsupervised clustering often guides the first bouts of training. One Japanese application uses a neural-fuzzy system to control a rolling mill that makes thin strips of steel or other alloys. Engineers first used the linear methods of optimal control on this problem and improved the milling process. A fuzzy system itself did not improve on this math-model approach until engineers added a neural system to guide the learning and improve the rules.[14]

Chapter 9 discussed the SAM or standard additive model of fuzzy systems. Supervised learning can tune all the parameters in the SAM system. It easily tunes the rule weights and the key values of the then-part sets. But it takes much more effort to tune the if-part fuzzy sets. These sets tend to account for most of the exponential rule explosion and for most of the gains in learning.[15]

These adaptive SAMs often work well on problems of nonlinear signal processing where there are few known math models of how signals behave. A key problem is filtering impulsive noise from a signal like the loud cracks and pops that you hear on a radio when lightning flashes nearby.

Engineers have tried to model this noise with the popular Gaussian bell curve made famous in IQ circles as the title of the controversial book *The*

Bell Curve.[16] There are in fact infinitely many bell curves (called alpha-stable curves).[17] All differ in kind from the Gaussian curve. These other bell curves give a better model of impulsive noise but almost none lead to a closed-form math model. A neural-fuzzy system lets a learning black box adapt to the impulsive nature of the noise to better filter it or to better predict a signal in its presence. Other systems can learn rules that describe this randomness.[18]

A geno-fuzzy system adds its search costs to the rule costs of the fuzzy system. The cooled random search often involves its own exponential complexity. Fuzzy rule patches can mutate slowly as the swarm of random balls climbs the hills and valleys of the system surface.[19]

This brute-force search tends to take far more crunch time than neural tuning takes. It also amounts to a type of supervised learning since a GA searches an error surface. Still more complex hybrids include the emerging new field of chaos engineering that ranges from problems in control and prediction to modern data encryption and communication.[20]

All hybrid fuzzy systems seek the old human goal of sound knowledge. This knowledge in turn improves how well a fuzzy system can model or control or approximate a system. It can also feed the old desire to predict the unpredictable.

Beating the Street

Sooner or later a neural or fuzzy engineer gets a big idea: Why can't neural or fuzzy systems predict fuzzy patterns in the stock market?

Markets produce streams of numbers or time-series data. This data can feed into a neural net or a neuro-fuzzy or a geno-fuzzy system. The smart black boxes can sift through the data and learn patterns and trends. The learned patterns can form rules and the rules can form a fuzzy system. Or experts can give the system rules to guide the sifting or learning process.

Markets also produce the age-old desire to beat the street and get rich quick or at least get rich slowly. This has spurred both innovation and a new level of financial speculation.

Financial analysts have applied new math tools almost as fast as they have appeared in the technical journals. These tools have their virtues. They replace raw guesses with precise equations and improve how one prices stock options or bonds or "futures" (contracts to buy or sell in the future an asset or commodity that one trades on an exchange). These tools have led to new ways to hedge risk in global markets because they show how to price the "derivatives" or buy-sell contracts based on foreign currencies or interest rates or commodities.

The world economy depends on the derivatives market. Indeed the currency market is the biggest market in the world. The contract value of derivatives was on the order of $20 trillion in 1994 and growing fast.[21] The stock market trades only a fraction of that dollar amount.

The key tool is a partial differential equation known as the Black-Scholes equation.[22] This equation stands to modern finance and derivatives trading much as Schroedinger's wave equation stands to quantum mechanics. The math is even more complex and has held up well in hundreds of formal tests.

Fischer Black and Myron Scholes published their math model in 1973 to value options on shares of common stock. They modeled stock prices as a random diffusion process that have a simple math form (a log-normal probability structure) in a market with no brokerage fees. Then an arbitrager can make money if the option does not obey the equation. Analysts have loosened some of these assumptions and applied the Black-Scholes equation to the full range of financial derivatives. Scholes and Robert Merton shared the Nobel Prize in economics in the fall of 1997 for their work on pricing derivatives. (Black died before the Swedish Academy had a chance to honor him.) A year later Scholes and Merton again appeared in the media when the Federal Reserve arranged an emergency bailout of their overleveraged hedge fund Long-Term Capital Management.[23]

The new math tools have also turned even some Swiss bankers into gamblers and currency speculators. Most large banks and credit firms now trade derivatives to hedge risks. They buy and sell the right to trade currencies or to cap interest rates at fixed rates for fixed lengths of time. Or they swap fixed-rate loans for loans with variable rates. Or they buy and sell "swaptions": options to swap such loans.

Derivatives have now largely replaced gold as the inflation hedge of first choice. This explains why central banks in Belgium and the Netherlands have been slowly selling off their gold reserves and many other central banks may follow suit. Central banks own about one-third of the world gold supply. Now even governments hedge and speculate in interest-rate derivatives. The residents of California's Orange County learned this when the county went bankrupt in late 1994 and lost about $2 billion in part because derivatives allowed the county to leverage its $7 billion into about $20 billion. Then a bad gamble on the direction of interest rates brought a

10% loss on the total leveraged amount and thus a real loss of $2 billion.[24] The media blamed the mysterious derivatives but the fault lay with the leverage and the gamble.

Derivatives can help smooth out fluctuations in interest rates and currency exchange ratios if you try to sell your corn or credit cards or electricity in a foreign country. They can also yield large profits or wipe you out at the speed of light. So much is at stake that there is a constant demand for new smart tools and decision aids.

Neural networks were among the first smart black boxes to find their way into modern finance because they are the most statistical in nature. You need only run the time-series data through the neural net and let it pick the values in a contract. An expert does not have to give advice or rules. Streams of past and current data can refine the contract values in bouts of supervised learning.[25]

Analysts have also begun to use fuzzy systems and genetic algorithms to price risk instruments. These systems range from user decision aids that help paint a risk profile of the user to GA search schemes that try to optimize values in futures contracts and stock options.[26] More humble fuzzy systems pick stocks based on rules of insider trading and interest rates and balance sheet ratios that involve assets and earnings and debt.

Hype and hope taint these systems as they taint the reasoned judgment of many financial analysts.

There is for instance scant evidence that "technical" or chart analysis can predict future stock prices. Markets are too efficient and already include most past information in current prices.[27] The charting idea is that a stock or gold price cannot rise too high above its "resistance" because then sellers would dump their supply and that will lower the price. And a stock or gold price cannot fall too far below its "support" because then buyers on the side lines would rush in and buy cheap and bid the price back up.

This chart view turns each price into a stable price equilibrium and it should apply to all prices of all things. But prices do rise far above their resistance levels and fall far below their support levels. The charts never show when these "breakthroughs" occur. They simply fail at their main task of predicting future prices.

Neural and fuzzy systems have added a new chic to technical analysis. They endow the price charts with the promise and mystery of machine intelligence and thus endow them with a new dose of hype and hope. But even the best neural-fuzzy system cannot see the future in past data if the future does not leave footprints in the past data.

Each process in the universe creates a stream of data. Your voice has grown a stream of sounds over the years. We could take the statistical average of that data trail and get a good feel for how your voice sounds when you speak calmly or under stress or in front of the mirror. Then we could build smart black boxes that pronounce words the same way you do. But the data trail would not tell us what you will say next no matter how much math or machine IQ we throw at it. The best we could ever do is place some odds on some speech outcomes.

Nothing dresses up a guess like stating it in math.

There are many places where neural-fuzzy systems can aid in finance. They can help assess a user's taste for risk. To what degree will you risk your capital to double it in five or ten years? To what degree do you prefer income to growth? They can learn to search for the kinds of investments you have made in the past. They can to some degree learn to pick investments as leading investors have picked them in the past.

But such smart systems do not offer a shortcut to wealth. There is not much chance that future smart systems will either. If everyone used ideal smart systems or agents to invest then markets would simply equilibrate faster. Market prices would more quickly reflect all available market information. Asset bargains would become much more rare or might even disappear.

Investors should also be wary of technology tools for the very newness that makes them attractive. Even experts in a field seldom fully understand a new advance. Fewer still know where it best applies. Engineers waited decades before they applied the math truths of information theory outside the "black" or secret projects of defense research. No one thought the results would first apply to gadgets that now scramble cable TV lines or route cellular phones or link satellites to computer screens.

The newness of technology makes it more an object of speculation than of investment. And yet that newness seems to seize the minds of many on Wall Street who bid up the stock of new public offerings or who buy a huge firm's stock just because the firm put out a new chip or Web browser or brain-based reuptake blocker.

Technology is a mere tool. Most such tools fade in a few months as new tools appear. Or the tool is no more than an equation and no one can patent or license math even though many try. Of course some patents can have great value. But many firms already own a batch of patents. Speed to market tends to count more than patent protection in the race for earnings. And a dozen firms with the same patents can fail or succeed in a dozen ways.

Betting on technology is dangerous. The computer industry gets something like 80% of its revenue from products that did not exist two years before. Betting in such a risky and changing market is best left to the experts.[28]

There is a math reason for this. It is almost impossible to predict the long-term future or "intrinsic" value of such fluid technologies. The basic formula of rational asset pricing sees the present or intrinsic value of an asset as its discounted future stream of earnings or profit. The formula looks at how the firm will pay off for all future years out to infinity.[29] It requires in practice an informed guess at how the firm will pay off for at least five years.

There are few firms for which we can guess well at their long-term future. Nebraska super-investor Warren Buffett admits this outright and so does not invest in technology at all.[30]

Betting on software is no safer.

Each new math advance spawns dozens of one-man and few-men outfits that sell the math in a software package. Students often sell software this way. This happened in the neural and fuzzy and genetic fields and in most others. Then larger firms put out their packages and in some cases a few firms pack the software onto a chip or computer board. The suppliers grow and the market rent soon falls to zero or nearly so.

The lesson of high tech in the marketplace remains one of basic economics: Those high-tech tools that succeed become mere commodities like corn or oil or memory chips.[31] Firms that produce commodities must compete largely on price. This helps the consumer but it eats away at the firms' profits and so eats away at investor returns.

Commodities give rise to so much risk that they lead to futures markets to hedge some of it. A person should not play with such derivatives unless they do not care if they lose their capital and maybe then some.

Think how long Isaac Newton would have lasted if he had run a firm that sold the calculus he had just invented. At first he would have earned some supreme consulting fees. But if the firm did well then his colleagues in the Royal Society and elsewhere would soon form like firms and then compete with him. Newton would have far less luck keeping a franchise on calculus than he had in fighting Wilhelm Gottfried Leibniz over who first invented it. There are no brand-name monopolies on math.

The big gains from technology come when the luster and hype have faded and the new knowledge has become old knowledge and diffused throughout the economy. The good news here is that each year high tech more quickly becomes low tech. The Information Society absorbs the science and the culture on its fringe at an ever faster rate.

Neural fuzzy systems will have arrived when they have all the mystery of a screwdriver.

Patterns of Mind

Neural fuzzy systems have done more than mix chemicals and pick stocks and tune rules. They have given us a richer picture of what we call mind. They paint a picture of fuzzy patterns stored in great webs of synapses.

The fuzzy patterns make up the conceptual anarchy. The patterns can be patterns of nouns like orange carrots or curve balls or white–hot coals. Or they can be patterns of verbs like kick or lie or risk. Studies show that when we think of a noun like a glass of beer we combine fuzzy sense patterns like cold and white foam and amber from many parts of our left brain.[32] Learning has distributed these fuzzy patterns across thousands or millions of neural webs.

This does not mean that fuzzy systems reside in brain tissue. There is little chance of that. Most wet neural nets have dense tangles of feedback loops and depend on dozens of factors that do not find their way into modern neural models. Rule explosion itself rules out any chance that brains consist of banks of fuzzy systems. Brains would have to be at least the size of large pumpkins if they housed enough fuzzy rules to carry out their tasks of motor and cognitive and sensory control.

But no one claims that real brains use neural fuzzy systems or geno-fuzzy systems or any other modern hybrid. We use these systems as tools to solve hard problems that we cannot yet solve as well or solve at all with standard tools.

A farmer does not assume "log–normal" odds when he picks a future price at which he will sell next year's wheat crop. He does not try to solve the Black-Scholes partial differential equation. He uses fuzzy rules of thumb to pick his future selling price and uses years of experience to form and tune those rules. Neural fuzzy systems let us absorb his wisdom and rules just by watching him buy and sell wheat futures. They do not force us to assume new equations to fit the still more complex equations of a math model even though neural systems have shown that they can learn even the complex patterns of the Black-Scholes equations.[33]

Still the neural fuzzy tools suggest how minds or brains might work at some level. They suggest a few tricks that brains might use in their own wet ways to store and process information. They store and match fuzzy patterns in parallel. They grow rules from experience. They form black boxes that do their best to approximate or to control the complex systems they watch.

The information engineer learns these mind tricks and searches for more.

The study of real brains might produce more mind tricks. So might the next paper in a math journal or the next study on why geese fly in a **V** or the next wireless scheme for routing phone calls. The information engineer does not care where the mind tricks come from. He seeks new tools for new minds and cares only how far those tools and tricks will stretch the minds of the future. The next chapter looks at this inquiry in terms of what our minds or future machine minds can say about the information structure of our universe. The final chapter goes further and looks at stretching minds in engineering rather than in science.

There is a good chance that future minds will use some form of neural or fuzzy or genetic systems. They will also use addition and multiplication and will structure data into sets or files. We do not know what future minds will use to reason or even whether they will reason in some way that we would recognize as reasoning.

Yet one bet seems safe to make: Future minds will have little need of three pounds of meat.

It from Fit

The world is my idea of it.

ARTHUR SCHOPENHAUER
The World as Will and Idea

By convention sweet and by convention bitter, by convention hot, by convention cold, by convention color: in reality there are only atoms and void.

DEMOCRITUS

Belief in this transcendental world taxes the strength of our faith hardly less than the doctrines of the early Fathers of the Church or the scholastic philosophers of the Middle Ages.

HERMANN WEYL
"Mathematics and Logic"
American Mathematical Monthly: Volume 53, 1946

Physical objects are postulated entities which round out and simplify our account of experience just as the introduction of irrational numbers simplifies our laws of arithmetic. The conceptual scheme of physical objects is a convenient myth, simpler than the literal truth and yet containing that literal truth as a scattered part.

WILLARD VAN ORMAN QUINE
From a Logical Point of View

Where were you when I laid the world's foundations? Tell Me if you understand. Who fixed its dimensions? Surely you know. Who stretched a measuring line over it? On what do its pillars rest?

Book of Job: 38:4–6

$$e^{i\pi} + 1 = 0.$$

> *Introductio in Analysin Infinitorium:* 1748
> LEONHARD EULER

*The bit count of the cosmos, however it is figured, is ten raised to a
very large power.*

> JOHN WHEELER
> "Information, Physics, Quantum: The Search for Links"
> *Complexity, Entropy, and the Physics of Information*

*God forbid that we should give out a dream of our own imagination
for a pattern of the world.*

> FRANCIS BACON
> *Novum Organum*

*The redundancy of a language is related to the existence of crossword
puzzles.*

> CLAUDE SHANNON
> "A Mathematical Theory of Communication"
> *Bell System Technical Journal:* volume 27, July 1948

*What peculiar privilege has this little agitation of the brain which we
call thought that we must thus make it the model of the whole universe?*

> DAVID HUME
> *Dialogues Concerning Natural Religion*

*The basic particles of our world correspond to the musical notes of the
superstring, the laws of physics correspond to the harmonies that these
notes obey, and the universe itself corresponds to a symphony of superstrings.*

> MICHIO KAKU
> *Strings, Conformal Fields, and Topology*

*The concept of "measurement" becomes so fuzzy on reflection that it
is quite surprising to have it appearing in physical theory at the most
fundamental level.*

> J. S. BELL
> *Speakable and Unspeakable in Quantum Mechanics*

*Almost all existing approaches to quantum gravity expect a modifica-
tion of the classical picture of space-time by introducing some sort of
fuzziness.*

D. V. AHLUWALIA
"Quantum Gravity: Testing Time for Theories"
Nature: volume 398, 18 March 1999

Physicists know that every equation is a lie.

GREGORY CHAITIN
Santa Fe Institute[1]

WHAT IS THE WORLD MADE OF?
That is an old question and it has had many answers in the last
3,000 years.

Thales of Greece said that the world is water. That was in 600 B.C. or so.
Thales helped found modern philosophy—and modern finance when
(Aristotle tells us in his *Politics*) he bought a forward contract that let him
control the next year's olive presses. Then Heraclitus said the world is fire.
Anaximenes said the world is air. Pythagoras said the world is math.

Parmenides was perhaps the first Greek philosopher to say that the earth
is a sphere. He said the world is a big unchanging ball of matter. He
said the world is a big unchanging ball of matter.
Democritus did not agree. He said the world of matter changes. The world
is made of infinitely many atoms that move through a sea of void.

Meanwhile mystics around the world said the world is love or spirit
or mind.

Western philosophers later recast these world views in the thick language of metaphysics. This language grew out of the "research" in the
Christian church of the Middle Ages.

French philosopher and military officer René Descartes said the world is
both mind and matter. The human brain's pineal gland connects the two.
Dutch Jewish philosopher and lens maker Baruch Spinoza said the world is
the pure substance God. German philosopher and mathematician Gottfried
Wilhelm Leibniz said the world is an infinite set of smart but small mathematical points or "monads." Anglican bishop George Berkeley said the
world is just God's thought of it and would go away if God blinked.

German philosopher Immanuel Kant said the world is the senseless goo
that excites our senses. He said we look at the goo through glasses colored
with the abstract notions of time and space and cause and effect.

German atheist philosopher Arthur Schopenhauer agreed with Kant. He said we sense the goo and then form our ideas about the world from what we sense. But he said we get one special peek at the naked goo itself. We see our own will and it does not depend on time or space or on cause and effect. So he said the world is will.

Modern physics agrees in part with the old Greek philosopher Thales of Miletus. It agrees that the world is a type of fluid but it does not agree that the fluid is water. The fluid is one of matter waves. So in the end it is a form of energy even though quantum mechanics says that up close energy exists only in discrete packets like bricks.

The world is a great store of energy.

Einstein showed how matter or mass converts to energy and vice versa. That is the lesson of his famous theorem that relates energy and mass and the speed of light ($E = mc^2$). Nuclear blasts later showed that Einstein had gotten the math right. At least it was right to a few decimal places and that was good enough for government work.

The energy view gives way to many math views of the world.

Most of these views cast the world as some type of fluid. The fluid view follows in most cases from the simplest rules of the differential calculus.

In 1873 James Clerk Maxwell cast the world as an electromagnetic fluid. The famous four Maxwell equations show how this charged fluid flows and how it can emit light waves. Einstein went further and cast the world as a curved fluid of energy and matter. The fluid even emits gravity waves at the speed of light. Quantum theorist Erwin Schroedinger used his own wave equation to cast the world as a matter fluid. He showed in the famous equation that bears his name that how the matter waves move in time depends on how matter concentrates in space.

A radical fluid view comes from physicist Frank Meno of the University of Pittsburgh. Meno casts the world as a compressible fluid made of tiny spinning gyroscopes or "gyrons." The flow of oblong gyrons gives electricity when it accelerates and gives magnetism when it rotates. A particle is a vortex in the gyron fluid whose sign (+ or −) depends on whether the vortex rotates to the right or left. Entropy comes from a type of stickiness or viscosity among the colliding gyrons. The whole theory reduces to standard equations of fluid mechanics. But such math simplicity comes at the high price of bringing back a form of Maxwell's ether.[2] Twentieth-century physics began in large part by dispensing with the "luminiferous ether" in favor of the "empty" vacuum of space.

More complex math models cast the energy fluid in higher dimensions than just the usual four of space and time. The German mathematician Theodor Kaluza showed in 1921 that adding a fifth dimension gives back Einstein's gravity fluid and Maxwell's electromagnetic fluid from the same set of equations. The extra dimension simplified the math and unified the science but did not seem to exist in the world.

Modern superstring theory goes well beyond this and works with 10 or even 26 dimensions. The extra dimensions may curl up on such a small scale that we do not see them. The superstring equations give back both the gravity fluid and the quantum matter fluid. Then the world consists of tiny loops of vibrating strings. Atomic particles act like musical notes that nature plays on the strings.[3] Einstein's gravity fluid is a type of low-energy version of superstring theory.

These energy world views lead to a new question: What is energy made of?

The math says only that we view the world as energy. We measure energy when we try to measure the world's goo. The goo reflects energy. That does not mean that the goo itself is energy.

So what is energy?

A new school of physics has put forth at least a partial answer and one ripe for the Information Age: Energy is a form of information. It comes from bit. The world is a big computer. Or at least the world computes or stores patterns of information.

This new view is the ultimate statement of the black and white or binary world view of science: Everything is a pile of 1s and 0s.

And it is a view that quickly gives way to gray.

The Mystery of Being

Holding an apple in your hand is at root a mystical experience.

How can that be? The apple lies in your hand. Your hand lies in space and sticks to your arm. Your arm sticks to you and you lie in space. These are "facts." A lot of them make up the complex "fact" that you hold the apple.

And you have direct experience of these facts. So their "proof" is but one step: Direct observation. You can call on science to back up the fact. You can measure your hand and the apple and the atoms that connect them.

So where is the mysticism?

It is in your head. The mysticism lies in your belief that the apple is in your hand. It lies in the belief that there is an apple and a hand that are part of the world.

It lies deeper still in the belief that there is a world *out there.*

The problem is that there is no way to prove the world is out there. Here we have to be careful with our verbs. We can *prove* only the empty tautologies of logic and math like "Red is red" or "1 + 1 = 2." All else is fuzzy and tentative.

You cannot prove a fact.

A proof assumes that some statements or premises are true. Then it derives a conclusion from them. You would have to assume at least one fact of the world to derive a fact of the world. But to assume that a fact is true is to reason in a circle. The point is to prove that a fact is true. Suppose you assume that the sky is blue. Then you can derive the fact that the sky is either blue or red. So what? The truth of the conclusion still depends on the truth of the assumption.

So logic does not help. That demotes the "proof" method down to the realm of evidence and witness and measurement. Formal proof gives way to mere suggestion. You can claim only that the evidence *suggests* that you hold the apple in your hand. Squeeze the apple hard enough and then juice runs over your fingers. This may suggest many things about muscles and pressure and carpets. Logically it proves nothing. The "law" of suggestion comes to no more than pointing. And pointing at pointing will not do.

You also cannot be sure whether you point. You cannot even be sure that you are awake or that your experience is real. It may be a dream or bad drug trip or a good ride through a virtual reality. Or it may be something more complex that a nano-surgeon programmed into a chip in your head.

Science can help only so far because of its fuzzy truth bounds. It can again get its math right to at most a few decimal places. Each decimal place comes at a higher and higher cost of test and measurement and logical consistency. Binary precision is not a free good. It is not even something we can achieve in the realm of facts.

This holds for a simple but deep reason: Science would have to get the math right to infinitely many decimal places to produce a 100% true fact. Then its claims would have the same binary status as those of math and logic. Even then you could not be sure that the logical fact matched something out there in the world. Someone or something would still have to point.

We have to accept that the small wet bundles of neural nets we call our brains did not evolve to prove theorems or to paint the world with math. We use these tools as best we can. Apes do not use these tools at all.

We use only the simplest of these math tools. And we use them only after we train for years and then only when we focus our mental effort in a way that we can seldom sustain for more than a few seconds. Most of us shake our heads in disbelief when we see the complex equations for *pi* and like terms that mathematicians Leonhard Euler and Ramanujan guessed at and then proved correct.[4] These guesses pierce our brain bounds and reach out a little into the great mathematical function space in the sky.

We have seen in the last two chapters how fuzzy and neural systems let us go a little further beyond these brain bounds. But even these and other brain boosters cannot hope to reach out and grab Kant's goo.

In the end we still take the apple and the world on faith. We believe in something that we cannot prove.

That is mysticism.

The apple reminds us of the mystery of being. Why is there an apple? Why is there a world? Why is there something rather than nothing? Why is there not just void? What question does the world answer? Why does the world have this ratio of electrons to photons rather than some other? Why does not the ratio change? Why is there this many atoms in the world and not one atom more or one atom less?

Thoughtful people have asked some form of these questions for thousands of years. They have put forth answers that have ranged from water to will to energy. Some of these answers have helped set up cults and religions and philosophies. Others have helped tear them down. But the answers have all *proven* the same thing: nothing.

The world remains a mystery.

So why not take it a step further? Why not think about what happens if you pierce the mystery and go beyond it? Suppose you could squeeze the world as easily as you can squeeze an apple. What happens?

What happens if you destroy the world?

The World in a Black Hole: It from Bit

Princeton physicist John Wheeler coined the term black hole in 1967.[5] But Wheeler had long before put forth a simple thought experiment. Suppose you throw a rock in a black hole. What happens to the conservation of energy? What happens to the information that the rock stores?

These are clever questions and it helps to review black holes before answering them.

Black holes act as a type of information prism in our universe. These strange objects offer a simple way to convert the world into information.

A black hole forms when a large star implodes under its own gravity. The star must be at least two and a half times as massive as our sun. A star that has less than about one and half times our sun's mass will in time collapse into a dead white star. In still more time it will go cold and dark. Stars with masses between these extremes will in time collapse into a dense neutron star or pulsar. These mass ratios assume that the stars have already lost mass through solar wind or an outright nova explosion. These old final stars end up as black holes or pulsars or cold dead embers.

The universe rewards the biggest and most massive stars and space objects for their size. It turns them into black holes. A key question is whether the universe itself will someday end in a black hole or in something like one.

Many physicists believe the intense X-ray source Cygnus X-1 is a black hole. It lies "only" 6,000 light years from us in space. If we shot a light beam at Cygnus X-1 then the light beam would take 6,000 years to get there. Massive black holes may lie at the center of many galaxies where they slowly devour nearby stars and other space matter. The center of galaxy NGC4258 may be such a massive black hole. A black hole 2.5 million times as massive as our sun may lie at the center of our own Milky Way galaxy.[6]

John Wheeler chose the name black hole because even light cannot escape from a black hole's intense gravity. The black hole can suffer quantum leaks and so emit radiation. It can leak itself out of existence but that may take eons of time. The black hole swallows all other matter if the matter gets close enough to it. A big black hole can even swallow small black holes and then merge with them to become a bigger black hole with even stronger gravity. A big black hole cannot break up or dissolve into smaller black holes. Its topology is stable and the hole can only grow.

A black hole's gravity bends or attracts light. If you shine a laser beam at a black hole then the beam would dive into the black hole's "event horizon" and leave this world for good. If you shine the laser beam at a far angle to the hole then the beam would bend toward the hole for a brief while and then move on past the black hole and return to a straight line. If you shine the laser beam at just the right angle (at exactly 1.5 times the radius of the black hole) then the beam would circle the black hole in a stable orbit just as Earth circles the Sun or as a satellite circles Earth.

The name black hole begins with Wheeler but the math concept of a black hole has a long history. British scientist John Michell showed in 1784 that a big enough star could act as a dark star and attract back its own light. He used only Newton's theory of gravity and the idea of a light beam as a stream of particles or photons. He showed that the star's gravity grew stronger as its surface area shrank and thus as its mass packed more densely. French scientist Pierre Simon Laplace published a like result in 1799.[7]

The modern theory of black holes begins with Einstein's general theory of relativity.

Einstein's math said that gravity is not an attractive force that acts among chunks of matter. Instead matter and energy curve space. Isaac Newton got the math right to a few decimal places when he described how the earth pulled the apple down from the tree and how at the same time the apple slightly pulled the earth up toward it. What Newton got wrong was the idea of attraction. One massive object no more attracts a second object than the Sun orbits the Earth.

The attraction is an illusion.

The Sun's gravity attracts a comet because the Sun acts like a massive rock placed on a large trampoline and the comet acts like a marble that rolls across the trampoline and rolls into the rock's gravity well. The closer the marble comes to the rock the more the marble curves in toward the rock. If you watched this from the ceiling then the trampoline surface would look flat and the rock would appear to attract the marble.

Matter warps the distance or metrical structure of the space-time continuum.

Karl Schwarzschild first solved Einstein's curvature equations of general relativity in 1915 just before he went off to fight and die in World War I. He assumed that a star mass was a perfect sphere and did not spin. Then the math says that black holes can occur where zero divides some terms in his solution and so where space-time points have infinite curvature.[8] Einstein and many other physicists worked hard to get rid of these "Schwarzschild singularities" but did not succeed. Further research showed instead that the singularities were part of our curved landscape.

New Zealand mathematician Roy Kerr solved Einstein's space-time equations for a spinning star in 1964. The spin creates a swirling vortex or gravity tornado in the space-time fabric around a black hole. Other physicists showed that a black hole could also have electric charge that flows straight out from the black hole. Still others showed that all large enough stars collapse to a spherical black hole no matter the shape of the star and no matter whether it has bumps on it or in it. The bumps turn into gravitational waves or ripples in curved space-time that travel at the speed of light.

The final round black hole has mass and spin and charge and nothing else. John Wheeler gave this a name too when he said in the late 1960s that "a black hole has no hair." The no-hair theorems say that one of the most mysterious objects in the universe is also one of the simplest to describe. It takes just the three values of mass and spin and electric charge.

Now suppose you throw the rock into such a no-hair black hole.

Have you just cheated the world out of some of its mass and energy? Have you broken the "law" of energy conservation that says the total energy in the world does not change?

Black holes feed on energy. With each bite their surface area grows larger and their gravity grows stronger. So does the black hole's belly balance what we lose from our world? How can we ever know if we cannot look inside a black hole?

Wheeler knew that this mass loss appeared to violate the second law of thermodynamics.[9]

This statistical law states that a system's entropy or disorder can only grow in time. It suggests the world may end in a cold "heat death." The world starts in the fires of the Big Bang with little or no entropy. Then it expands out like a balloon with less and less average order and more and more entropy. This continues until heat death occurs unless there is enough matter to draw the world back into a gravitational collapse.

Throwing the rock into the black hole appears to break the second law of thermodynamics. It appears to reduce some of the disorder of the world. The same problem would occur if we fed a black hole a swarm of colliding molecules. Their disorder or entropy would vanish down the hole.

Wheeler's student Jacob Bekenstein found a way out of this problem. He thought he saw entropy in the surface area of a black hole. Stephen Hawking had shown that the hole's surface area could only grow in time. This was his so-called *area-increase theorem*.[10] Bekenstein and Hawking each pursued this idea. In time they proved what we now call the Bekenstein-Hawking equations.

Bekenstein was right. The surface area of a black hole does measure its entropy or its information content. It does more than measure the entropy. It *is* the entropy. The entropy of a black hole is exactly one-fourth its area or event horizon. So the second law of thermodynamics holds after all. The entropy growth in the black hole more than makes up for the entropy lost from the rock or from the swarm of molecules. So disorder still grows.

This makes sense on its face because we do not know what lies in a black hole. We are uncertain about that region of space-time. We have no idea what type of matter fell into the black hole and made it expand. It could have been star dust or a stack of newspapers. Any of a huge number of atomic particles could have led to the mass in the black hole. We could say the same thing about the large number of possible alphabet letters that appeared in a newspaper burning in a trash can.

Entropy measures this huge number.

Quantum theory says that this number is finite. Mass comes in small quantal units just as letters in the alphabet come in discrete units. The English alphabet has 26 letters while Chinese characters number in the thousands. If mass were not quantized then an infinite smear of infinitesimal mass points could have made up the mass of the black hole. So quantum structure is key to the Bekenstein-Hawking result.

You can compute an object's black-hole entropy with a pocket calculator. You need to know the number of possible quantum states that could make up the object's mass. String theory has shown new ways to count these quantal microstates.[11] But you can estimate the number of states without it. Then just take the logarithm of this number. This is a much smaller number but still a huge number. It gives what Wheeler calls the "bit count" of the object. The number cashes the object into a pile of 1s and 0s.

So the black hole acts as a type of bit machine. It chews matter into literal bits. Feed the black hole and it grows and thus its bit count grows. The black hole turns things or "its" into bits or binary yes–no units of information.

This leads to a natural question: What if we throw the whole universe into a black hole? Will we get a finite number or an infinite number of bits? How many bits are we worth?

The answer is about 10^{120} bits or units of information.[12]

So the bit count of the cosmos is on the order of 10 multiplied by itself 120 times. This is a huge number when we compare it to the number 10^{87} or so of all subatomic particles in the known universe. So even the smallest particles are supercharged with information. This bit-count number is on the same order as the number of all possible chess games. So we can map the world into all the chess games that some super-alien could play. Mathematician Roger Penrose uses the same equations to arrive at the like bit count of 10^{123}.[13]

The world's bit count is a big number but still a finite number. Why is that? Why are we worth an exact number of bits and not one bit more or less? Who or what picked this number of bits for our world rather than any of the infinitely many other numbers? What broke the symmetry? These are again modern versions of the ancient question of why is there something rather than nothing.

Meanwhile John Wheeler has found yet another cryptic name to describe this binary world view: *It from bit*. Wheeler even conceives a black hole as more of a bit ball than a black ball in empty space:

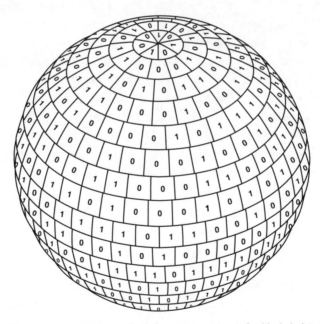

It from bit. Binary units or bits give the information content of a black hole's surface area or event horizon.[14]

Matter comes from information. Things come from binary facts. Its come from bits: "Every physical quantity, every it, derives its ultimate significance from bits, binary yes-or-no indications, a conclusion which we epitomize in the phrase, *it from bit*."[15]

Wheeler's it-from-bit thesis does not explain why there is something rather than nothing. The thesis recasts the something into a new thing. The world becomes at root a pile of bits rather than a swarm of particles that undulate in a sea of frothing quantum vacuum. This is a bold world view and one that has already led to a new research program and a wedding of physics and information science.

But is the world so perfectly black and white?

The World in a Fuzzy Cube: It from Fit

It from bit rests on an old assumption. It assumes that science facts have the same logical status as math facts. It assumes that we or God or the Black Hole Mind can get the facts right to 0 or 1 precision just as if they were statements of math.

There are at once two problems with this idea.

The first problem is that all logically true statements imply one another. This is an artifact of how we define the if-then term in logic. The whole if-then statement is true if both the if-part statement is true and the then-part statement is true. The statement "If red is red then blue is blue" is true because red is red and because blue is blue. Those same logical facts mean that the converse statement "If blue is blue then red is red" is also true. The if-then structure suggests that the if-parts cause or spill over into the then-parts. The one empty tautology pushes into the other and vice versa. Logical truth pays no heed to this suggestion. It simply says that logically true statements imply one another.

So we have to ask a question of the black hole: Does it know or possess a truth of mathematics? Is any statement of the form "1 + 1 = 2" part of its bit count? The quantum realm obeys such laws. If you add one quark to a second quark you get two quarks. The black hole should contain the structure of at least one such truth in its mammoth bit count for any chunk of matter.

But this leads to a new question: Does the black hole know all the infinitely many math truths that are logically equivalent to the first truth? How can it possess one of these truths and not the entire logical chain to which it belongs? How can the black hole break such symmetry? Where can the hole put all these truths? How can a black hole of finite mass and thus finite bit count contain infinitely many math truths?

The finite bit count does not square with the infinitude of math truths. This problem stems from the common but sweeping assumption that facts have the same binary truth status as math truths. And that assumption stems from little more than the old habit of making it.

The second problem is the one we have looked at many times in this book. No one has found a binary fact of the world. No one has gotten the science right to more than a few decimal places. Maybe no one ever will. Maybe even a God could not produce a binary fact. Or maybe a God is just the thing or force that can produce binary facts.

Perhaps the tightest result we have is a prediction of general relativity about how pairs of dense neutron stars or pulsars orbit each other. We may have gotten that right to as many as 14 decimal places.[16] Better clocks may let us get it right to a few more decimal points.

Big deal: A few more decimal places buys nothing in terms of logic. They are no better than no decimal places when we compare them with the infinite decimals we need for binary truth.

Fuzzy truth is a simple way to deal with these and other problems. We do not say "The gas molecule is in this room right now" with 100% or 0% truth. We say it with some truth value between these extremes. We might say it is 80% true that the gas molecule is in this room right now.

We replace the *bit* value with a *fit* value or fuzzy unit.

This will look to many as no more than putting odds on whether the gas molecule is in the room. This is the random view or the probability view that has for some time held something of a monopoly on models of uncertainty. It uses the same numbers between 0% and 100% but often uses them in distinct ways and always means something binary with them.

The random view says that it is 80% likely that the gas molecule is in the room or not. It still sees the event as a binary event. It just places a bet on it. And it always places the 0% bet on the purely fuzzy claim that the gas molecule is both in and not in the room at the same time. It always places the 100% bet on the either-or event that the gas molecule is either in the room or not at the same time. Fuzzy truth never takes on these two 0-1 values in those cases.[17]

The random view keeps the old binary view of Aristotle and classical mathematics. That assumption is not one of logic but one *about* logic. Wheeler's it-from-bit thesis just reaps the harvest.

Fuzzy truth does two things in one stroke. A fuzzy unit or fit combines both the logic and the uncertainty of a statement in the same value. So a fit value like 80% means that the statement is only true to degree 80% and yet that itself is a vague state of affairs. The uncertainty does not sit on top of the world picture. It is part of that picture.

Fuzzy truth also reminds us that binary truth is rare and extreme. We should not expect to achieve it for free. We should expect that it will cost and cost dearly.

One fuzzy model says just this. It says we cannot achieve binary certainty in this world. The facts of the world do not come in a pile of 0s and 1s. Those extreme values would require infinite mass or energy because something would have to fluctuate infinitely fast.

The fuzzy view begins with something called an H number. An H number is just the entropy or bit count of some uncertain state of the world. Scientists often use the H symbol to refer to entropy. Atomic theorist Ludwig Boltzmann proved in an 1872 memoir what we have come to know as the H theorem: Thermodynamic entropy only grows in time.[18] This was an early attempt to prove the second law of thermodynamics from basic math concepts.

Claude Shannon also used the H symbol to describe entropy in his pioneering work on information theory at Bell Laboratories in the 1940s. It had the same math form as the Boltzmann entropy and it measured the same property. It measured the randomness of an entire probability distribution that describes a binary world. This was a natural outgrowth of Shannon's famous MIT master's thesis in 1938 that first showed how to tie 0 and 1 to whether an electric circuit was closed or open.[19] This thesis helped launch the binary digital age.

The entropy H does not just measure how random a swarm of molecules is. It measures the randomness in its probability description. The entropy is highest if all events have the same odds as in the throw of a fair pair of dice. This leads to the strange result that the most informative book or film is one where all words or images occur at random. The entropy is zero if all events are certain and thus have no randomness.

The entropy H describes much of the worlds of matter and information. But where does *it* come from?

I explored this question from the point of view of Wheeler's it-from-bit thesis. Shannon and others showed that the math form of H comes from the unique structure of the logarithm (which alone turns products into sums). That came to more of a description than an explanation. It did not show how to eliminate H in favor of something simpler.

I viewed H as what physicists call a potential. It assigns a number to each state of a system. Here the state is a probability description or a list of numbers that add up to one. Then it follows that H measures the compressibility of a fluid (a gradient field). The surprise is that the fluid is the abstract "fluid" of fuzzy mutual entropy.[20]

Fuzzy mutual entropy measures how close a fuzzy description of the world is to its own opposite. It has no random analogue in general. Compressibility measures how hard it is to shrink a given volume of a gas or fluid. Scuba tanks contain compressed air and so airlines will not let you take a filled tank on an airplane for fear that they might explode in flight. Water is harder to compress. Children learn that a balloon full of water tends to burst if you try to compress it. The incompressibility of water gives a lake its surface. Steam is compressible and so does not have such a sharp surface.

The fuzzy fluid leads in turn to a type of wave equation. The wave shows how the Shannon entropy H fluctuates in time. It has the form of a so-called reaction-diffusion equation. These equations occur in all branches of science and engineering. They show how a term changes in time depends on how it changes in space. Shroedinger's wave equation has this form. So do most models of how things diffuse. These models range from how a gene for blue-eyedness can spread through a human population to how a cloud of insects can disperse in the sky to how rabies can spread through foxes in Europe or through raccoons in North America.[21]

The fuzzy wave equation assumes only that information is conserved. The total amount is fixed and we do not create or destroy information. Some form of the wave equation would still apply if information were conserved only locally or in small regions of system space. The space itself is a fuzzy cube of high dimension. It has as many dimensions as there are objects or even subatomic particles of interest. Each point in the cube describes the uncertainty or fuzz of each object or particle. The Shannon entropy H changes at each point in this cube and defines a wave. One result is that the entropy H can only grow in time in the spirit of the second law of thermodynamics.

A deeper result is that the entropy changes slowest at the fuzzy cube midpoint of maximum fuzz. That is the only point in the cube where the fuzzy description equals its own opposite. The Shannon entropy wave grows faster and faster away from the cube midpoint and near its skin. The skin or surface of the fuzzy cube is the only place where a 0 or 1 appears in the system description. The wave equation shows that the entropy H changes infinitely fast if and only if it touches the cube's skin.[22] This is impossible in a universe with finite bounds on velocity like the speed of light.

So there are no 0-1 bits.

There are only fits between these extreme values. It would take too much speed or energy to achieve complete binary certainty. Binary truth may hold for the math schemes we work with. It cannot hold for statements about the world we work in. The result is not it from bit. The result is it from *fit*.

Is the World a Computer?

So is the world made of fuzzy fluid?

That may be as plausible as a world made of 0s and 1s. It remains no more than a math abstraction and it is not the type of math scheme that one can easily test. We could throw stones or entire planets into black holes and not be sure whether the final information structure was binary or fuzzy. But the fuzzy view still throws into doubt the old belief that binary truth comes for free.

It also raises the question whether existence itself can admit degrees. This might make the most sense in the quantum world where the uncertainty principle can forbid exact values. String theorists have been quick to observe that how many dimensions the world has depends on tests and data and not on the dictates of common sense. String math demands 10 dimensions to avoid inconsistencies. Only data can say. The same holds for whether fits or bits give a better model of the micro-world.[23]

There are other schemes that see the world as information.[24] Physicist Frank Tipler has used general relativity and the Bekenstein-Hawking equations to argue that the world is information and will end someday in a final heavenlike "omega point" if the world is closed and so if it falls back on itself in a Big Crunch.[25] The omega point would replay our lives and variations of our lives in a type of cosmic simulation.

Physicist Lee Smolin has taken these ideas to a bold new level. Worlds collapse under their own gravity in Big Crunches that give birth to new worlds with slightly different initial conditions. This forms an evolving labyrinth where worlds grow and collapse and where worm holes or space-time tubes connect them. Black holes and Big Crunches breed these new worlds. Darwin's natural selection favors those worlds that have the most black-hole offspring. Our world may just be one such hole-rich world.[26]

These and other info-world views raise a new question. If the world acts like a computer then what about the converse? Why can't a computer or a chip act like a world? This leads the discussion away from physics and into engineering and into the realm of what we can build and what we may someday experience.

It leads to heaven in a chip.

Part Three

Fuzzy

Digital

Culture

Fuzzy Digital Culture

Man is a species that invents its own responses. It is out of this unique ability to invent, to improvise his responses that cultures are born.

ASHLEY MONTAGU
Culture and the Evolution of Man

The barbarians are the brooms which sweep the historical stage clear of the debris of a dead civilization.

ARNOLD TOYNBEE
A Study of History

The information superhighway is a revolution that in years to come will transcend newspapers, radio, and television as an information source. Therefore I think this is the time to put some restrictions on it.

SENATOR JAMES EXON (1996)

The empires of the future are empires of the mind.

WINSTON CHURCHILL
Speech given on 16 September 1943

O UR DIGITAL CULTURE is not a culture of black and white. It is an ever more diverse culture of fact and opinion and art and science and all those fuzzy patterns we call ideas. The digital culture offers a fresh glimpse at the full multidimensional spectrum of the human experience cast at the level of ideas.

Many social critics and policymakers point with scorn at some of these ideas and at the frequency with which these ideas propagate and recycle and transform into new ideas or cultural trends. Social critics and policymakers have always done that. Now they point at the digital culture's level of chat and gossip and rumor and libel and pornography and gore games and gambling and encryption and advertising and cult paranoia and trash journalism and whatever else they think springs from the baser side of human nature. They know such ideas can challenge or erode or even over-turn current systems of teaching or morals or religion or politics.

But the digital medium only transmits the supply and demand of our ideas. It does not create the ideas. It provides the infrastructure that lets us transmit them. The digital medium houses this rich marketplace of ideas at a lower cost and with more options than have the other information medi-ums of the past. It transmits ideas more efficiently than have words vibrat-ing in the air or the less transient media of clay tablets or books or pamphlets or newspapers or state-monitored radio or network TV channels.

We create the ideas from the world around us and the genes within us. We create them from our own experiences and our dark hominid hardware.

The Internet is the emblem of the digital culture.

The Internet rose from the wired and wireless communication paths that circle the planet's surface and skies. It rose quickly and while few seemed to notice. There were on the order of 50 million users on the Internet by the late 1990s and the number of users tends to double each year or two. So too grow Internet congestion and brownouts.

The Internet is a modern example of what the late Nobel laureate econ-omist Friedrich von Hayek called a "spontaneous order" or an evolved social system like the English language or the price system. These systems grow or evolve by action and not by design. Some observers have in this spirit called the Internet the accidental superhighway. The Internet is a new institution or global cultural equilibrium that has evolved largely on its own. No one planned it. That scares many people and state planners and adds vigor to their efforts to control it.

The Internet caught all governments by surprise and caught some of them in the act. Governments have moved swiftly to censor the Internet and reduce its flow of information. Asian governments have felt the most threatened. You can go to prison for 15 years in Myanmar (Burma) if you own a network computer that the state does not approve. China and Singapore have tried to force Internet data to flow through their state-run choke points and filters. China sentenced Web-page designer Lin Hai to two years in prison in early 1999 for giving 30,000 e-mail addresses to a pro-freedom group in the United States. This was China's first prison sen-tence for Internet activity. Such cyberphobic techniques and laws have produced only mixed results though they surely have helped suppress some measure of political dissent and technical creativity.

Where will this digital culture take us?

The three chapters that follow explore that question. They cannot hope to answer it but they can explore it from new angles.

Chapter 13 looks at the nature of art and how smart systems or machines can create more art. Many persons see art as the highest form of human culture. This chapter sees art in the more neutral terms of mathematical function approximation.

Chapter 14 looks at ways you can search the growing number of Internet and other digital databases without commercial firms and governments searching you in the process. It calls for a digital rights act to let us code our own bits in our way.

The final chapter 15 explores that great digital threshold when chips replace brains. Here culture and science and politics as we have known them fuse and transform into new systems and processes that we can scarcely guess at. The guesses are hypotheses that a digital world will someday test.

That digital world may have the power to do more than test such hypotheses. It may have the power to invent new worlds where the hypotheses are true.

Smart Art

Beauty is the purgation of superfluities.

MICHELANGELO

Science, like art, is not a copy of nature but a re-creating of her.

JACOB BRONOWSKI
Science and Human Values

I am absorbed in the blooming fruit trees, pink peach trees, yellow-white pear trees. My brush stroke has no system at all. I hit the canvas with irregular touches of brush which I leave as they are: patches of thickly laid-on color, spots of canvas left uncovered, here and there portions that are left absolutely unfinished, repetitions, savageries.

VINCENT VAN GOGH
Letter to Emile Bernard (April 1888)

Modern art: a revolt against the imitation of reality in the name of the autonomous laws of art.

MILAN KUNDERA
Testaments Betrayed

An arbitrary playing with the means of art without proper knowledge of the end is in all art the fundamental characteristic of bungling.

ARTHUR SCHOPENHAUER
The World as Will and Idea

Abstract Art: A product of the untalented sold by the unprincipled to the utterly bewildered.

AL CAPP

*So long as artistry exists there is no need of theory or logic to direct
the painter's action. The inner voice of the soul tells him what forms
he needs.*

WASSILY KANDINSKY
Concerning the Spiritual in Art

*The power of subtle orchestration is a secret impossible to transmit. The
composer who possesses this secret should value it highly and never
debase it to the level of a mere collection of formulas learned by heart.*

NIKOLAI RIMSKY-KORSAKOV
Principles of Orchestration

*Only truly meaningful polyphony can disclose the loftiest tone-miracles
of the orchestra.*

RICHARD STRAUSS
Treatise on Instrumentation

*Beauty is subject to laws and rules that depend on the nature of human
intelligence. The difficulty consists in the fact that these laws and rules,
on whose fulfillment beauty depends and by which it must be judged,
are not consciously present to the mind either of the artist who creates
the work or the observer who contemplates it. Art works with design but
the work ought to have the appearance of being undesigned and must be
judged on that ground. Art creates as imagination pictures, regularly
without conscious law, designedly without conscious aim.*

HERMANN L. F. HELMHOLTZ
On the Sensations of Tone as a Physiological Basis for the Theory of Music

*Slowly I discovered the secret of my art. It consists of a meditation on
nature, on the expression of a dream which is inspired by reality.*

HENRI MATISSE
Matisse: The Artist Speaks

The artist's ideal is to produce a crystal-clear reflection of his own self.

M. C. ESCHER
On Being a Graphic Artist

What a strange illusion it is to suppose that beauty is goodness.

LEO TOLSTOY
The Kreutzer Sonata

Art is a quick indicator of status, which—in human as in animal societies—is a key to acquiring food, land, and sex partners.

JARED DIAMOND
The Third Chimpanzee: The Evolution and Future of the Human Animal

There is no female Mozart because there is no female Jack the Ripper.

CAMILLE PAGLIA
Sexual Personae

WHAT IS ART?
Philosophers have tried to answer that question for thousands of years. Plato and Aristotle saw art as the imitation of life or ideas. The great 19th-century German scientist Hermann Helmholtz saw all music as an outgrowth of the human voice and to prove it he wrote a seminal book on the periodic nature of musical tones and how the inner ear perceives them. Others have seen art as how the mind plays or how it tries to express or transmit emotion. Philosopher George Santayana saw art as "pleasure objectified."

Philosophers are not the only ones who ponder the nature of art. Every child who has ever smeared finger-paints on paper has wondered if what she has created is art. Some social critics have claimed that drawings in restrooms and spray-can graffiti on building walls and billboard signs count not just as art but as some of its most vibrant current examples. Censors in all cultures still try to define art and draw a line between it and non-art. And now the computer has invaded art both as a tool for the creation of art and as the newest and perhaps most avant-garde creator of art.[1]

But what is it? What is the thing or process we call art?

Most archaeologists and cultural anthropologists agree that art is some form of *symbol representation* and that artists engage in a symboling activity. They also agree that the first hard evidence for art dates back about 50,000 years when humans likely first ventured beyond Africa and battled the Ice Age world to the north.[2] Art is surely symbol representation. But is all symbol representation art? That would include the equations of science and the typed entries in the *Federal Register.*

There may be only one definition of art that most people will accept. And most people will accept that answer because it transfers the burden of definition from art to those who create it. That definition is this: Art is what an artist does.

But then what is an artist?

This chapter explores a math answer to that question. The answer may not comport with all of our notions of artists of the past or present. But it offers insight into those artists who will create much if not most of the commercial art and perhaps fine art in the future: smart machines.

Smart Art: Artists as Functions in Art Space

What does an artist do?

An artist expresses himself. He signs or paints or writes or dances or sculpts. He expresses himself in some medium. The artist expresses more than just the hormonal whim of the moment even though that whim may affect his art. He expresses his ideas and the stings of life that have produced them.

So an artist is a system or a mapping: *An artist maps experience to symbols.*

The artist converts life inputs to symbol outputs. The symbols can be colors or musical tones or words or dance twists or any other medium.

You may express yourself in a shout when you stub your toe. In most cases the shout does not come from a set of structured symbols. So most of us would not see the shout as art. In the same way we tend not to see as art a cat's random walk on a piano keyboard or the shivering of a dancer who has the flu. The artist must map experience to symbols for the result to count as art.

Life is the grist for the artist's symbol mill.

This function view reflects what abstract painter Wassily Kandinsky meant when he said that the "inner desire of the artist expresses itself in different ways." To each his own function.

And the art function changes or adapts as the artist's brain adapts to new inputs. The artist looks for new symbol ways to express himself and for new inputs to excite the expressing. Sampled input drives creative output. Here the best creative advice comes from behaviorist psychologist B. F. Skinner: Vary your stimuli.

The function view of art also opens the door to computer automation. It opens the door to that species of machine intelligence we called *model-free* function approximation in chapters 9 and 10. Smart black boxes or fuzzy gray boxes map inputs to outputs. No one guesses at a math model of the throughput process or what engineers sometimes call the "plant."

Rules or neural synapses learn the input–output structure and endow the black box with machine intuition.

Human experts seem to use some mix of rules and neural associations. The poultry expert holds the new chick to the light and then puts the chick in the hen pile or rooster pile. She has no idea of the equations that define the nonlinear boundary between hens and roosters in some chick pattern space of stunningly high dimension. She classifies chick sex by trained intuition even though she may state some incomplete reasons why she classified this chick as a rooster and that chick as a hen.

She does it *by feel*. And so does the artist.

Neural networks can learn some of an artist's intuitions by training on the artist's symbol patterns. Neural nets can better and better approximate the artist's function with more and more training patterns.[3] The same holds for fuzzy systems that use neural-like learning schemes to tune the fuzzy patterns in their if-then rules. We have seen that both neural and fuzzy systems are universal function approximators. So in theory they can model any art system if they use enough neurons or rules. They can do this because we can view artists as input-output functions from experience to symbols.

Model-free learning systems grow search balls in art space.

A work of old art defines a point in some huge art space of paintings or novel plots or song melodies. The art space of paintings contains all possible paintings. This space exists in the same sense that numbers exist. It is a formal abstraction. It exists in the sense that we can capture it in symbols and use formal rules to manipulate the symbols. It is the continuous extension of what many have called a universal library.[4]

Leonardo Da Vinci's Mona Lisa is one point in the art space of painting. Next to it are millions or even infinitely many paintings that differ only slightly from the Mona Lisa. We would see most of these paintings as the same. Further away would be even more versions where the nose is short or long or where Mona has Asian eyes or three and a half eyes.

The artist picks a point in such an art space with each new work of art. The copycat picks a point nearby. The great acts of artistic creativity pick not just a point in an art space. They define a whole new art space or a new region of an art space. Vincent Van Gogh did this for the impressionists and Pablo Picasso did it for the cubists. Graphic artist M. C. Escher carved out an unnamed niche with his works based on the mathematics of symmetry and curvature.

An artist can start an artistic movement or revolution when he or she or it defines a new art space.

Machine art starts as a form of copying. It samples a work of art and picks a nearby art point or a set of nearby art points. Learning goes further and generalizes. Learning forms a ball around the original art point. Artists can guide the learning process by seeding art space with ball centers. Art spaces are so vast that a learning system has little chance of picking a good starting point on its own. That will surely change as machine intelligence progresses.

Consider a neural-fuzzy system of adaptive if-then rules. The rules can come from a real artist and capture some of her rules of thumb. Or statistical clustering schemes can find the first set of rules if they sample enough art works and average them. Then neural learning schemes can both tune the rules in new directions and teach the fuzzy rules to store abstract fuzzy patterns of art objects like landscapes or new 3-D perspectives or facial smiles or frowns.

A neural-fuzzy system might encode and compress the old art points in a set of fuzzy rules. Then noise or chaos might vary the rules and search for new minima of some aesthetic cost function. Genetic algorithms offer one way to do this. Neural learning first helps define a bumpy surface over the art space of paintings or piano sonatas or dance sequences. Hill peaks are good or beautiful and hill valleys are bad or ugly. Neural learning sculpts this surface with each new art point it samples.

Then genetic algorithms can try to find the highest peaks on the beauty surface. They turn loose hundreds or thousands of random search agents that act like balls that bounce at random on the surface. The bad search balls die. Some of the better search balls climb higher and higher and then mate to form hybrid search balls that replace the dead ones. The breeding swarm tends to find higher and higher peaks and so give a better and better rule structure.

The first goal of a smart art system is to copy the masters.

All artists fall into creative ruts that define their style. Smart systems try to learn these ruts as fuzzy patterns or rules: If the shading is light on the left side of an object then make shading very dark on the right side. If the brass group plays in near unison then combine the woodwinds with the upper strings. If the hero asks for a big favor then the answer is usually no.

A good painting example is Van Gogh's staccato brush stroke. Most aspiring painters paint studies or exercises in the style of Van Gogh and many of the other masters. The students' brains quickly learn many of the rules and patterns and hunches that let them approximate Van Gogh's sharp style. The very sharpness of an artist's style or "voice" makes it easy to copy. And the simplest way to copy is to learn or abstract the fuzzy rules that produce the sharpness.

The ultimate goal is to go beyond the old masters and create new masters. That may be a long way off. We would have to search vast and distant regions of old and new art spaces. We might stumble on novel art masters or master patterns only after we searched these huge spaces purely at random. But all these efforts require that we find new and efficient ways to define art objects as numerical points (or vectors) in art spaces. Naive schemes do not compress the art works and so can lead to art spaces of prohibitively high dimension.

So the next goal of smart art is simply to extend the masters.

We want to plant the masters as seeds in art space. We want to grow balls centered at their seeds. We want more than 49 Fabergé eggs and want miles more of medieval stained glass. Such culture farming lies within our grasp and has already begun.

There is a lot of pleasure and commerce left in the old brand names.

The Sound of Math

Why should Beethoven's symphonies have stopped at nine?

The old brain answer is that Beethoven's brain died when he was still sketching his tenth symphony. His brain was the goose that laid the golden eggs. Nature and nurture made the eggs and then in time nature killed the goose. We will have to make due with those nine grand eggs for the next thousand or million years.

This view does not even hold up in meat terms.

Someday we could in theory go to Beethoven's grave in Vienna and scrape the DNA from his bones and decayed flesh. Or we could use the tufts of Beethoven's hair that have sold at auction. Then we could re-gene Beethoven clones in the spirit of cloning dinosaurs from far less complete DNA in the film *Jurassic Park*. We could train the little Beethovens on his old works. They could play the works on the piano or synthesizer. Or they could conduct the works before fawning crowds. Rich philanthropists could commission new string quartets and concertos and perhaps engage the young brothers in competitions. Beethoven's works would only grow in number and perhaps grow in quality.

Many groups would of course protest this "genetic engineering." Some would try to pass laws to stop it. Others might try to save the human race by bombing the dead man's tomb. These vocal efforts would supply a like demand from the press for quick displays of virtue. But there would be strong incentive for some form of the Beethovens to supply the worldwide demand for more Beethoven. The franchise would be too profitable not to exist. It would slake too many thirsts.

There would be more Beethoven.

Smart systems do not need to scrape the old master's bones. They can watch and learn. They can train on art patterns. They can learn the key features of the art patterns and learn the rules that produce those features.

And smart systems can in principle learn whatever skills musicians acquire as music training reshapes their brains. Brain scans have shown that musicians have a larger left planum temporal region of their brain than non-musicians have. Playing a musical instrument also improves verbal memory.[5]

Music makes a good test domain. Simple rules form tones out of frequencies and form chords out of tones. Composers work with broad rules of harmony to move from chord to chord. Many also use explicit or implicit rules of melody: If the melody line walks down in steps then it leaps up many tones. If the melody line leaps down twice in a row then it walks up three steps. Composers from Mozart to John Williams also group chords and themes in larger chunks or other characteristic patterns.[6]

Music makes a good test domain for a deeper reason: It is the art form closest to math. The frequencies that make up tones have an exact math form. A tone is energy that vibrates in time. An octave above a tone doubles the tone's frequency and so vibrates twice as fast. An octave below halves the frequency and vibrates twice as slowly. The ancient Greeks explored the simplest tones in terms of ratios of whole numbers. Most Western music still depends on these simple tones and scales. Eastern and some modern music often use much more complex math schemes.

Music is the sound of math.

We could in theory write Beethoven's Ninth Symphony as one long differential equation. Then the music would unfold as a type of sonic wave in a sea of silence.

Music symbols have evolved as a crude shorthand for this math. The symbols compress a lot of information about time and energy. Music symbols allow one to compose and play music without math. Beethoven was so bad at math that he often had to add up long columns of numbers because he could not multiply well.

The key problem with music symbols is learning to think in terms of them.

Many people simply play the piano or guitar by ear rather than paying the ante of learning to read the alien-like symbols of music. Most people cannot read music at all. Painting and dancing do not impose such start-up costs. Anyone can paint or dance to some degree. Music is not so user-friendly.

I struggled as a teenager to read orchestral scores just as I later struggled to learn ever more complex math. I had gotten a recording of Beethoven's nine symphonies as a Christmas present and set out to understand these masterpieces. So I bought a copy of the orchestral score and tried to follow along in the score. I would get up an hour early before school and sit with headphones on and each day play one movement of one symphony over and over on the stereo.

The first task was just trying to keep up. The music might pass through a page of notes in a few seconds when the full orchestra played. The eye and brain had to follow a dozen or so lines or "voices" at once. They had to parallel process in real time. Some voices ran on the same line or staff to save space. The woodwinds almost always doubled this way. Sometimes the first violins or violas would double up and so would the brass instruments. The Ninth Symphony went further and added a chorus and vocal soloists and new percussion instruments. I would race to keep up and to find my place when I fell behind.

Then reading score became easy. I could keep up with the music and start to see how Beethoven grouped the instruments to express themes and see how he chose his chord patterns. I formed a background intuition of how Beethoven balanced his sonic landscapes. Soon I could open to any page of the score and quickly put the needle on the vinyl LP to locate the recorded sound. I could in time follow the much more complex orchestral scores of Richard Wagner's operas. Some of these scores require that you turn the score sidewise to allow for the expanded orchestra. The end of *Das Rheingold* has so many harp parts that they have to go in an appendix.

The next advance came when I could open any page of the score and hear the music in my head. Only then had I truly paid the ante and learned to read score. I could for the first time grasp the smallest part of what a conductor sees when she looks at a score and maps it to thousands of commands for the orchestra.

Once I could read score I could think in score. So I began to compose orchestral scores of my own. I carried music notebooks with me at school and worked on scores during study hall. I played parts of these scores on the piano when I got home. I found at first that my written scores often differed from what came out of the piano. More practice and study narrowed that gap between idea and sound and in time narrowed the gap between piano sketch and orchestral score. By age eighteen I was rewriting my first symphony.

The start-up costs had taken at least a thousand hours of focused mental effort.

Few people have that much extra time. Yet most people are full of song. We hum and whistle and play out themes and variations in our heads. And few people have the patience to spend the minutes or hours it takes to notate just a few seconds of orchestral music. The orchestral punctuation of one final chord involves dozens of tones spread out across the orchestra.

This helps explain why more of us do not compose sonatas or symphonies or rock ballads.

Surely there is a real demand for such new art. Computer MIDI systems help some musicians supply their personal demand. These systems map piano key strokes to score symbols and can let one mix sound tracks at will. But these systems are not yet smart. They still require some knowledge of musical notation. And of course they do not map our moods to full scores. They cannot take much of a hint from the fuzzy musical patterns we try to whistle or hum.

Modern computer music has largely ignored this demand. Modern composers have often based their music on models of formal grammars or on math structures like Markov chains that move the music at random from note to note or from chord to chord.[7] The schemes are clever and often involve a high level of technical skill. But the result often sounds like what one hears on a walk through a video arcade. Friedrich von Schelling said last century that "architecture is frozen music." One hates to picture the gray bombed-out structures that would come from "freezing" rap music or atonal serial music.

This is a problem with much of modern art: It tastes bad. And the more the artist relies on formal structures like Arnold Schoenberg's twelve-tone rows or Kandinsky's color grammars the worse it tends to taste.

We want the artist to do something both that we cannot do and that comports with our fuzzy notions of beauty. We find that in the works of

Shakespeare or Bach or Michelangelo. These artists worked with finely honed intuitions or neural associative memories. They did not work with explicit rules of construction.

The worst three-act screenplay uses the same rough rules of drama that Shakespeare used. And the kind of rules that Bach could state were for the most part the negative rules of counterpoint: Avoid parallel voices in unison or perfect fifths. None of the great artists of the past left enough written notes to allow someone to create as they did. They mapped experience to symbols by feel.

The great artists showed their artistry as they searched largely unstructured regions of their art space or as they bent or broke the working rules of their art forms. Most engineers would find it easier to program a robot or other smart system to sculpt twisted bronze modern-art pieces for posh shopping malls than to program such a system to sculpt marble statues in the style of Michelangelo. It is one thing to give a computer rules but quite another to give it intuition. Neural and fuzzy systems have some chance at building such intuitions.

Neural musical research has tried to learn from example.

Neural pioneer Teuvo Kohonen of Helsinki University has built a neural musical grammar system that learns the rough patterns of sampled styles from medieval madrigals to J. S. Bach's two-part and three-part inventions to modern jazz and Latin improvisation.[8] Kohonen's neural music has often played in the background of the welcoming party of neural conferences.

Kohonen's new Bach music sounds like Bach and yet it wanders in and out of new regions of music space that sound little like Bach. Such smart music systems may not at first sell many compact discs. They may in time help a film composer or film director fill in a background score or help each of us create new works at will. We may want to hear background dinner or workout music that comes in the fuzzy mix of 30% Felix Mendelssohn and 40% Franz Schubert and 30% John Phillip Sousa. Or we may want to take out the 30% Schubert and put in the tune we hum as we exercise or shower.

Other researchers have combined chaotic search schemes with neural-fuzzy systems to arrange music in chosen styles.[9] Chaos varies themes. Chaos acts as a type of controlled randomness that lets the user control how much the music wanders from the seed score. These smart-art schemes are also first steps and there will be many more as information machines grow faster and smarter.

The real advances will come when machines create their own symbol schemes from their own intuitive rules. Then they can map their own virtual experiences to their own symbol spaces. Then they too can create by feel.

In the meantime we may have to make due with smart *karaoke* machines.

Art for Computer's Sake

Smart art faces a legal challenge: Who owns it?

Property rights in art have always been fuzzy. One painter or composer or screenwriter can slightly or grossly copy another. Very few of these cases come to court. And now rich patrons can buy the digital rights to many of the masterpieces that hang in museums. Art vendors will soon sell limited digital rights to art just as software vendors now sell limited digital rights to their software programs. Digital "watermarking" may help bring secure property rights in cyberspace.[10] Firms already buy or rent the digital right to use dead actors or cartoon figures in beer or cigarette ads or music videos.

But what happens if someone broadcasts a new painting or new novel on the Internet? Or suppose you turn an old novel into a new "branching story" that unfolds in directions at key plot twists. The hero takes the bribe this time at the end of the first act. The rest of the story has nothing to do with the rest of the old story. Who owns what and who sues whom? A fuzzy Coase Theorem may be too fuzzy to help.

Smart art adds to the legal complexity.

Suppose a graduate student creates a new software program that sings as Barbra Streisand sings or that writes novels as Tom Clancy or John Grisham write novels. Can the student sell new artworks in these brand-name styles and not infringe the rights of the living artists or their estates?

This may seem a moot question for now. How realistic are such systems? When will smart system write a book in the style of a preferred author? The answer is one already has.

The romance novel *Just This Once* appeared in August of 1993. The novel from Birch Lane Press had a new type of subtitle: "A novel written by a computer programmed to think like the world's best-selling author as told to Scott French." The best-selling author in question was Jacqueline Susann and she had been dead for almost 20 years.[11]

Scott French had studied artificial intelligence and he had read Susann's *The Valley of the Dolls* and other novels. He came up with thousands of rules that he felt captured much of her writing style and plotting techniques. French called the program HAL. He used HAL to create a story or symbol sequence that he later edited into the novel's final form. He mapped his experience of Jacqueline Susann's writings to new symbols.

A new smart-software market has also emerged for would-be screen-writers. The software programs began as script formatters. They helped the writer put her story into the play-like notation of screenplays and teleplays. Then rule-based systems appeared. These programs capture the lessons of screenwriting workshops in a series of story questions. Other programs suggest plot ideas or character traits. More advanced systems help the user structure stories into three acts and then structure the acts.

Vendors have sold tens of thousands of these programs. Rumor has it that some harried TV writers come up with new teleplays by running the *TV Guide* description of old TV reruns through such smart software. This would amount to mapping one art space to another.

Smart software has even replaced its first human writer. Nebraska's *Humphrey Democrat* newspaper fired its sportswriter and replaced him with a $100 software program called Sportswriter.[12] The program maps sports facts into the sentences and paragraphs that make up a brief sports column. It maps sports experiences to text symbols.

The legal issues will likely grow as smart art becomes more common and more profitable. Art to date has been a poor investment on average. We have data on the risky business of owning art prints. The average real rate of return is only about 1.51%.[13] You are better off owning Treasury bonds. They yield a higher and safer rate of return and you can sell them at will.

Smart art programs will someday be standard features in office and household multimedia packages. You may want one wall swimming with new paintings in the style of Dalí or Rembrandt while new music plays in the style of Verdi or the Beatles. Researchers at MIT and elsewhere are already working on mood detectors that can further smarten your walls or car and perhaps guide the smart art.[14] Other software research in artificial intelligence lets the user shape the plot and characters in interactive fiction.[15]

Other systems may let us use digital actors to make our own TV shows or films. Why watch the old films of the 20th century when you can use them and their actors to seed your own artworks that you star in?

Today we can control simple digital actors as little more than moving skeletons in a computer animation.[16] That will surely change as computers become cheaper and more powerful and as art systems become smarter. Future systems will let us seamlessly augment reality or transform old art-works into new ones. The same product demand that today drives the markets for film and TV will push technology to supply our desire to play with Mickey Mouse in a cartoon or save a damsel in distress in an action film or make love to pop stars in a blue film.

Advanced multimedia systems offer opportunities for copyright infringement in all the sense modalities.

The march of smart art may also change the science that led to it. Scientists act much as artists do. They both map experience to symbols. They both engage in symboling. And they both trade in innovation: Science reveals structure in the world. Art adds structure to the world.

So why can't smart systems displace or augment scientists?

The answer is that they can and will.

Neural and fuzzy black boxes are just the first step. They learn complex mappings from inputs to outputs that we often cannot understand. Adaptive fuzzy systems are more artlike in this sense because they convert their training data into a pile of if-then rules. Neural systems just learn the abstract mapping from training data to desired outputs such as 0s or 1s.

More structured systems may help scientists look at a pool of data and map it to a set of symbolic hypotheses. We think of time in seconds and patterns in terms of what we can see in our mind's eye. More objective systems might look for time patterns on the scale of quantum fluctuations or all the way up to geological or cosmic changes. These smart systems could cast the patterns in terms of formal math conjectures. AI software systems have shown for decades that they can find and prove simple theorems in logic and math.

The digital age will have crossed into a new frontier when a machine first discovers a new law of physics or sociology or information theory. That frontier will be truth for machine's sake. We will have given birth to what roboticist Hans Moravec calls our mind children.[17] It will be like that moment parents face when for the first time their child comes up with a good idea that they did not see. But the impact will be thousands or millions of times greater.

A like frontier awaits smart art: Art for computer's sake.

We could appreciate the 10th or 100th Beethoven-like symphony that a smart-art system composed. But the system could soon search regions of new art spaces that we could not grasp.

And a smart machine's eye could see paintings or sculptures in 4 dimensions or in 4,000 dimensions. The smart-art system could abstract our fuzzy notions of beauty from our tiny number of art samples in low dimensions. It could search out or "create" beauty in art spaces of ever higher dimension and complexity.

The human ear cannot hope to grasp a fugue in 20 musical voices or in 1,000 voices. It takes a trained ear just to follow counterpoint with 3 or 4 independent voices. Mozart capped the last movement of his last symphony with a brief fugal passage in 5 voices. I struggled in the finale of my own first symphony to keep 7 voices going for a few seconds. I am sure composers could on paper double that number of voices but I doubt they could know what the joint effect sounds like. That does not mean there are no wonderful passages of music with 10 or 20 or more voices. There should be an infinitude of them.

So why should minds find beauty only in the low dimensions that our brains can grasp? Why should the range of notes stop outside of what our brains can hear? Why should the notes move no faster than our brains can follow?

We can try to confine the smart-art systems of the future to our regions of art space. That will work at first just because the first smart-art systems will be simple and will do little more than what we or our preferred artists tell them. More advanced systems will search deeper and wider and will soon move further away from our regions of art space. They will create for themselves even as they at first create for us.

But why confine artistic creativity of any sort?

That has seldom worked in our past though many governments and religions have tried to do it. If computer art exceeds our brains then maybe we should let it. Maybe we should expand our brains.

Or maybe we should get new ones.

1 4

Secret Agents

The wise man proportions his belief to the evidence.

DAVID HUME
An Inquiry Concerning Human Understanding

A habit of basing convictions upon evidence, and of giving to them only that degree of certainty which the evidence warrants, would, if it became general, cure most of the ills from which this world is suffering.

BERTRAND RUSSELL
Why I Am Not a Christian

I killed people in war, challenged them to duels with the purpose of killing them, and lost at cards. I squandered the fruits of the peasants' toil and then had them executed. I was a fornicator and a cheat. Lying, stealing, promiscuity of every kind, drunkenness, violence, murder—there was not a crime I did not commit. Yet in spite of it all I was praised. And my colleagues considered me and still do consider me a relatively moral man.

LEO TOLSTOY
Confession

Never do anything against conscience—even if the state demands it.

ALBERT EINSTEIN
Albert Einstein: Philosopher-Scientist

Are we going after their [the Jews'] tax returns? . . . I can only hope that we are frankly doing a little persecuting. . . . What about the rich Jews? The IRS is full of Jews, Bob [Haldeman]. Go after them like a son of a bitch!

PRESIDENT RICHARD M. NIXON
White House tapes: 8 and 14 September 1971

Woody Allen: "Do you believe in God?"
Diane Keaton: "Well I believe that there is somebody out there who
watches over us."
Woody Allen: "Unfortunately it's the government."

Sleeper

To be governed is to be watched, inspected, spied upon, directed, law-
driven, numbered, regulated, enrolled, indoctrinated, preached at, con-
trolled, checked, estimated, valued, censured, commanded by creatures
who have neither the right nor the wisdom nor the virtue to do so. To
be governed is to be at every operation, at every transaction noted, reg-
istered, counted, taxed, stamped, measured, assessed, licensed, autho-
rized, admonished, prevented, forbidden, reformed, corrected, punished.

PIERRE JOSEPH PROUDHON
General Idea of Revolution in the Nineteenth Century

You never know what is enough unless you know what is more than
enough.

WILLIAM BLAKE
The Marriage of Heaven and Hell

Most of us would trust GM, IBM, or AT&T currency more readily
than that of many developing nations because the "currency" repre-
sented by those companies is more likely to remain convertible. As soon
as currency becomes bits (dutifully encrypted) its reach is unlimited.

NICHOLAS NEGROPONTE
Wired: 4.11

I call on Congress to pass the requirement of a V-chip in TV sets so
parents can screen out programs they believe are inappropriate for their
children.

PRESIDENT BILL CLINTON
23 January 1996

If in personal affairs, where all the conditions of the case were known to
me, I have so often miscalculated, how much oftener shall I miscalculate
in political affairs, where the conditions are too numerous, too wide-

*spread, too complex, too obscure to be understood? Here, doubtless, is a
social evil and there a desideratum; and were I sure of doing no mis-
chief I would forthwith try to cure the one and achieve the other. But
when I remember how many of my private schemes have miscarried;
how speculations have failed, agents proved dishonest, marriage been a
disappointment; how I did but pauperize the relative I sought to help;
how my carefully governed son has turned out worse than most children; how the thing I desperately strove against as a misfortune did me
immense good; how while the objects I ardently pursued brought me little happiness when gained, most of my pleasures have come from unexpected sources; when I recall these and hosts of like facts, I am struck
with the incompetence of my intellect to prescribe for society.*

HERBERT SPENCER
The Man Versus the State

Ask yourself whether you are happy and you cease to be so.

JOHN STUART MILL
Autobiography

WHOM DO YOU TRUST?
Most of us trust our family and friends to varying degrees. We
may trust the legal or medical advice we pay for even though we tend not
to get the same advice if we pay other lawyers or doctors. We place less
trust in the claims of commercial firms or state agencies. We doubt their
motives and question their track records.

We also have to discount the claims of scientists. We have to factor into
their claims at least the narrowness of their knowledge or aims and the
effect of their state or private funding. Scientists seldom know the real
extent of their expertise or the fuzziness of its borders. And few scientists
pursue pure research in the sense of unfunded research. Most seek funds
that pay for system overhead and teaching relief and staff and graduate student salaries as well as funds that pay for software and computer time and
lab equipment.

We may trust ourselves by default but that offers no help when we need
help. And we have to face up to our own complex motives and mixed track
records.

The question of whom to trust in the digital age soon comes down to what to trust. What we have to trust or not trust is data. Facts and opinion and nonsense make up the data content that lies in the vast databases of the world and cyberspace.

Most data has the form of pulses of light or electricity. So we can transfer huge amounts of it at the speed of light. This massive data transfer drives the digital age and swells our personal information power.

But such power comes at a double cost. The first cost is a geometric increase in search complexity. There is too much data for us to fully search it and we seldom have concrete search criteria. No one solves tangles of equations to match up their wants and budget with what the local food market offers. Optimal search is a lost goal that each day grows more distant. The best we can achieve is some trade-off between a good search and a fast search.

The second cost is a compound loss of privacy. More people and more firms and state agencies can search our data. They can read some or all of our vital statistics that describe where we work or whom we speak to or how much we borrow or which drugs we buy. It no longer takes an errant president to invade our privacy and leak or "go after" our tax returns or medical files. A new-hire data clerk can do it with a few keystrokes.

The problem compounds because we leave part of ourselves behind as we search databases. Our Internet click trails or "clickstreams" persist in cyberspace long after we log off. The more we search the more we reveal. Our clickstreams can feed complex programs that fit us into corporate or government "profiles."[1] Using Internet TVs and cybercash and Internet phones or I-phones will grow still more detailed clickstreams.

We need someone or something we can trust to smartly search databases for us. We need them to discreetly do our digital bidding and to shield us as best they can from the prying eyes of private firms and state agencies around the world.

We need a good agent.

Your Agent Will See You Now

You start each day with a trip to the bathroom. Your blood sugar is low while your bladder is full and your mouth swarms with bacteria. This is an ideal time to measure your bio-variables and collect data and make a few testable predictions.

It is an ideal time to see your software agent.

Small personal sensors can process this data and send it for analysis to your household computer. The computer also stores your medical history and might send some of the data over the Internet for further analysis or to store it or match it against like samples in other databases. The results could help you choose your breakfast or choose the right dose of medication to take with your breakfast.

The results could also help dissolve the powerful medical monopoly.

Physician groups have passed laws that restrict the supply of those who can practice medicine and thus compete with them. Nurses can compete most directly with physicians. Physicians have passed licensing and certification laws that outlaw or limit competition from nurses and from others who offer substitute health care such as midwives and chiropractors and herbalists.

Physicians passed these laws at least in part for our own safety whether or not we wanted them to choose our safety levels for us. Computer experts could make the same argument for restricting who can make or buy or sell software or computers. Our bits can be just as important to us as can be many of our molecules.

Many of these restricted nonphysicians offer comparable services to physicians.[2] A study in the *New England Journal of Medicine* found that Americans visit these nonphysicians more often than they visit physicians.[3] Many people simply prefer the risk/cost ratio of nonphysicians at least for basic health care and the treatment of some ailments. And the government does not subsidize the training of nonphysicians as it does most physicians.

Physician groups also limit competition by restricting which medical schools get and maintain accreditation and who can attend them. More than half the medical schools in the United States are state owned.[4] A reduced supply of medical services increases health care costs and physician incomes while it gives the rest of us fewer choices.

The Internet will help erode the medical monopoly.

Physician groups have long since passed laws that restrict out-of-state competitors. The Internet will allow people to seek medical services not just out of their state but out of their country. Why not pay a credit-card fee and send your health data and questions to professionals in Sweden or India or China? The Food and Drug Administration and other state agencies have only a slim chance of stopping such binary trades. The chances will get slimmer if we can securely encrypt our digital communications.

Now suppose you had your own medical software agent.

The agent is a set of software programs that can search and learn. The agent knows far more about your health and medical history than you do. It knows the values of all your bio-values and biorhythms each day for the past few months or years. The agent knows the food mixes you like and treatments you do not like and which you can afford. It searches thousands of databases for you and bargains with other agents on your behalf so that it can better understand your data and offer you more treatment choices at better rates.

And you can trust it.

Your medical agent has only your interest at heart because it has no other interests. It wants only to give you the best medical assistance at the best price. Its judgment does not depend on blood sugar or fatigue or concerns about personal income or student loans. And by then it may not have to worry about drug cops who watch how it prescribes Schedule II painkillers.

The first medical agents may face the same legal threat of malpractice as physicians now face. Physician groups will no doubt demand that medical agents meet safety and legal requirements that they will set and monitor. This will reduce competition from agents at first. But it is a practical if self-serving way to deal with a real problem: Whom do you sue if your agent poisons you?

Physicians alone will not control the answer. Computer markets are too efficient for that.

The first medical agents will likely be smart diagnostic tools that the consumer can buy for about the cost of a medical checkup. These tools will be little more than home blood and urine tests with chips in them. Consumers will use these smart tools at their own risk subject to product warranties. But their price will fall and their skills will grow and more and more people will use them. Cheaper and smarter systems will in time cross the fuzzy line from diagnosis to prescription and on to active treatment.

The poor may benefit the most as both the supply of and demand for cheap smart medical agents swell and help shape health care in the digital age. The poor will be able to afford the personalized and high-quality health care that now only the better-off can afford. And the world teems with poor people who need much better health care than what the current closed system of human experts provides.

Medical agents will give the medical monopoly a run for its money.

Legal agents will do the same for the legal monopoly as will other agents for the talent pools of accounting and investing and education. Why not tailor each student's education as we would tailor his medical or financial counseling? Cable and satellite systems already let many students shop for courses at colleges and universities far from home and far from one another. Intelligent agents will go beyond this and act as on-site super-tutors.

Agents will be the great equalizers of the digital age.

Secret Search Agents

Intelligent agents are the focus of intense research in multimedia engineering.[5]

Intelligent agents come in many forms. They all come in software even if that software sits in a chip and the chip sits in a dashboard or headphone or robot arm. Their variety stems from how researchers endow the agents with intelligence. Much of this "intelligence" just describes how agent programs quickly check logic cases or crunch numbers or match objects when they search a database.

Most agents act as database filters. Data pours into the agent and only some of it pours back out. The agent's search has in effect filtered out some of the data and found or mapped in the rest. The agent may map a database into just one of these objects or map it into a subset of the objects. A medical agent does this when it searches a database of drug prices and finds the cheapest drug for its master's ailment.

The agent acts as a proper subset filter if it finds more than one best match. Then the agent needs a scheme for "conflict resolution" so that it can then rank the objects in the preferred subset or find the best of the best.

Internet browsers are filter agents. The user types a key word and then the browser searches its database of Web sites for matching key words. The key words "Rembrandt paintings" may produce only a few hundred sites or "hits" while the key word "Art" may produce hundreds of thousands or millions of sites.

Most browsers try to rank these sites because so many matches are dumb. The key words "Rembrandt paintings" might call up all the posted e-mail messages of people whose real name or Net pseudo-name ends in Rembrandt and call up all the posted messages that respond to these messages. Dumb search is brute-force or exhaustive search. It relies on speed and data access while smart search prunes the database with knowledge.

Knowledge is the essence of a search agent's intelligence: *More knowledge implies less search.*

Suppose you want to find the listed phone number of someone in the United States. You walk into a room and find a huge pile of all the current U.S. phone books. If you had no knowledge then you would have to search through all the phone books and call all the numbers. If you know the city the person lives in then you could prune the search to just one phone book. If you know the person's last name starts with a *G* then you could prune the search within that phone book to just those names that start with G. If you know only the spelling of their last name then you could prune the search to just those names and so on.

Knowledge shrinks the effective search space. The hard part is capturing that knowledge in symbols or numbers so that a computer can use it.

Many agents in artificial intelligence combine binary logic trees with interactive software programs.[6] AI agents already help buyers shop at on-line record stores for bargain compact discs.

AI logic trees may have the form of expert systems of binary if-then rules. Or they may define parsing trees of language grammar. Or they may not be pure trees because they have closed logic loops in them as in a semantic or causal network of links and concept nodes. The software programs come in great variety but most operate on symbols or strings of text. Their knowledge tends to be explicit and rest more in words than in numbers.

Neural and fuzzy techniques offer a related but different way to increase an agent's machine intelligence. These techniques may have a user interface that works with natural language but the techniques themselves work with numbers rather than with logic symbols or text strings.

Neural networks let the agents learn fuzzy patterns or concepts from imperfect samples. This type of knowledge is intuitive. The user does not give the agent rules or try to define the fuzzy patterns in words. The user just shows the neural net examples of the patterns by training it on the objects in the database. Then the neural network can give a fast parallel match of new objects to the old objects stored in the database.

Suppose an architect sketches a new building and asks a neural agent to find the most similar buildings out of all the buildings stored in an image database of buildings. The neural network pulls the matched building pattern out of its distributed associative memory no matter how many images the database contains. The neural network does not compare the sketched building to the stored buildings one at a time just as our brains would not do that if we tried to answer the architect's question from memory. We would just somehow come up with the answer.

Computer scientists call this type of search *content addressability*.[7] The neural network does not look up the label or address of stored patterns as would a standard serial search scheme. Instead it searches for the closest building patterns based on how much the stored patterns match the content or description of the building sketch. This type of search requires an extra scheme to resolve the conflict among the best matches.

This neural search works well so long as the neural network has enough neurons and synapses and enough time and computer power to learn or absorb all the building images into its associative memory. Then the neural network can keep the database patterns spread out in its associative memory and avoid too much crosstalk or interference among the stored patterns.

The challenge is to load up a neural network with a large number of building patterns that are similar to one another or close in some metrical sense. Then the neural network may need extra synapses and thousands or millions more training cycles to learn the fine distinctions among the patterns. A human expert would also need to spend more time studying the similar building patterns if she wants to be able to tell them apart from memory.

Agents can use fuzzy systems in at least two broad ways. The first way is to define the fuzziness or degree of match of the associative search.[8] The building sketch acts as a point in the building space or the set or database of all stored building patterns. Then fuzzy search schemes can define fuzzy balls centered at the new building sketch. The ball is a mathematical fiction because the building space consists of a discrete set of building patterns.

A small ball defines a precise search. A 90% search ball contains only those building patterns that match the building sketch or search key at least to degree 90%. A 20% search ball is much larger than the 90% search ball and would tend to contain many more building patterns. A 20% search ball defines a much broader search since building patterns can more easily satisfy its membership requirement. It contains all the building patterns in the 90% search ball and likely a lot more. The building-sketch point defines the center of an infinite number of nested fuzzy search balls in the database.

The fuzzy match itself can depend on a fuzzy measure of equality. Fuzzy equality measures how much two building images equal or resemble each other. Two office buildings may resemble each other more than either office building resembles a library or church. How much two building images equal each other depends on how much each contains the other or how much each building is a subset of the other. This reduces fuzzy equality to a measure of fuzzy mutual subsethood or mutual containment.[9] Then the search agent combines the math of fuzzy sets with the math of neural learning. The agent can adapt its degree of match to best suit the task at hand or to best reflect its master's tastes.

The second way an agent can use fuzzy systems is to use fuzzy rules to help build its knowledge base and to reason from input matches to search outputs. Fuzzy rules fall somewhere between the explicit word rules and symbols of AI and the learned numerical input-output associations of neural networks.

The user or agent can state the fuzzy rules in words but the rules have a numerical form as we saw in chapter 9. The fuzzy rule "If the building is rectangular then the user much likes the building" defines a fuzzy rule patch that associates the fuzzy set of much-liked buildings with the fuzzy set of rectangular buildings. Neural learning can tune these fuzzy sets to match what the user means by rectangular and much-liked and so define the user's niche in the conceptual anarchy.

Fuzzy rules can help a user teach a search agent the user's tastes. An agent needs guidance when it searches for its master. Fuzzy systems can house and structure this guidance.

The most basic kind of guidance is what the master likes and dislikes. An agent needs to know these tastes when it screens and perhaps answers its master's phone calls or e-mails messages or when it patches together its master's morning newspaper from all the papers and news sources of the world. The agent must learn a profile of its master. Economists call this profile a preference map or utility function.[10]

A good profile lets the agent rank each object in the database from those least liked to most liked. Then the agent would know that the user likes a gothic building more than she likes a modern building. This is an *ordinal* ranking.

A better profile lets the agent say how much more its master likes a gothic building than a modern black-mirrored building. The agent could assign value or utility numbers to the objects in the database. This is a *cardinal* or numerical ranking.

Cardinal rankings are easier for neural and fuzzy systems to work with but they are harder to get from the user. The agent may have to ask the user hundreds of questions of the form "How much do you like this building?" or "Do you like this building two or three times as much as you liked the last building?" Or the agent may have to infer or guess at these numbers as it watches the user pick out buildings she likes or deletes what she dislikes.

The query process may prove tedious and time-consuming. That may still be the case if the agent has a speech-recognition interface that lets the user speak to the computer agent rather than feed it key strokes and mouse clicks. Future interface systems may lighten this burden with retinal displays or even supersensitive brain wave detectors.[11]

Adaptive fuzzy systems can speed up the session of questions and answers or at least squeeze more rule structure out of it. Neural learning grows and tunes the fuzzy sets and rules. This can give a quick fuzzy approximation to the user's profile or preference map. The map itself is a bumpy surface of hills and valleys defined on some database or pattern space of high dimension. The user likes the hill patterns and dislikes the valley patterns. The hill height or valley depth measures how much the user likes or dislikes these patterns.

The trouble as always is that no one knows what these surfaces look like.

Economists have assumed for decades simply that these utility surfaces exist and that market buyers and sellers hide them in their brains. But economists have not been so bold as to guess at the equations that define these surfaces.

Neural-fuzzy agents can guess for us. The guesses will be far from perfect but they will tend to improve as the agent system absorbs more question-answer pairs and as the interface technology improves.

The agent tries to patch the bumps of the preference map.

The agent forms its first fuzzy rules at or near the first samples the user gives it. The user may click through a stream of building images on a computer screen and state whether or how much she likes or dislikes each image. The agent views these answers as some of the bumps on the user's underlying but unknown profile. The agent covers these bumps and fills in around them with other rule patches. Constant neural tuning shifts these rule patches to best fit the mounting set of data. The agent's fuzzy system forms its own evolving bumpy surface that better and better approximates the user's profile.[12]

The deepest problem with the neural-fuzzy approach is how it represents the patterns it tries to model. An image consists of thousands of tiny picture elements or pixels. The dumbest approach is to model each pixel as it changes many times per second. Computers are not fast enough or powerful enough for neural-fuzzy systems to do this efficiently.

So the agent system needs a front-end system that compresses each image into contours or energy measures or even color blocks.[13] There is a vast literature of such techniques in the field of pattern recognition. Smart agents will likely combine these with many techniques from AI and neural networks and fuzzy systems and genetic search algorithms and dozens of other technologies.

Future agents will do more than our digital bidding. They may be our closest friends. They will be living diaries and companions with which we share our deepest secrets and seek our most trusted advice.

They will be our secret agents.

Imagine an agent that had the moral wisdom of an Albert Einstein or Bertrand Russell or Leo Tolstoy or any of the great thinkers of the past. Smart systems will absorb their wisdom by absorbing the rule structure and concept patterns of their written works as with the smart art systems discussed in the previous chapter.

Some users may want a hybrid agent. Its wisdom might come from a fuzzy mix such as 30% Einstein and 40% Buddha and 30% Benjamin Franklin. Others may want to draw on the more familiar characters of pop culture like Elvis Presley or Isaac Asimov or John Wayne. Software firms will have to license many of these character patterns from their estates or from the book publishers or film studios that control their digital rights.

For my part I would like a personal agent based on the written works of the 19th-century philosopher and economist and profoundly nice guy John Stuart Mill.[14]

Mill had one of the deepest and most wide-ranging minds ever captured in print. He mastered Latin and Greek at a young age and went on to help found modern economics and the utilitarian social philosophy of seeking the greatest good for the greatest number of persons. Many in British society saw him as a sage in his own lifetime even if they did not reelect him for a second term in Parliament. Mill's book *Methods of Logic* has been a classic on what we now call the philosophy of science since he published it in 1843. His short book *On Liberty* remains perhaps the clearest statement of the need for personal freedom and social tolerance and still inspires thinkers of all political stripes.

Others may find Mill too stuffy and verbose. Or some may want to combine Mill with dozens or hundreds of others of the great thinkers of the past. They may want to graft together more and more intelligent agents to form intelligent super-agents.

This impulse to multiply agents beyond necessity may prove irresistible. It may offer so much knowledge and search power for so little cost that few will be content with the docs-in-a-chip that grace their smart cards or ear pads.

Cybergreed may crash cyberspace.

Agent Explosion and Smart Markets

Not all information researchers like intelligent agents.

Some feel that we will come to rely too much on agents or that working with agents will make us treat people more like machines. Virtual reality pioneer Jaron Lanier shares this concern:

> Intelligent agents stink. I am concerned that people will gradually, and perhaps not even consciously, adjust their lives to make agents appear to be smart. If an agent seems smart, it might really mean that people have dumbed themselves down to make their lives more easily representable by their agents' simple database design.[15]

Smarter agents can address these and like concerns. But their increased search power and intelligence can pose new ones.

Perhaps the greatest threat that cheap smart agents pose is a cyber-version of the curse of dimensionality: exponential agent explosion.

Your personal agent will need many agents to perform its tasks. It will need to bargain with other persons' agents for some tasks and services. And it will need to control its own slave galley of task agents for hundreds or thousands of search and processing tasks. Your personal agent will be a hierarchical system of agents or what MIT's Marvin Minsky has called a society of mind.[16]

There may be no sure way to keep agents from making more agents. These new agents can perform dedicated tasks and can make still more agents. This can quickly lead to an exponential explosion of agents. Where will they all fit on the Internet?

Many researchers have predicted that the Internet will crash. More users attract more users and their bit streams can swamp the wired and wireless paths of cyberspace. This has already caused delays and brownouts.

Agent explosion may prove more serious. We can quickly create more agents to exceed the capacity of any information system. Net firms or governments may try to restrict the number of agents per user but that may prove hard to enforce. But users may still not have much incentive to restrict their agent budget. And if other people restrict their agent budget then any given user can secure more relative information gain if he exceeds his budget.

Cyberspace is a bit-based tragedy of the commons.

Some form of the Coase Theorem from chapter 6 may help change that. Right now ownership is far too fuzzy in cyberspace. If we somehow assign binary property rights to all the bit-stream pathways then users would have to be more frugal in how they used them. The Invisible Hand of the Coase Theorem may keep the total number of software agents in balance as well as secure some type of Pareto optimum for Internet users. An army of agents may simply cost most users too much to use.

Micro-pennies will also help.

A perfectly efficient market would charge some slight fee for each e-mail message or search query we send. And you might want to pay an insurance firm an even tinier fee to insure each bit stream. Chopping the digital cent into micro-pennies or even nano-pennies permits this fine-grained book-keeping.

Micro-currencies would also help firms and graduate students earn income on the Net. And micro-fees would reduce the number and extent of e-mail broadcasts that clog the Internet. Some Net-heads will protest these tiny fees since cyberspace seems to give us something for nothing. But firms will still charge them.

The Coase Theorem also suggests that cyberspace will see vastly more light-speed auctions. Cyberspace may transform from a massive bit-based common to a massive smart market of complex and shifting agent equilibria.

The Coase Theorem promises Pareto optimality or a market outcome such that even God cannot make an agent better off unless He makes some other agent worse off. The agents can just as well be software agents as Hollywood agents. This stable equilibrium demands that property rights are well defined or binary and that there are zero transaction costs of bargaining. That way the agents can trade and swap all objects of interest until they can trade or swap no more. Micro-currencies can help reduce transaction costs while helping us assign binary property rights to all Net actions and transactions.

Light-speed auctions will extend the reach of the Coase Theorem. The Coase Theorem carries more force as more agents buy and sell more things. Most of us do not hold auctions when we trade atoms or bits among ourselves. Fast bit trades will let more of us or our agents trade quickly and cheaply and this favors the auction format.

We can often gain more for ourselves if we hold an auction for our wares than if we just negotiate for them. But atom auctions can be hard to hold because of the cost of finding bidders and the atom-based cost of shopping around our goods or services. Economists Jeremy Bulow and Paul Klemperer have shown that auctions beat negotiations under fairly broad conditions: "The value of negotiating skill is small relative to the value of additional competition."[17] This holds if the agents in the auction place their bids independently of one another as they are likely to do in high-speed worldwide auctions for small bit tasks.

Idea markets and e-money (electronic credit) have helped form the first and simplest smart bit markets. Economist Robin Hanson invented idea markets or idea futures while still a graduate student at Caltech:

> Imagine a betting pool on disputed science questions where the current odds are treated as the current intellectual consensus. For example, people might bet on whether cold fusion will be used to produce power by the year 2020. Right now the odds would be fairly low—say 20-to-1. But as the results of new research became known and if more people became convinced that cold fusion worked, the odds would rise. And if cold fusion became a reality by 2020, those early supporters would make a bundle. Such betting markets would become "idea futures" markets—like corn futures markets except you'd bet on the future settlement of a scientific controversy instead of the future price of corn. The system could increase the public's interest and role in science. And betting odds could serve as a scientific barometer to guide mass media and public policy.[18]

There are trial idea futures markets on a few Web sites but so far they do not involve real money.[19] If they did then some governments might object that these innovative bit markets are no more than the old numbers racket in digital disguise. They might view idea futures as agent gambling.

But even cash-free versions of idea futures can provoke a more informed debate on public issues. These debates might offer some guidance on how to structure fuzzy tax forms and which funding options to list on them.

E-money or digital cash can give idea futures real financial punch.

Electronic money is the latest way to transfer credit from one place or person or institution to another. Credit transfer has always been a matter of moving bits rather than atoms. We have transferred credit this way for hundreds of years from IOUs and loan notes to personal checks and money-grams. E-money just lets us transfer credit over the Internet or over smaller personal or institutional communication systems. The Net has already spawned e-money banks with digerati names like First Virtual Holdings or DigiCash. E-money is an ideal tool for Net trading in micro-currencies.

The multitrillion-dollar derivatives market runs largely on e-money. So too does the vast London-centered market that trades in Euro-dollars or the foreign exchange of American dollars. The Euro-dollar market has the further distinction of being the world's largest stateless market. The U.S. government does not control or regulate the Euro-dollar market though it would like to.

The Euro-dollar market is a crude precursor to future private currencies. Nobel laureate economist Friedrich von Hayek long ago proposed private currencies as an efficient way to stabilize the world currency markets and help dampen a country's efforts to inflate its own currency. This has spawned a vast literature on what economists call free banking.[20]

Hayek surprised many of his European colleagues in the 1970s when he said they should give up their old goal of a unified European currency. Why not pay your bills or receive payments in Swiss francs or British pounds or Japanese yen if you prefer? Let the currencies compete and let consumers choose. Users would shift away from weak currencies and shift toward stronger ones.

World markets would punish or reward countries based on how well their central banks managed their currencies. Bond and stock mutual funds already treat governments this way and do so at the speed of light. Much of the money that flowed out of Mexico during its late-1994 peso crisis was mutual fund e-money.

Few economists took Hayek seriously in the 1970s. The transaction costs were too high for the French citizen to trade in so many currencies with each purchase. E-money on the Net has since reduced those costs to the point where trading in competing currency is practical for at least some large wealth transfers. Smart agents will make it practical for all of us. Yet the 11 countries of the European Monetary Union moved in the opposite direction and imposed the new monopolistic euro currency on themselves at the start of 1999.

Private currencies can still come next.

Each of us can in theory start our own currency. We can tie our private currency to a basket of goods or issue a pure fiat currency like *Monopoly* money or the U.S. dollar. But we would have to convince the money markets that we will not inflate or deflate at will and that we can always convert back and forth between our currency and theirs. There is not much chance of that.

Large firms or banks might establish such a reputation. It might take them years to do it but some large firms or banks may have the time and profit expectation to try. What could it hurt to let them? Your agent can always work out the e-money details with their agents.

The digital culture always pushes for more bit flow. E-money and competitive currencies show just two ways how the digital culture can challenge the concentrations of power that still make up much of our older atom culture. The digital culture pushes for more information decentralization and more of the "creative destruction" that makes up innovation. Culture clashes are inevitable.

The largest clash so far involves secure communication.

Last Call: A Digital Rights Act

Privacy may be doomed in the digital age.

We want to speak and spend freely and securely whether we give out our credit card number over the counter or over the telephone or over the Internet. We want our agents to work for us in private and to protect our transmissions and our digital signatures.

Governments have a different view.

Many governments around the world have passed laws to limit this digital privacy. They have passed vague laws that limit speech and censor content and that give the state the legal right to tap phones and data lines or to keep users and their agents from encrypting their bit streams.

Governments have of course done this for our own good and in the name of national security or to protect children or to fight drug dealers and terrorists. And they have done so despite existing laws that outlaw the same behavior in the world of atoms.

The United States has taken the lead in many of these anti-privacy efforts.

The United States has outlawed data encryption software for American firms and citizens even while other firms and citizens can buy the same software in Europe or parts of Asia. It has given law enforcement ever more powers of search and seizure as it has fought a failed war on drugs in its second attempt at substance prohibition. And it has outlawed broad forms of Internet pornography even though these efforts have so far largely snagged in the courts. It and other governments have made no major legal efforts to secure or promote digital privacy.

I had the chance to witness a small part of one of these anti-privacy legal efforts in the fall of 1994. I suspect what I saw has occurred and will occur again many times around the world. The experience gives some insight into how the state can quietly pass legislation that restricts digital civil liberties when like efforts to restrict like liberties at the level of spoken or written speech might lead to protests or riots.

This digital privacy issue dealt with the Federal Bureau of Investigation's rush to tap digital telephones as more and more phone systems switch from analog to digital. The tale requires some history of recent FBI efforts to prevent private encryption in digital systems.

The FBI has pursued digital wiretap legislation since at least 1992 when it helped prepare a General Accounting Office briefing to the House Subcommittee on Telecommunications. The briefing described what the FBI saw as its need to restrict digital communications so that the FBI and other state agencies can always tap into them:

> The FBI considers wiretapping an essential information gathering tool in fighting crime. The federal government and 37 states have statutes governing wiretapping.
>
> The FBI now has the technical ability required to wiretap certain technologies, such as analog voice communications carried over public networks' copper wire. However, since 1988, the FBI has become increasingly aware of the potential loss of wiretapping capability due to the rapid deployment of new technologies, such

as cellular and integrated voice and data services, and the emergence of new technologies such as Personal Communication Services, satellites, and Personal Communication Numbers.

In response to the rapidly changing technology, the FBI prepared two legislative proposals in April and May 1992. The May proposal replaced the April proposal. According to the FBI, these proposals are intended to maintain the same level of wiretapping capability for new telecommunications technology that it has with technologies such as older analog communications using copper wire.[21]

The FBI had cast its proposed 1992 wiretap bill as a way to fight crime. FBI chief of investigative technology James Kallstrom described the FBI's concerns: "We see a rocky road ahead for law enforcement because the [new digital] technology hasn't been designed with the correct feature packages."[22] Those "correct features" were just those that would keep us or our agents from sending secure messages by phone or fax or e-mail or personal wireless or any other digital system.

But the real issue was not about technology. It was and remains about the First Amendment right to say what we want to say in our own way. We just speak a lot more now with bit strings of 1s and 0s than we did in the past.

New telephone systems have raised this old concern about free speech. The new digital phones convert what we say into long bit strings. We can securely say something in our own way if we encrypt those 1s and 0s with the latest smart software. This software relies on a large and growing literature of encryption math.[23]

The FBI wanted and still wants to be sure it can crack the binary codes we use. It wanted to limit the codes we can use and wanted to be sure it can listen to the codes it lets us use. The FBI did not and does not want us to speak in a language it cannot hear or understand.

The growing shift from analog to digital communication systems has eroded some of the FBI's power. Legal wiretaps are a key part of that power.

Congress approved wiretaps in 1968 that stem from a court order. This was in response to a 1967 Supreme Court decision that made all wiretapping illegal. To get a court order for a wiretap the FBI must show "probable cause" of a crime or a threat to national security. "Probable cause" can now mean in practice just that the FBI asks for a wiretap. That stems in large part from the extended search and seizure and asset-forfeiture powers that the Reagan and Bush and Clinton administrations and the Supreme Court gave the FBI and police and other state agencies. They gave them these new powers to help them wage the cold war and the war on drugs.

Phones were analog in 1968 when AT&T still held a monopoly on phone lines in the United States. Many phones are still analog. You speak into the phone and it turns your speech into smooth changes in current that flow over wires to some other phone or microwave link or satellite link. The FBI can easily tap a phone or fax call if it can intercept the analog signal.

Most new communication systems are digital. And each phone company uses its own digital schemes and devices to send and receive digital messages. The FBI can crack our bit codes of 1s and 0s only if it has direct access to them. It has such access for our wireless phones and faxes since these transmit bits through the air. The FBI and police can tap a cellular phone without a warrant. The FBI and other state and defense agencies can scan some if not all of our calls at random and look for key words and case leads. These random scans occur far more often than most people realize.

But the FBI did not get what it wanted. It did not get to install the "correct features" to crack our codes. Industry complaints and privacy concerns helped kill its 1992 wiretap bill.

Then the Clinton administration tried to do much the same thing for the FBI with its failed "Clipper Chip" proposal in early 1994. Civil libertarians focused media attention on the clipper chip and helped kill it too. That scheme would have built code breakers right into new American computers and phones and modems and satellites.

But a new version of the FBI's 1992 wiretap bill quietly emerged and became the 1994 wiretap law now passed into law. This law did not ban private encryption as some of the earlier FBI efforts had tried to do. That battle still continues. But the 1994 law did give the FBI and drug agents and other law enforcers the legal right to scan our digital phone and fax calls when they feel they need to.

The 1994 FBI wiretap bill bypassed the home phone and went straight to the phone companies.

The FBI had claimed that it failed to carry out several court-ordered wiretaps in 1993 because of poor line access. The new wiretap bill fixed that and more. The bill required the FBI to pay the phone companies $500 million to retrofit the phone lines as the FBI saw fit. The phone companies had to install devices that let the FBI and other state agencies "plug in" and listen to digital talk. Thus did taxpayers pay for a new loss of liberty.

The 1994 FBI wiretap bill was not the omnibus telecommunications reform bill that passed in February of 1996 even though its sponsors called it a telecommunications reform bill before it passed. The 1996 omnibus bill's Communication Decency Act outlawed much sex talk and pictures on the Internet. Courts have challenged and largely defanged the Decency Act but some form of it or other laws like it may well pass in the future.

The 1996 telecommunications bill also forced TV firms to build a violence censor or "V-chip" into their new TV sets. This was part of the legacy of the failed clipper chip. President Bill Clinton cited the V-chip law as one of his key achievements when he ran for reelection in 1996. Passage of the V-chip law also helped convince TV stations to rate the content of their TV programs in terms of sex and violence.

The 1994 FBI wiretap bill passed in the eleventh hour of the 103rd Congress in October. A student brought it to my attention a month or so before it passed. The *Wall Street Journal* and few other press sources had briefly mentioned the effort but there was no major discussion of it in the media. The reporters I spoke to thought the issue was "too technical" for most Americans.

The House first passed such an FBI-backed measure of Democrat Don Edwards in a late-night voice vote. Democrat Senator Patrick Leahy had pushed a similar measure through the Senate Judiciary Committee with a 16-to-1 vote. Then he got the Senate to pass his version of the Edwards bill moments before the 103rd Congress broke for recess.

I called Senator Leahy's staff a few hours before the FBI wiretap bill passed in the Senate. A young male staffer told me that what he called the "telecommunications reform" bill had stalled. The bill still had two holds on it. But it might come to the floor at any minute.

I told him that I was writing an op-ed piece on the wiretap bill for the *Los Angeles Times*.[24] I said I wanted to give the bill's sponsors a chance to explain the bill and I asked to speak to someone in charge. He told me to hold and then put a young lawyer on the phone.

The lawyer said she could not fax me the wiretap bill because it contained too many hundreds of pages. She also called it the "telecommunications reform" bill and refused to call it the "wiretap bill" even though the press and cyber-Netizens called it that. I asked her to summarize the bill's main points but slipped and called it the "wiretap bill." She said that showed my bias and ignorance and she promptly hung up.

The bill passed later that night.

I called back a few weeks later to see if President Clinton had signed the bill because I could find no mention of it in the media. I spoke to the same Leahy staffer with whom I had spoken the first time. I asked him what had come of the "telecommunications reform" bill.

He paused at first as if he did not understand. Then he spoke: "Oh! You mean the *wiretap* bill? Yeah. Clinton signed it last month."

The FBI got its wiretap bill but it still has not banned private encryption. The government has even relaxed some of its encryption export laws to help software makers prevent digital piracy. But the FBI has made other digital inroads in the wake of the 1995 Oklahoma bombing and concerns over terrorism. It still seeks more wiretap authority.[25]

The FBI has also won the right to snoop our digital credit reports. Congress quietly passed the Intelligence Authorization Act in January of 1996. This law lets the FBI raid key parts of your digital credit reports without first asking a judge. So it bypasses a key step of due process. The FBI needs only suspect that you know a spy or terrorist or know someone who does. The FBI does not have to let you know that it is watching you or that it has watched you. The law makes credit firms keep the FBI searches secret from you.

The FBI needs a court order or a subpoena to search your whole report but there is little need for them to do that. The new law lets the FBI see your past and present addresses and your work history. It also lets the FBI see your lenders and your other financial contacts. The FBI may well choose to feed these digital facts into their own neural networks or other smart statistical programs and so compute or update your profile. The FBI can create its own secret software agents to watch ours.

The fight for digital privacy will last well into the 21st century.

At root the FBI still wants us to use codes that it or the National Security Agency can crack. The NSA can crack codes in many ways with its twelve underground acres of computers at its Fort Meade compound in Maryland. The NSA also has little or no oversight as is true of many state intelligence agencies. Even members of Congress were surprised to learn in 1995 that those who run the U.S. spy satellite program at the supersecret National Reconnaissance Office had quietly but legally squirreled away more than $1 billion of unspent funds.[26]

The FBI wants us to give up some or all of our digital privacy so it and other secret agencies can do their jobs better and search for the suspected criminals of the day. But we need that encryption-based digital privacy so we can protect our bit streams from thieves and criminals and from the powerful state agencies who search for them.

We need a Digital Rights Act.

We need a constitutional amendment or at least a federal law that ensures the *right to code our own bits in our own way.* One sentence can help secure this digital freedom and set an example for the rest of the world. It would not take a Thomas Jefferson to write it.

A Digital Rights Act would extend the First Amendment to the Information Age. The law would cost zero tax dollars and a bold politician might make a new career out of passing it. Then we would not have to trust the goodwill of unchecked state agencies that buy computers by the ton and that swap databases at the speed of light.

The hard technology problem is how to design intelligent digital agents that let us share our secrets with them.

Research in the information sciences will wrestle with this problem and address it in new ways that combine science and markets. Our agents may prove smart enough to find new secure schemes to encode our bit streams. Or our agents may have to create dummy bit streams that help deceive snooping hackers and credit bureaus and state agencies.

The political problem presents a greater challenge. We must find ways to make sure that our agents keep the secrets we share with them.

Or else the state will find the ways for us.

Heaven in a Chip

My body and my will are one.

> ARTHUR SCHOPENHAUER
> *The World as Will and Idea*

The most lively thought still is inferior to the dullest sensation.

> DAVID HUME
> *An Inquiry Concerning Human Understanding*

Every moment of our life belongs to the present for only a moment. Then it belongs forever to the past. Every evening we are poorer by a day. Our existence has no foundation on which to rest except the transient present. Thus its form is essentially unceasing motion *without any possibility of that* repose *which we continually strive after. It resembles the course of a man running down a mountain who would fall over if he tried to stop and can stay on his feet only by running on.*

> ARTHUR SCHOPENHAUER
> *Essays and Aphorisms*

When we are, death is not. When death is, we are not. Death is nothing to us.

> EPICURUS
> *Letter to Menoeceus and Principal Doctrines*

PRIEST: But surely you believe something awaits us after this life? DYING MAN: What other, my friend? Nothingness. It has never held terrors for me. In it I see only what is consoling and unpreten-

tious. All the other theories are of pride's composition. This one alone is of reason's.

> MARQUIS DE SADE
> *Dialogue Between a Priest and a Dying Man*

I do not believe that any man fears to be dead but only the stroke of death.

> FRANCIS BACON
> *An Essay on Death*

I'm not afraid to die. I just don't want to be there when it happens.

> WOODY ALLEN

You are a bundle of self-centered fears, hopes, greeds, jealousies, and self-conceit, all doomed to death.

> C. S. LEWIS
> *Mere Christianity: Let's Pretend*

The clergy of every established church constitute a great incorporation.

> ADAM SMITH
> *The Wealth of Nations*

My objection to organized religion is that it tends to use the name of God in vain. So far as religion is testable it seems to be false.

> SIR KARL POPPER[1]

Religion is based primarily and mainly upon fear. It is partly the terror of the unknown and partly the wish to feel that you have a kind of elder brother who will stand by you in all your troubles and disputes. Fear is the basis of the whole thing—fear of the mysterious, fear of defeat, fear of death.

> BERTRAND RUSSELL
> *Why I Am Not a Christian*

The old men he had known when a boy had known old men before them. They did not count. They were episodes. They had passed away like clouds from a summer sky. He also was an episode and would pass away. Nature did not care. To life she set one task and gave one law. To perpetuate was the task of life. Its law was death.

> JACK LONDON
> *The Law of Life*

We begin in the madness of carnal desire and the transport of voluptuousness. We end in the dissolution of all our parts and the musty stench of corpses. And the road from the one to the other goes, in regard to our well-being and enjoyment of life, steadily downhill: happily dreaming childhood, exultant youth, toil-filled years of manhood, infirm and often wretched old age, the torment of the last illness, and finally the throes of death.

> ARTHUR SCHOPENHAUER
> *Essays and Aphorisms*

Death is an imposition on the human race and no longer acceptable.

> ALAN HARRINGTON
> *The Immortalist*

It is impossible to imagine the height to which may be carried, in a thousand years, the power of man over matter. All diseases may by sure means be prevented or cured, not excepting even that of old age, and our lives lengthened at pleasure even beyond the antediluvian standard [beyond 1000 years].

> BENJAMIN FRANKLIN
> Letter to Joseph Priestly: 8 February 1780

Death is no more natural or inevitable than smallpox or diphtheria. Death is a disease and as susceptible to cure as any other disease. Over the eons, man's powerlessness to prevent death has led him to force it from the forefront of his mind, for his own psychological health, and to accept it unquestioningly as the unavoidable termination. But with the advance of science, this is no longer necessary—or desirable.

> STANLEY KUBRICK
> *Playboy* Interview: 1968

The price of computing has dropped by more than a factor of 100,000 during the past 20 years.

> BILL GATES
> CEO and Chairman of Microsoft
> *The Wall Street Journal*
> 16 March 1995

The definition of "Moore's Law" has come to refer to almost anything related to the semiconductor industry that when plotted on semi-log paper approximates a straight line.

> GORDON E. MOORE
> Chairman Emeritus of Intel Corporation
> "Lithography and the Future of Moore's Law"

We back up computers. Why not ourselves? Two-year recordings of our corpus callosum signals may provide a good record.

> MARVIN MINSKY[2]

Mind and spirit are both composed of matter.

> LUCRETIUS
> *On the Nature of the Universe*

Your brain is a material object. The behavior of material objects is described by the laws of physics. The laws of physics can be modeled on a computer. Therefore the behavior of your brain can be modeled on a computer. Q. E. D.

> RALPH C. MERKLE
> "Uploading: Transferring Consciousness from Brain to Computer"
> *Extropy:* volume 5, number 1, 1993

God is what mind becomes when it has passed beyond the scale of our comprehension.

> FREEMAN DYSON
> *Infinite in All Directions*

Is biology destiny?
This question has haunted and delighted parents since the dawn of time. Children combine the genetic hardware of their parents into their own wet hardware. So children follow to some degree in their parents' gene prints if not in their footprints. This keeps some of the parent genes swimming in the gene pool while it constrains what the children can be and what they do.

It also constrains their minds.

Mind resides in neural software or "wetware." Culture and the environment can nurture and shape this wetware. A kitten or a child will grow up blind in one eye if you tape that eye closed for a few weeks after birth.[3] The tape shuts out the environment. The tape shuts out its information-rich light stimuli. The neural pathways and circuits need those outside signals to grow and structure themselves.

All signals shape our wetware to some degree because wetware is plastic. All sights and smells and pains leave their mark in wetware. They all change how neurons fire and how hormones flow and how synapses release their chemical neurotransmitters. These wetware changes feed on themselves. That leads to more changes as the tangles of neurons and synapses race to find a new dynamic balance or equilibrium. And so we learn from experience whether we want to or not.

Thinking shapes our wetware on a larger scale.

Thinking is a type of global self-programming. Thoughts swirl around as nonlinear resonances in our neural networks and perhaps in our entire brain.

These fleeting patterns of resonance are not static. They learn or adapt and evolve into new patterns of resonance. They combine new signals from the world with old signals and thoughts stored in wetware memory.[4] This creates global mental equilibria of exquisite complexity. They too feed on themselves and create new equilibria or new thoughts. The resonance patterns leave the electromagnetic footprints we call brain waves and produce the ghostly flux of awareness we call consciousness.

Our genes constrain it all. It does not matter how hard or how well we think. We still think in the same old meat.

And meat dies.

The meat consists of cells that nature has programmed to die. Biochemists call this programmed cell death.[5] All multicellular creatures share this dark gift from nature and so do many single-cell life forms and slime molds. It may have helped cell groups quickly weed out weak cells. But then nature had to evolve an anti-death chemical response so that winners could suppress their suicide programs long enough to let them pass on their genetic data.

Our genes condemn us to old age and disease.[6] Cell damage builds up as we age and our cells do an ever poorer job of fixing the damage. A cell's countdown to death begins the first time it divides. Each of the cell's 50 or so divisions snips off a bit more of the cell's fixed DNA telomerase tape. The cell dies when the tape runs out.

The cell damage piles up into the fuzzy pattern of dry skin and light bones and fading senses and failing organs and rotting teeth that we call old age. These weaker cells fail to suppress our toxic genes as well as they once did. So gene-based diseases like cancer of the breast or prostate or brain can destroy our cells en masse.

Our genetic ancestors did not care if we died slowly in a parts decay or more quickly but in racking pain. They did not even care if we died. They just cared that we did not die before the mean age of reproduction. The viruses and the cancers and the wolves could have us after that. These killers weeded out life-forms and gene schemes that bred too late. So our genetic ancestors programmed our genes by survival default.

Our ancestors programmed our genes with a supreme biological commandment: Breed before you die.

Our culture has built on this biological imperative as it too has evolved in the last few thousand years. Science and markets help us control nature and secure the tools and wealth we need to survive and reproduce. Language has evolved as a tool to convey thought in the service of our battles with nature and with one another. Systems of morals and laws have evolved that favor the family while less child-friendly systems have decayed and dropped off the cultural tree.

Religious myths and churches have also evolved to support the family. The drive to gain more members has given competing religions and churches a lifelike structure of their own. States or power monopolies grew both out of early tribal social structures that favored the family and later tribal efforts to conquer or defend against competing tribes.[7] The drive for conquest and for more subjects has given states their own evolving lifelike structure. Art has evolved as a symbolic byproduct of all these social quests that try to secure ever more means of production and reproduction.

This cultural focus on breeding and dying raises a simple question but one we seldom ask: Why do we accept death so passively?

Why do we not rank each new idea or action by how well it helps us conquer death? We would not after all need to breed if we did not die and if we could adapt our own genes to fit our changing world. We could keep our own place in the flowing gene river rather than have a shrinking fractional claim on future places. Death has an infinite opportunity cost.

So why is killing death not our supreme goal in society?

The answer has to do with how we guess at the progress of technology. Most people expect science to further extend the human lifespan. But few people think that science will extend life to hundreds or thousands of years in the near or distant future.

Most people are *deathists*. They think death is as natural and inevitable as taxes.

That helps explain why fewer than a thousand of us have so far signed up for cryonic suspension in liquid nitrogen upon our death. Cryonics involves many gambles about the future of society and molecular engineering and it is not cheap. But this "deanimation" technique of today freezing and someday rebuilding cells remains the only known technique that has even the slightest scientific chance of beating death.

Embalming and cremation offer no chance. Prayer and a deathbed wish for an afterlife may offer comfort to those who make them. But faith does not count as scientific evidence no matter how intense or how fearful the emotion or how many of us share it.

The rest of the human race has chosen by default to avoid the experimental group of cryonicists and stay in the control group of deathists. Time will sort the winners from the losers in this ongoing social experiment.

Meanwhile culture reflects the deathist view.

Party jokes tell how we might barter with Saint Peter at the gates of heaven. Street slang tells us that life is hard and then we die. Art laments death in sad songs and bleak paintings and tragic tales and has done so since the days of cave dwelling. Religions spin tales of life beyond death in all cultures and in all ages. Religion creates in this way a "moral hazard": its quick promise of an afterlife lessens our resolve to conquer death.

Religions may be the purest offspring of deathism. Most religions promise rewards in heaven or punishments in hell even though no religion has produced one atom of scientific evidence to support these super-sensory and even fantastic claims.[8] The passion that psychologist William James called the *will to believe* runs deeper than our brain-based belief in science.[9]

The will to believe taps into our genes and hormones. These wet structures ground our deepest intuitions about ethics and authority because they encode the harsh survival strategies of our dead ancestors. There may be no genes that urge us to believe in God or to salute up the chain of command. But there surely are no genes that urge us to suspend our judgment when we lack evidence for a claim. The heart quickly fills in what the head leaves blank.

We should not expect the digital culture to blindly accept deathism. The digital culture sees the world as flows of information or bit streams of 1s and 0s. It sees life forms as information machines that store and process bit streams. That puts brains on the same bit-based footing as computers.

We do not lose forever our software or data files if our computer stops or crashes or falls apart. We repair the hardware or transfer the bit-based software to new hardware. We tune the information machine or replace it with a new one. Why not do the same with our minds? Why not repair the meat with better meat or replace the meat with something more durable? Why tie mind and mood to blood sugar?

Our brains lack a key feature of computers and software: They have *no backup.*

We have no backup for our most precious biological and mental data. Each second our memory decays a little more and we forget a little more of our life. We barely remember our lives a few days ago let alone a few years ago. Our neural wetware lets our past slip away and slip much faster than we may realize. The very structure of how our synapses learn new patterns consigns us to exponential memory decay.[10]

We live our physical and mental lives like careless musicians who never write down the notes they play and who soon forget what notes they have played. This is the greatest danger of living in meat and we can thank our genetic ancestors for it.

Dumb and blind evolution has forced us to live in machines that have no backup.

Should we just accept that and continue to sleep and have strokes and die? The brain is one of the great marvels of science but it is also one of the great fiascoes of engineering. We would never design a tooth or an eye or a mind as nature has. So how might things change if we design and replace our own mind parts? What if we could back up our meat machines?

From Brainloaf to Chipnet

We can explore what may come next for brains in the digital age by look-ing at how brains cooperate: Why is it so hard to work together as a soci-ety or a team or a family? Why do all symphonies have just one author?

The meat answer is that brains do not communicate well. Brains do not have direct access to one another.

Skulls get in the way.

When we talk or write or gesture our brain sends signals through com-plex networks of nerves and muscles. Some version of these signals then disturbs the air or paper or keyboard.

The receiving brain must reverse this process. It must convert the noisy sense data into electrical impulses and chemical packets. Then it must pass these signals through layer after layer of neural filters. Only then can the brain process these signals in the same coin as brain talk.

Now consider a thought experiment.

Think of a big meatloaf in the center of a room that has a lot of people in it. The people make the meatloaf. Each person takes out his brain and throws it on the floor as if it were a loaf of bread. The loaves combine to form one large meatloaf or brainloaf.

Each person has his own chunk of the brainloaf. That grounds his iden-tity in the space-time continuum. But now all the brains can talk in the same neural language just as the two halves of each brain talk to each other. The brains can receive the same flux of sensor signals. Or they can divide that labor among themselves.

What would it be like to live or think in a brainloaf?

At first there would be what William James called the "blooming buzzing confusion" of all the brain talk. Soon you would get used to that just as you get used to the background chatter at a party or in a mall.

And at first the parallel talk and thought would tax you. You would get used to that too as your neural muscles stretched and tuned themselves to meet the new processing demands. Dissent and debate could reach new heights. But think of the art and science the brainloaf might produce. There would be joint symphonies.

There would be something new in the brainloaf: global thought. This is a meat version of "collective consciousness." The brainloaf would form one huge neural network or dynamical system. So it would tend to cool down or equilibrate when inputs stimulate it. The whole brainloaf would resonate.

The points of resonance would define the group thoughts just as the modes of neural-net resonance in our brains define our much simpler thoughts.[11]

The brainloaf dynamical system would swirl into complex patterns of resonance. These patterns would include chaotic or aperiodic attractors. The chaos thoughts would bring novelty into the brainloaf. They would create new information. Or the chaos thoughts may just be fun to think for their own sake.

This is not science fiction. It is a problem of engineering design. We can in theory replace the brainloaf with a "chipnet" of dozens or thousands of digital souls.

The human brain stores about 10^{18} bits of essential information or a billion billion bits. The brain processes those bits at about 10^{16} bits per second. These numbers are not exact. But they are close to within a power of 10 or so.[12] The human body itself contains on the order of 10^{28} or ten billion billion billion atoms.

How big a chip would you need to replace your brain? What size chip do we need to *upload*?

Right now it would be the size of a house or an office building. A few years ago it would have been the size of a skyscraper.

Soon it will be the size of a sugar cube or even smaller.

This stems from Moore's law as do so many velocity effects in the digital age. Recall that this observed "law" or tendency states that the density of circuits on a computer chip doubles every two years or faster. Moore's geometric law has held up for over 20 years in the chip race and now acts as a type of optimistic information antidote to Malthus's grim law of human population growth.

We can assume that Moore's law will hold for at least the next decade. It may hold longer than that despite the quantum and other device limits in its future.[13] Some form of Moore's law of circuit doubling and shrinking may go on to hold for nanoprocessors and quantum computers.[14]

Moore's law implies that by the year 2020 or so your brain will in principle fit in a chip the size of a sugar cube. The year may be as soon as 2010 if chip densities keep doubling every 18 months. Or it may be closer to 2030 if Moore's law slows down.

The exact year or even decade does not matter. Society will soon have to face a stark and yet exciting fact: Computer chips will have the raw processing power of the human brain. Then chips will forever exceed the old human benchmarks. This will not mean that we can then just pull out our brain and plug in a chip. We will also need breakthroughs and steady progress at the level of the chip-neuron interface. These results may take an extra decade or they may lead to new breakthroughs in chip design that speed up the entire process.

Young people today may see the chip-brain day come to pass. If they do not see it then their children will. Surely their grandchildren will see it. Those who cannot wait until then to upload can cryonically suspend their synaptic topknots as a crude form of backup.

But what would those who upload see? What happens to the musiclike patterns of the self when they transfer from one instrument to another and then merge with the other patterns in a chipnet orchestra?

What happens to the *I* in *me*?

Chipping Away at Your Brain

Suppose you can load your brain into a chip.

Where do you go? Do you die first? Do you stay in the brain or in the chip or in both?

The mind seems to split or bifurcate in the same way that some cells split and grow as new cells. How can you be sure that it is still you in the chip? Suppose we make a hundred copies of the chip brain. Which one are you?

The problem is that the mind in the chip does not depend on the mind in the brain. The brain can live on or die. That does not affect the chip. So how can your sense of you jump from brain to chip? How does it survive the transition from meat to non-meat?

This is a digital version of the old mind-body split of Western philosophy. French philosopher René Descartes thought mind and body met in the pineal gland in your head. He still had the same problem at death as we have in the chip transfer. His soul lived in the brain and then had to hop from there to a hall in heaven or a cell in hell. Our mind has to hop from meat to chip and the chasm seems just as great.

One way to do it is to go to sleep in your brain and then wake up in the chip. In some sense you die each night when you fall asleep. You are you when you wake because of the continuity of your memory and dreams.

But how do you know that the you in the old brain did not face death? You don't know. So why grow a new you somewhere else if the old you still dies? The whole idea is not to die.

Fuzziness shows a way out.

The fuzzy world view sees even consciousness as a matter of degree. You face this when you slowly fall asleep or wake up or when a drink or bump on the head diminishes your capacity. And you face this sense of changing in the continuum of consciousness when your thoughts soar from the sight of flashing red police lights or from a shot of caffeine or smart drugs.

The idea is that you go from thing to not-thing in small steps just as day passes to night in degrees. There are no binary hops from thing to not-thing or from brain to not-brain or from not-chip to chip. The old idea of replacing your brain with a chip makes a binary hop. That is the source of the bifurcation.

So go from brain to chip in small steps. Chip away at it in degrees.

Picture this: The nano-surgeons open your skull and you are wide awake. The nano-surgeons may be humans or robot arms that guide tiny nano-robots that unstack and stack the molecules of your skull and brain.

You might wear virtual reality goggles and fly as a golden eagle over the Rocky Mountains during the procedure. Or you might watch yourself in a TV sitcom or rock video. Or you might watch the nano-operation from a number of angles in the operating room.

At first the nano-surgeons cut out a small gray chunk of your brain. Say it comes to 1% of your brain. You do not feel it or notice any change in your perceptions or memory.

You are still you.

Then they replace your brain chunk with a tiny nanochip wrapped in sponge and studded with nano-sensor tendrils. This chiplet has the same rough input-output ports as the brain chunk had. The chiplet mimics the old brain chunk as a modern neural net might. But the chiplet runs a million times faster than the old gray matter ran. It can store and process more data than all the rest of your brain combined. So you might notice a boost in your cognitive skills as if you just drank a glass of iced tea laced with choline or norepinephrine.

The nano-surgeons next cut out a second brain chunk and replace it with a chiplet wrapped in a like shell of sensor-control ports. A neural net has also tuned this chiplet too to act as the second brain chunk did. The chiplet sends the same inputs to the same outputs. Again there need be no break in the continuity of your consciousness. And again you may find your mind racing a little faster and better.

You might find it racing a lot faster and better. The key is that you find your mind that way and you never lose it. You never lose you.

Then the nano-surgeons cut out a third brain chunk and replace it. Then they replace a fourth chunk and so on until they finish the job.

Now you are not just in the chip. You are the chip or the net of chiplets. You can pass in the same smooth way from the chiplet net to a master chip or a network of master chips.

The nano-scalpel or laser or particle beam need never have broken your train of thought. You did not die but the meat did die. And a better nano-surgeon might reverse the steps and rebuild the brain a chunk at a time. That mind would have had a big break in consciousness. But that mind is not you.

The fuzzy point is that you can stay awake the whole time on your path to chip pseudo-immortality. You shade smoothly from meat to non-meat and avoid the mental bifurcation that could come from a wholesale swapping of brain with chip.

The electro-chemical cloud of patterns you call you need not die or dissolve if you code the patterns in a new medium. You can get out of the meat car before it wrecks. And in time they all wreck.

No backup ensures that.

Heaven in a Chip

So suppose you wake up in a chip.

Now you can test the many claims that philosophers of mind have made for centuries and still make about mind and matter.[15] Or you might want to direct your information engineering toward one philosophical school of thought or another. The point is not so much that a chip mind would let you answer old questions of mind as that it would let you bypass them and explore new ones.

You might want to follow Leibniz and synchronize your chip mind with the slowly moving meat machine that houses it and establish a mind–meat harmony. You might want to follow Descartes and keep the two worlds separate. Or you might take the extreme reductionists and physicalists at their word and ignore them and the whole issue of what you are in philosophical terms.

You are in digital terms a stream of bits. You are a pattern of patterns in a bit stream. And God help us if (by then) we do not have reliable digital encryption.

If any philosopher wins then perhaps it may be the 19th-century philosopher Arthur Schopenhauer. Schopenhauer saw the world as will. Our will is all we ever know directly. It is not a filtered sensation or a recalled thought or memory. It is the raw thing itself. Now the world is your idea of it. And your body and your will are one. They are information patterns in the bit stream.

So suppose you have uploaded successfully. Your brain has switched from meat to silicon but you are the same mind.

At least you are at first.

The old memories are still there but now you do not access them the same way. They are not just vivid when you recall them. They are as intense as when you first lived them. And you can edit them as if they were dreams you controlled.

Your memory is just one small database that you can access at the speed of light. You can command armies of intelligent agents and search thousands or millions of databases and knowledge networks. And you can sense all stored knowledge of art and science and news and history much as you now scan a newspaper.

And you can feel and act and do it alone or with thousands of other chip souls.

In a brain you see an apple and squeeze it in your brain. The hand touches the apple and the eyes see its reflection. Those signals feed to the brain. Then command signals feed back down from the brain to the hand and eyes.

In a chip you do the same thing but you do more of it and faster. You have more types of sense data. You can see all parts of the spectrum and hear the subsonic and supersonic. You can dampen or amplify pain and pleasure signals and weave new patterns of emotion or feeling and shut them off when you tire of them. You can edit thoughts and memories at will. You can relive any past memory or modified memory as vividly as you live the current moment. And all or some of this you can share with others in the chipnet or chiploaf.

Your subjective sense of time also changes. Slow neural time can accelerate to ultrafast *nanotime.*

In the chip time passes a million or billion times more slowly if you want it to. The time it now takes you to read this chapter may then seem like years. A good crystal chip could last for thousands or millions of years or until it fell into a star or a black hole. In the best case it could last most of the billions of years until the universe falls in on itself and ends in the Big Crunch or until it peters out in a cosmic heat death. Multiply that by the millions or billions of new subjective seconds per old second and you get an engineering approximation of eternity.

That long life in a chip might be as close as we can come to heaven in a universe made of matter and energy and information. It can be heaven or hell in a chip. You can choose which. Bit streams have no more purpose than do the blind genetic pathways of Darwinian evolution.[16] The will must impose on a bit stream any value or purpose it has even if the will is itself part of a bit stream.

So government need not be destiny.

There will be no need to work or to be governed unless you want to. There will be no sickness or pain or death unless you want to play with them. The virtual will be real and the real will be virtual in the same stream of electrons and photons. The economics of growth would have little material constraint in a chip.[17] Pure thought could create the old utopian worlds of Plato or Sir Thomas More or Karl Marx or it could create new ones.

Here the bit-based digital culture can compete with the old atom-based religions. It does not have to but in time it surely will. Adam Smith first discussed the market for religions in *The Wealth of Nations*. He argued that competition would improve religions or rather religious firms. The evidence seems to support his view.[18] So let the real competition begin.

Believers may also transfer their belief to a chip's bit streams. The most devout believer could postpone his imagined spiritual destiny if he uploads to a chip for a few nano-millennia. Chip super-minds may have deep new insights about the nature of God and first causes. Human brains have complex religious insights that fly brains and dog brains do not have and that flies and dogs could not fit into their minds if they had such insights. The trend of growing insight will surely continue as we build ever more powerful minds.

Believers might also shape their chip minds to match and extend their visions of heaven. But others can explore heaven in their own bits and in their own way. Here the real competition begins.

Religion holds no monopoly on the concept of heaven. Whoever said it did? That would be like restricting flying to birds or restricting electricity to lightning. Heaven is too wonderful a concept to restrict it to the fearful imaginings of long-dead and pre-scientific men and women.

And religion has failed to explain the mechanics of souls.

Religion has failed to explain how you could sense heaven or hell with no body even if believers could produce evidence for such places. The soul somehow takes you from here to there. But what can a soul do when it gets there? How does it feel or think or act in heaven? A senseless soul cannot taste grapes or hear harps or even see light. Atoms and bits would pass right through the proposed vapor.

A digital soul or mind does not have these problems. A digital mind has a physical identity because it is a unique pattern of patterns in a bit stream. And it can sense other patterns at least as well as a meat mind can. If in our brains we can have the mental taste state of eating a red-flame grape then we can have that same mental taste state in our chip mind if we feed it the same sensory signals. Both instruments can play the same music. But the digital instrument can play the music at greater intensity and can vary and edit it in incalculably more ways.

Religions have also cast heaven as an ethical absurdity. Where is the justice in granting someone infinite pleasure in exchange for finite pain or for finite good deeds or worship?

Benjamin Franklin spiked this wishful longing for an eternal free lunch in his letter of 6 June 1753 to Joseph Huey: "By heaven we understand a state of happiness, infinite in degree, and eternal in duration. I can do nothing to deserve such rewards: He that for giving a drink of water to a thirsty person should expect to be paid with a good plantation would be modest in his demands compared with those who think they deserve heaven for the little good they do on Earth."

This heavenly deal may be the greatest sucker strategy of all time. It is surely one of the most cynical swaps of present and future value: Pay now and collect after you die. Most major religions have promised such an eternal free lunch in an alleged hereafter in exchange for obedience and for donations and often for outright mind control in the here and now. The extreme version offers infinite paradise in exchange for a quick murder-suicide. But justice is measure for measure. That alone argues for term limits in heaven in direct proportion to one's good deeds in life. No finite sum of finite good deeds adds up to an infinite reward. So once again the religious view of heaven contradicts its own assumptions.

Heaven in a chip also avoids this problem. We do not use a wheel or a toothbrush or a lightbulb or any other product of engineering because we deserve to use it. We use it because it is at hand and because we can afford it. Our best heaven in a chip will always be only an engineering approximation. The finite structure of the universe imposes the ultimate term limits on any heaven in a chip. And the promise of our own digital heaven creates no moral hazard. That promise can only increase our resolve to conquer death and so spur more research and innovation in that direction.

Will it be worth it? Is the digital quasi-immortality of any designer heaven worth the time and energy we would have to exchange for it? The pyramids of Egypt show how deeply the human heart craves even primitive visions of immortality. Heaven in a chip can supply much of this demand and likely do so for less than the cost of the average computer of the day.

Heaven in a chip does more than solve the practical problems of religious views of heaven. It completes the rival world view of science.

The creation myth gives way to the Big Bang or to a whole sequence of Big Bangs if the universe oscillates or bifurcates into an infinite labyrinth of baby universes and black holes and chip-like bit pools. Divine law gives way to the binary laws of math and the fuzzy laws of science. The soul gives way to complex patterns of information processing in feedback neural circuits made of flesh or silicon or light or plasma or of any other form of structured energy.

The resurrection myth gives way to cryonics and cell repair with nano-computers and nano-robots or to the gentle sleep that takes you from brain to chip. And Gods give way to our digital superminds and all the worlds they can imagine.

Why live in a heaven that you have created at will and then take orders from an alien despot? Why take orders from anyone or any government or anything else? Why live in your own heaven and not *be* God? Why give up the best seat in the house? Each chip soul may have a chance to explore such questions.

So is biology destiny?

Biology may have been destiny for the 100 billion or so humans who came before us on this planet. We owe them our genes. But we pay them little or no mind just as our heirs will pay us little or no mind when we pass. Biology is not destiny for the minds that will follow us.

Biology was never more than tendency. It was just nature's first quick and dirty way to compute with meat.

Chips are destiny.

Notes

1. INTRODUCTION: CREEPING FUZZINESS

1. An article in the 22 March 1998 *New York Times* announced the use of fuzzy logic in the Volkswagen New Beetle: "The New Beetle has a five-speed manual transmission (the old Beetle had four), or an automatic that uses 'fuzzy logic' to adapt to the driver's style." For technical details see Schroeder, M., Petersen, J., Klawonn, F., and Kruse, R., "Two Paradigms of Automotive Fuzzy Logic Applications," in *Applications of Fuzzy Logic—Towards High Machine Intelligence Quotient Systems,* ed. M. Jamshidi, Prentice Hall, 1997.

   The GM Saturn also uses a fuzzy system in its automatic transmission: Legg, G., "Transmission's Fuzzy Logic Keeps You on Track," *Electronic Design News,* 23 December 1993, 60–63. The fuzzy controller helps the automatic transmission downshift as if it were an expert driving a stick shift. The fuzzy system consists of 5 rules and takes up only 500 bytes of memory in the Saturn's Motorola 68HC11 chip: "Six fuzzy variables serve as inputs to the rules: grade, speed, throttle position, brake-application time, brake application with high deceleration, and coasting with acceleration. Each rule involves some combination of fuzzy variables such as 'if speed is low and grade is negative and brake-application time is long, then downshift,' (In other words: 'You're riding the brake going down a hill, and it would be better to use a lower gear.') The fuzzy controller evaluates the rule set four times each second to determine if it needs to shift gears."

2. Egusa, Y., Akahori, H., Morimura, A., and Wakami, N., "An Application of Fuzzy Set Theory for an Electronic Video Camera Image Stabilizer," *IEEE Transactions on Fuzzy Systems* 3, no. 3 (August 1995): 351–56.

3. Von Altrock, C., "Recent Successful Fuzzy Logic Applications in Industrial Automation," *Proceedings of the Fifth IEEE International Conference on Fuzzy Systems (FUZZ-96),* vol. 3, 1845–51, New Orleans, September 1996. For more details on the German applications see Von Altrock, C., *Fuzzy Logic and Neuro-Fuzzy Applications Explained,* Prentice Hall, 1995.

4. There is a long history of researchers in many countries applying fuzzy systems to the control of nuclear power plants: Bubak, M., Moscinski, J., and Jewulski, J., "Fuzzy-logic Approach to HTR Nuclear Power Plant Model Control," *Annals of Nuclear Energy* 10, no. 9 (1983): 467–71; Akin, H. L., and Altin, V., "Rule-Based Fuzzy Logic Controller for a PWR-type Nuclear Power Plant," *IEEE Transactions on Nuclear Science* 38, no. 2 (1991): 883–90; Kuan, C. C., Lin, C., and Hsu, C. C., "Fuzzy Logic Control of a Steam Generator Water Level in Pressurized Water Reactors," *Nuclear Technology* 100, no. 1 (1992): 125–34; Iijima, T., Nakajima, Y., and Nishiwaki, Y., "Application of Fuzzy Logic Control Systems for Reactor Feed-Water Control," *Fuzzy Sets and Systems* 74, no. 1 (1995): 61–72.

   Nuclear power plants already use standard controllers with mixed results. Researchers benchmarked a fuzzy controller against a standard controller at Belgium's BR1 nuclear reactor. This research reactor serves as an international reactor for nuclear calibration. The initial fuzzy experiment was favorable but the research continues. Ruan, D., and van der Wal, A. J., "Controlling the Power Output of a Nuclear Reactor with Fuzzy Logic," *Information Sciences* 110, no. 3 (October 1998): 151–77.

5. "Because the language of fuzzy logic control is closer to that of engineers and technicians than that of programmable logic control, fuzzy logic control is beginning to replace programmable logic control in Brazil's oil industry and elsewhere. Typical fuzzy logic rules are of the type 'IF pressure is "high" and level is "low" THEN pressure valve aperture is 'medium' and level valve aperture is "low," ' which is intuitively obvious to the plant operator. Such partial and fuzzy reasoning support decision making for two reasons. First, in fuzzy logic systems, actual pieces of data are matched against typical expert knowledge about the behavior of process variables in abnormal conditions. Second, some information may be

missing in crucial moments when decision making is called for." Da Rocha, A. F., Morooka, C. K., and Alegre, L., "Smart Oil Recovery," *IEEE Spectrum,* July 1996, 48–51.

Brazil's Carajas Railway carries iron ore on a single track 892 kilometers long: "Fuzzy logic techniques are being used primarily for analyzing train movement and for aiding the operators to make the best possible decisions. For instance, a badly delayed unloaded train arriving at a station will get a priority over a loaded train arriving at the same station ahead of its schedule from the opposite direction, even though usually the loaded train has a higher priority." Vieira, P., and Gomide, F., "Computer-Aided Train Dispatch," *IEEE Spectrum,* July 1996, 51–53.

6. De Ru, W. G., and Eloff, J. H. P., "Enhanced Password Authentication through Fuzzy Logic," *IEEE Expert: Intelligent Systems and Their Applications,* November 1997, 38–45.

7. Many Japanese firms hold patents on fuzzy elevator systems. The United States has granted a patent (number 4,244,450) to Hitachi for its "Group Supervisory System of Elevator Cars." For details of a typical fuzzy elevator system see Kim, C., Seong, K. A., and Lee-Kwang, H., "Design and Implementation of a Fuzzy Elevator Group Control System," *IEEE Transactions on Systems, Man, and Cybernetics* 28, no. 3 (May 1998): 277–87.

For business and investment applications see Cox, E. D., *Fuzzy Logic for Business and Industry,* Charles River Media, 1995. *Electronic Engineering Times* also ran a special issue on embedded fuzzy systems in its 29 July 1996 issue.

Omron sells an e–mail spam filter that uses fuzzy logic "artificial intelligence": "If junk mail is your biggest bugaboo, you might want to unleash a full–fledged spam assassin. Omron Advanced Systems sells a $25 companion program for Eudora and Microsoft's Outlook 97 called MailJail (www.mailjail.com). The software relies on artificial intelligence to sift through the headers, addresses, and texts of all incoming messages. By identifying known spammers, or by picking up key phrases ("make money from home," "free offer"), the program can detect and dispose of junk." Baig, E., and Dunkin, A., "Taming the E-Mail Monster," *BusinessWeek,* 2 March 1998.

8. The following list includes a fair sampling of the major fuzzy textbooks published in recent years: Kandel, A., *Fuzzy Mathematical Techniques with Applications,* Addison-Wesley, 1986; Pal, S. K., and Dutta Majumder, D. K., *Fuzzy Mathematical Approach to Pattern Recognition,* Wiley, 1986; Klir, G. J., and Folger, T. A., *Fuzzy Sets, Uncertainty, and Information,* Prentice Hall, 1988; Miyamoto, S., *Fuzzy Sets in Information Retrieval and Cluster Analysis,* Kluwer, 1990; Kosko, B., *Neural Networks and Fuzzy Systems: A Dynamical Systems Approach to Machine Intelligence,* Prentice Hall, 1991; Zimmermann, H. J., *Fuzzy Set Theory—and Its Application,* 2d edition, Kluwer, 1991; Terano, T., Asai, K., Sugeno, M., *Fuzzy Systems Theory and Its Applications,* Academic Press, 1992; Driankov, D., Hellendoorn, H., and Reinfrank, *An Introduction to Fuzzy Control,* Springer-Verlag, 1993; Yager, R. R., and Filev, D. P., *Essentials of Fuzzy Modeling and Control,* Wiley, 1994; Klir, G. J. and Yuan, B., *Fuzzy Sets and Fuzzy Logic: Theory and Applications,* Prentice Hall, 1995; Pedrycz, W., *Fuzzy Sets Engineering,* CRC Press, 1995; Ross, T. J., *Fuzzy Logic with Engineering Applications,* McGraw-Hill, 1995; Kosko, B., *Fuzzy Engineering,* Prentice Hall, 1996; Lin, C.-T., and Lee, C. S. G., *Neural Fuzzy Systems,* Prentice Hall, 1996; Petry, F. E., *Fuzzy Databases: Principles and Applications,* Kluwer, 1996; Jamshidi, M., *Applications of Fuzzy Logic—Towards High Machine Intelligence Quotient Systems,* Prentice Hall, 1997; Jang, J.-S., R., Sun, C.-T., and Mizutani, E., *Neuro-Fuzzy and Soft Computing,* Prentice Hall, 1997; Rouvray, D. H., *Fuzzy Logic in Chemistry,* Academic Press, 1997; Wang, L.-X., *A Course in Fuzzy Systems and Control,* Prentice Hall, 1997; Berkan, R. C., and Trubatch, S. L., *Fuzzy Systems Design Principles: Building IF-THEN Rule Bases,* IEEE Press, 1997; Klir, G. J., St. Clair, U. H., and Yuan, B., *Fuzzy Set Theory: Foundations and Applications,* Prentice Hall, 1997; Cox, E., *The Fuzzy Systems Handbook,* 2d edition, Academic Press, 1999; Yen, J., and Langari, R., *Fuzzy Logic: Intelligence, Control, and Information,* Prentice Hall, 1999.

9. Discussed in McNeill, D., and Freiberger, P., *Fuzzy Logic,* Simon & Schuster, 1993, and in Kosko, B., *Fuzzy Thinking,* Hyperion, 1993.

10. Kosko, B., and Isaka, S., "Fuzzy Logic," *Scientific American,* July 1993, 76–81.

11. Arthur C. Clarke may have been the first writer to have mentioned a fuzzy set in a mainstream science fiction novel: "But the boundary of the M-Set [a fractal] is *fuzzy*—it contains infinite detail." Clarke, A. C., *The Ghost from the Grand Banks,* Bantam Spectra, 1990, p. 127.

This is right in spirit but not in logic. Fractals are math objects. They have exact black and white boundaries even if those boundaries have jagged edges of exquisite detail.

MIT artificial-intelligence pioneer Marvin Minsky may be the first to have mentioned fuzzy logic as a technology in its own right in a science-fiction novel: "I prefer not to assign an unconditional probability to a situation with so many contingencies. For this kind of situation it is more appropriate to estimate by using fuzzy distributions rather than deceptively precise-seeming numbers. But plausibility summaries on a scale of one to one hundred can be provided if you insist." Harrison, H., and Minsky, M., *The Turing Option*, Warner Books, paperback, October 1993, p. 210. Machine-man Data has spoken of his fuzzy logic on *Star Trek: The Next Generation* as have some of the evil scientists on Showtime's series *The Outer Limits*. And the villain of the 1994 Tom Clancy action film *Clear and Present Danger* assures us: "The world is gray!"

12. This converse relation holds only for denumerable (finite or countable) sets. Suppose X is any space of objects. Then $A \subset X$ is a fuzzy or multivalued subset if and only if A has the *set function* $a:X \to [0, 1]$. A is a binary set if and only if the set function a maps to two values: $A:X \to \{0, 1\}$. So it is true that we can write $a(x) = 1$ if $x \in A$ and write $a(x) = 0$ if $x \notin A$. But this informal talk does not imply that we can enumerate all the elements in a binary set by writing even a countably infinite number of 1s and 0s. The unit interval $[0, 1]$ is a binary set. It is not denumerable because it contains as many elements as there are real numbers.

13. Let $a:X \to [0, 1]$ define fuzzy subset $A \subset X$ and let $a'(x) = 1 - a(x)$ define the set complement $A^c \subset X$. Then the fuzzy sets A and A^c intersect just in case there is at least one object $x \in X$ such that $a(x) = a'(x) = 1 - a(x)$. So $a(x) = \frac{1}{2}$. This proves that if a fuzzy set fully intersects its complement set then the two set functions coincide at the midpoint value. It does not prove that all fuzzy sets intersect their complements. The constant fuzzy set function $a(x) = .9$ for all x has complement set function $a'(x) = .1$ for all x. Then the sets A and A^c are fuzzy and have non-empty intersection. But the fuzzy set intersection $A \cap A^c \neq \emptyset$ does not have the "full" 100% set value. Objects belong to this set to at most the degree 50%. In this case if $x \subset A \cap A^c \neq \emptyset$ then $a_{A \cap A^c}(x) = .1$ (because $a_{A \cap B}(x) = \min(a(x), b(x))$).

14. Suppose we want an adaptive fuzzy system $F:R^n \to R$ to approximate a test function or approximand $f:R^n \to R$ as closely as possible in the sense of minimizing the mean squared error between them ($\|f - F\|^2$). Then the jth scalar "sinc" function

$$a_j(x) = \frac{\sin\left(\dfrac{x - m_j}{d_j}\right)}{\dfrac{x - m_j}{d_j}}$$

with center m_j and dispersion d_j often gives the best performance for mean-squared function approximation even though this generalized set function can take on negative values: Mitaim, S., and Kosko, B., "What Is the Best Shape for a Fuzzy Set in Function Approximation?" *Proceedings of the IEEE International Conference on Fuzzy Systems (FUZZ-96)*, vol. 2, September 1996, 1237–43; Mitaim, S., and Kosko, B., "Adaptive Joint Fuzzy Sets for Function Approximation," *Proceedings of the IEEE 1997 International Conference on Neural Networks (ICNN-97)*, vol. 1, July 1997, 537–42. See also note 15 in chapter 10 below.

15. "The epistemology of fuzzy systems is that of a conceptual anarchy. We all use the same terms but do not mean the same things by them. The set of circles or other math objects may not be fuzzy or relative in the language of math but they may be both in a brain-based mind or neural network. Outside of math all patterns are fuzzy and relative to at least some degree. The humble sets of *dirty wash water* and *cool air* give a first glimpse of our conceptual anarchy. The fuzz and relativity grow as we move from sensory terms like *blue* or *large* or *cool* to the more abstract terms of head and heart like *similar* or *smart* or *fear* and on up to the broad terms like *season* or *progress* or *war* that describe complex physical or social processes. The same newspaper prints all the fuzzy terms in the same clear type as it states the 'facts' of the world. It is a surprise that we can communicate at all since no two persons mean the same thing with what they think or say. Fuzzy systems maintain a common rule structure while each user tunes the sets and finds his niche in the conceptual anarchy. This turns the anarchy

into a search space for user-friendliness. Each person digs his own niche in the fuzzy state space." Kosko, B., *Fuzzy Engineering,* Prentice Hall, 1996, p. *xx.*

16. See note 17 of chapter 6 below for details on the formal structure of the probability of both binary and fuzzy events. A quick proof that fuzziness differs from probability comes from the fact that the probability of a contradiction is always zero: $P(A \cap A^c) = P(\emptyset) = 0$ for all binary sets $A \subset X$. The dual extremal result holds for the probability of "either-or": $P(A \cup A^c) = P(X) = 1$. The fuzziness measure F in the next note shows that these extremal results never hold for fuzzy sets:

$$0 < F(A) = F(A \cap A^c) = F(A \cup A^c) = F(A^c) < 1$$

if and only if A is a fuzzy set (iff at least one element $x \in X$ belongs to A to nonbinary degree: If $a(x) = Degree(x \in A)$ then $0 < a(x) < 1$). The result $F(A) = 0$ holds iff A is a non-fuzzy or binary set. Still we can equate each set function or "membership" value $a(x)$ with its own discrete conditional probability density function that depends on x:

$$a(x) = p(A \mid X = x) \text{ and } a^c(x) = 1 - a(x) = p(A^c \mid X = x).$$

17. Let fuzzy subset $A \subset X$ have the fuzzy cardinality or count $c(A) = \sum_{i=1}^{n} a_i$, $c(A) = \sum_{i=1}^{\infty} a_i < \infty$, or $c(A) = \int_{R_n} a(x) \, dx < \infty$ for a discrete fuzzy set $A = (a_1, \dots, a_n) \in [0, 1]^n$ or for a real fuzzy subset $A \subset R^n$ with arbitrary integrable joint set function $a{:}R^n \to [0, 1]$. Then we can show from the geometry of fuzzy cubes that the fuzziness F of A is the ratio of counted violations of the "laws" of noncontradiction to excluded middle: $F(A) = c(A \cap A^c)/c(A \cup A^c)$ where we define intersection and union pointwise with minimum and maximum $(a \cap a^c \, (x) = \min(a(x), 1 - a(x))$ and $a \cap a^c \, (x) = \max(a(x), 1 - a(x)))$ and define set complement with order reversal $(a^c(x) = 1 - a(x))$. This is the "fuzzy entropy" theorem. Hence $F(A) = 0$ iff A is binary and $F(A) = 1$ iff $A = A^c$. An l^p-version of the Pythagorean theorem gives the measure of sub-sethood or partial set containment as another ratio of counts:

$$S(A, B) = Degree(A \subset B) = \frac{c(A \cap B)}{c(A)}$$

This is the subsethood theorem. It follows that we can eliminate the fuzziness of A in favor of a unique whole-in-the-part relation of subsethood: $F(A) = S(A \cup A^c, A \cap A^c)$.

The same Pythagorean structure leads from first principles to a measure E of fuzzy equivalence between two sets as a different ratio of counts:

$$E(A, B) = Degree(A = B) = \frac{c(A \cap B)}{c(A \cup B)}$$

The choice $B = A^c$ gives at once the "yin-yang" theorem that the text mentions: $F(A) = E(A, A^c)$.

So the fuzziness of a set A is exactly the degree to which A equals its own opposite or complement A^c. Note that F reaches its maximum $F(A) = 1$ if and only if $A = A^c$ holds with 100% equality and thus if and only if (for a discrete fuzzy set) A lies at the midpoint of the fuzzy cube $[0, 1]^n$. We can further eliminate fuzzy equality E in favor of subsethood S:

$$E(A, B) = \frac{S(A, B) \, S(B, A)}{S(A, B) + S(B, A) - S(A, B) \, S(B, A)}$$

for non-empty fuzzy sets A and B. This shows that for any fuzzy set A and its complement A^c there is a unique coincidence of the three distinct concepts of fuzziness and subsethood and equality: $F(A) = S(A \cup A^c, A \cap A^c) = E(A, A^c)$

Each of these equalities holds only trivially if A is a binary set since then each measure gives the value zero. $(A \cup A^c = X$ and $A \cap A^c = \emptyset$ if and only if A is binary. Then $F(A) =$

$S(A \cup A^c, A \cap A^c) = S(X, \varnothing) = Degree(X \subset \varnothing) = 0$ if X is not empty. This holds in a fuzzy cube of n dimensions if and only if A is one of the 2^n vertices of the cube.)

Subsethood emerges as the fundamental concept in fuzzy theory. Counting or cardinality reduces to it because $c(A) = nS(X, A)$ from the subsethood theorem. This is another whole-in-the-part result. Elementhood or fuzzy membership also reduces to subsethood because $a(x) = S(\{x\}, A)$ from the subsethood theorem. For details see Kosko, B., "Fuzzy Entropy and Conditioning," *Information Sciences* 40, (1986): 165–74; Kosko, B., *Neural Networks and Fuzzy Systems*, Prentice Hall, 1991; Kosko, B., *Fuzzy Engineering*, Prentice Hall, 1996.

18. Suppose there are n trials and each trial results in a success or a failure. Suppose n_A of these trials are successes and $n - n_A$ are failures. Then X is the sample space of n trials. So $A \subset X$ is the binary subset of successes and A^c is the subset of failures. We can describe the success set A as a bit vector of length n. It contains a 1 in the ith slot if the ith trial is a success and contains a 0 in the ith slot if the ith trial is a failure. The relative frequency f_A is just the success ratio: $f_A = n_A/n$. Now the subsethood theorem in note 17 above gives

$$S(X, A) = \frac{c(X \cap A)}{c(X)} = \frac{c(A)}{n} = \frac{n_A}{n} = f_A$$

since $A \subset X$ holds even if A is fuzzy. So the purely "random" concept of relative frequency reduces to fuzzy subsethood. Frequency probability is just the "whole in the part" or the partial inclusion of the sample space X in the event A. This relation cannot occur if the subsethood operator is binary. This result still holds in the more general case when A defines a fuzzy subset of successes.

The subsethood theorem gives back many of the defining axioms of abstract conditional probability. This suggests that subsethood is the more fundamental notion. See Kosko, B., "Fuzziness Versus Probability," *International Journal of General Systems* 19, no. 2-3 (1990): 211–40. This paper led to a formal debate on fuzziness versus probability that takes up the February 1994 issue of the *IEEE Transactions on Fuzzy Systems*.

19. Legg, G., "Transmission's Fuzzy Logic Keeps You on Track," *Electronic Design News*, 23 December 1993, 60–63. See also note 1 above.

20. Whitehead, A. N., and B. Russell, *Principia Mathematica*, 2d edition, Cambridge University Press, 1927. George Boole and Gottlob Frege also helped lay the groundwork for modern (binary) symbolic logic. But logic changed little for over 2,000 years since Aristotle wrote the first book on the subject. Aristotle's *Organon* contains his collected writings on logic. Typical binary quotes come from *De Interpretatione*: "Everything must either be or not be, whether in the present or in the future" and "There is nothing between asserting [A] and denying [not-A]." Aristotle did make brief mention of degrees of truth but his followers over the centuries largely ignored these remarks. Aristotle made his binary stance clear in his "most certain of all" principles in his *Metaphysics*. He claims there in section 1005b that "the same attribute cannot at the same time belong and not belong to the same subject in the same respect." He further claims in section 1011b that "there cannot be an intermediate between contradictories, but of one subject we must either affirm or deny any one predicate." Hence A or *not-A*. For details see Lane, R., "Peirce's 'Entanglement' with the Principles of Excluded Middle and Contradiction," *Transactions of the Charles S. Peirce Society* 33, no. 3, Summer 1997.

21. Russell, B., *The Philosophy of Logical Atomism*, Open Court, 1985.

22. "So far as the laws of mathematics refer to reality, they are not certain. And so far as they are certain, they do not refer to reality," Einstein, A., "Physics and Reality," *Journal of the Franklin Institute*, 1936. Positivist philosopher Friedrich Waismann states this quote in his essay "How I See Philosophy" in the anthology *Logical Positivism*, A. J., Ayer, editor, The Free Press, 1959, p. 360.

23. Russell, B., "Vagueness," *Australian Journal of Psychology and Philosophy* 1 (1923): 84–92. Modern philosophy defines vagueness as what we now call fuzziness: "Words like 'smart,' 'tall,' and 'fat' are vague since in most contexts of use there is no bright line separating them from 'not smart,' 'not tall,' and 'not fat.' Vagueness needs to be distinguished from ambiguity, which is a property of a word or phrase with two distinct meanings. Whereas 'is

drunk' is vague, 'is at the bank [river? commercial?]' is ambiguous," from the "Vagueness" entry in the *Oxford Companion to Philosophy,* T. Honderich, editor, Oxford University Press, 1995, p. 894.

24. Lukasiewicz, J., *Selected Works,* Borkowski, editor, *Studies in Logic and the Foundations of Mathematics,* North Holland, 1970; "Aristotle on the Law of Contradiction," *Articles on Aristotle,* vol. 3, *Metaphysics,* J. Barnes, M. Schofield, and R. Sorabji, editors, St. Martin's Press, 1979, pp. 50–62.

25. The unity sum holds because the truth value $t(not\text{-}A)$ of statement $not\text{-}A$ is one minus the truth value of A: $t(not\text{-}A) = 1 - t(A)$. The connectives AND and OR lead to the further truth-value equality $t(A) + t(not\text{-}A) = t(A \ AND \ not\text{-}A) + t(A \ OR \ not\text{-}A)$. This holds for an infinitude of operators that include the popular *min* and *max* operators as boundary cases: $t(A \ AND \ B) = \min(t(A), t(B))$ and $t(A \ OR \ B) = \max(t(A), t(B))$.

This gives the fundamental theorem on the "laws" of noncontradiction and excluded middle for a multivalued logic: $t(A \ AND \ not\text{-}A) + t(A \ OR \ not\text{-}A) = 1$. Then binary logic leads to a logical equivalence of noncontradiction and excluded middle: $t(A \ AND \ not\text{-}A) = 0$ if and only if $t(A \ OR \ not\text{-}A) = 1$. These two truth values take on numbers in the open interval $(0, 1)$ for multivalued logics.

26. Black, M., "Vagueness: An Exercise in Logical Analysis," *Philosophy of Science* 4 (1937): 427–55. Black returned to fuzziness in his paper "Reasoning with Loose Concepts," *Dialogue* 2 (1963): 1–12.

27. *Logical Positivism,* editor, A. J. Ayer, New York: The Free Press, 1959. For extensions and criticisms of the logical positivism of the early 1930s see Popper, K. R., *The Logic of Scientific Discovery,* Harper & Row, 1959; Hempel, C. G., *Aspects of Scientific Explanation and Other Essays in the Philosophy of Science,* The Free Press, 1965; Nagel, E., *The Structure of Science,* Hackett, 1979; Quine, W. V. O., *Theories and Things,* Harvard University Press, 1981; Quine, W. V. O., *Pursuit of Truth,* Harvard University Press, 1992.

28. Harvard philosopher Willard Van Orman Quine sees the cost of bivalence as high but ultimately worth it in "What Price Bivalence?" *Journal of Philosophy* 78 (February 1991): 90–95: "If the term 'table' is to be reconciled with bivalence, we must posit an exact demarcation, exact to the last molecule, even though we cannot specify it. We must hold that there are physical objects, coincident except for one molecule, such that one is a table and the other is not."

Quine proceeds in the same article to define a mountain: " 'Mountain' affords a rich example, for there is the vagueness of acceptable altitude, the vagueness of boundary at the base, and the consequent indecision as to when to count two summits as two mountains and when as one. Possible stipulations are as follows. Leaving foreign planets conveniently aside, we may define a mountain as any region of the earth's surface such that (a) the boundary is of uniform altitude, (b) the highest point, or one of them, is at an inclination of at least ten degrees above every boundary point and twenty degrees above some, and is at least a thousand feet above them, and (c) the region is part of no other region fulfilling (a) and (b). (Theorem: the boundary of a mountain is the outermost contour line that lies wholly within ten degrees of steepness from the summit and partly within twenty.)" Quine never explains why a hill becomes a mountain if it crosses his thousand-foot threshold by a single micron.

29. Zadeh, L. A., "Fuzzy Sets," *Information and Control* 8 (1965): 338–53. Zadeh published his collected papers in *Fuzzy Sets and Applications: Selected Papers,* ed. R. R. Yager, S. Ovchinnikov, R. M. Tong, and H. T. Nguyen, Wiley-Interscience, 1987.

Zadeh's seminal paper "Fuzzy Sets" mentions multivalued logic only in a footnote: "If the values of [the degrees of membership] $f_A(x)$ are interpreted as truth values, the latter case [where the truth values lie in the unit interval $[0, 1]$] corresponds to a multivalued logic with a continuum of truth values in the interval $[0, 1]$" (p. 339). The paper does not reference the prior vague work of Lukasiewicz or Black. Yet the paper uses the *min* and *max* operators of Lukasiewicz to define the AND operation of set intersection and the OR operation of set union as well as the negation operator $1 - x$ to define set complement.

Fuzzy information pioneer George Klir explains the historical context of Max Black's fundamental 1937 paper on vagueness and reprints Black's paper in a special issue of the journal that Klir edits: "Ideas do not emerge in a vacuum. They develop from other ideas through a process that is still ill-understood. To trace ideas that contributed to the emergence of the conception of fuzzy sets would be a difficult, if not impossible, task. By and large, mathematics, science, and engineering were not interested in the notion of fuzziness or vagueness. It was taken for granted that vagueness is unscientific, and thus undesirable, and that it should be avoided by all means. Precision and sharpness were the orders of the day. Nevertheless, there were some lonely voices that recognized and discussed the importance of imprecise boundaries of sets or concepts. The term *vagueness* was usually used in these early discussions rather than the modern term *fuzziness*, which was coined by Zadeh.

"The best discussion of vagueness, which, at least in my opinion, has not been surpassed yet, was published in 1937 by Max Black, a well-known American philosopher [see note 26 above]. This classic paper is so important, historically and conceptually, that I decided to reprint it in this Special Issue. This republication is also intended as a tribute to Max Black, who passed away in August 1989," Klir, G. J., "An Introduction to the Special Issue on a Quarter-Century of Fuzzy Systems," *International Journal of General Systems* 17, no. 2 (June 1990): 89–93. Fay Zadeh (Lotfi's wife) has written a personal history that deals with some of the academic sociology of these events: Zadeh, F., *My Life and Travels with the Father of Fuzzy Logic,* TSI Press, 1998.

30. The first dedicated international fuzzy journal was North Holland's *Fuzzy Sets and Systems.* It began in 1978 and still publishes the most frequently of all fuzzy journals. Wiley began the *International Journal of Intelligent Systems* in 1986. North Holland followed with the *International Journal of Approximate Reasoning* in 1987. The Institute of Electrical and Electronics Engineers (IEEE) began the *IEEE Transactions on Fuzzy Systems* in 1993. Many other journals now count fuzzy systems among their publishing topics.

31. Holmblad, L. P., and Ostergaard, J. J., "Control of a Cement Kiln by Fuzzy Logic," *Fuzzy Information and Decision Processes,* ed. M. M. Gupta and E. Sanchez, North Holland, 1983, pp. 389–99.

32. Mamdani published many papers on rule-based fuzzy systems in the early and mid-1970s: Mamdani, E. H., and Assilian, S., "An Experiment in Linguistic Synthesis with a Fuzzy Logic Controller," *International Journal of Man-Machine Studies* 7 (1977): 1–13. Mamdani, E. H., "Application of Fuzzy Logic to Approximate Reasoning Using Linguistic Synthesis," *IEEE Transactions on Computers* 26, no. 12 (December 1977): 1182–91.

33. LIFE members published many of their research results in *Applied Research in Fuzzy Technology: Three Years of Research at the Laboratory for International Fuzzy Engineering (LIFE),* Yokohama, Japan, ed. A. L. Ralescu, Kluwer, 1994.

34. Hitachi had designed and tested their fuzzy system for years in simulations. The first paper on the subject appeared in 1983 from Yasunobu, S., Miyamoto, S., and Ihara, H., "Fuzzy Control for Automatic Train Operation System," *Proceedings of the 4th IFAC/IFIP/IFORS International Conference on Control in Transportation System,* Baden-Baden, Germany, 1983, 33–39.

35. Schwartz, D. G., and Klir, G. J., "Fuzzy Logic Flowers in Japan," *IEEE Spectrum,* July 1992, 32–35.

36. Perry, T. S., "Profile: Lotfi A. Zadeh," *IEEE Spectrum* 32, no. 6 (June 1995): 32–35.

37. Suppose X is a finite set: $X = \{x_1, \ldots, x_n\}$. Then we can define a fuzzy subset $A \subset X$ algebraically as the multivalued indicator function $a:X \to [0, 1]$. We can also define A geometrically as the fuzzy-unit or fit-vector point

$$A = (a(x_1), \ldots, a(x_n)) = (a_1, \ldots, a_n) \in I^n = [0, 1]^n$$

in the fuzzy n-cube or n-dimensional unit hypercube $I^n = [0, 1]^n$. We can also define fuzzy subsets of arbitrary sets X as points in function spaces. Most of the structure of the fuzzy theorems on finite fuzzy cubes carries over to the general continuous case (and beyond) because all results hold pointwise. Fuzzy engineers tend to work with continuous fuzzy sets when they approximate functions. The theorems on universal (uniform) function approximation still hold for discrete fuzzy sets if the discretized set contains enough samples. For details see Kosko, B., "Fuzzy Systems as Universal Approximators," *IEEE Transactions on Computers* 43, no. 11 (November 1994): 1329–33. Kosko, B., *Neural Networks and Fuzzy Systems,* Prentice Hall, 1991; Kosko, B., *Fuzzy Engineering,* Prentice Hall, 1996. The results are far less conclusive when one attempts to approximate an arbitrary *feedback* or dynamical system. Few feedback fuzzy systems are stable: Kosko, B., "Global Stability of Generalized Additive Fuzzy Systems," *IEEE Transactions on Systems, Man, and Cybernetics* 29, no. 3 (August 1998): 441–52.

3. LEFT AND RIGHT AND NEITHER: THE FUZZY POLITICAL SQUARE

1. "The legal system is part of the system of *social control*. In the broadest sense, this may be *the* function of the legal system; everything else is, in a way, secondary or subordinate. To put it another way, the legal system is concerned with controlling behavior. It is a kind of traffic cop. It tells people what to do and not to do, and it backs up its directives with force." Friedman, L. M., *American Law,* W. W. Norton, 1998.

2. Social scientists and critics have often pointed to the trade-off between state and social power: "If we look beneath the surface of our public affairs, we can discern one fundamental fact, namely: a great redistribution of power between society and the State . . . an increase of State power and a corresponding decrease of social power," from the essay "Social Power vs. State Power," in Nock, A. J., *On Doing the Right Thing and Other Essays,* Harper & Row, 1928.

3. Nobel laureate economist Kenneth Arrow defines a dictator as such a taste tyrant in his "impossibility theorem" in *Social Choice and Individual Values,* 2d edition, Yale University Press, 1970. The theorem states that there is no good way for society to vote on all issues. Arrow lists five intuitive axioms that most economists feel any such voting scheme should obey (such as the axiom of transitivity of preferences: If all the people prefer option B to option A and if they all prefer option C to B then they all prefer C to A). Non-dictatorship is one of the five axioms: It should not be the case that society prefers option A to B if and only if some person X prefers A to B. Then the impossibility theorem states that any four of the five axioms contradicts the fifth. Ideal democracy is impossible.

   Arrow's theorem has spawned a family of like results for voting and other nonmarket attempts to structure society. These theorems are uniformly negative: "Three widely shared objectives—collective rationality, decisiveness, and equality of power—stand in irreconcilable conflict. . . . There is little comfort here for those designing ideal procedures for collective choice. . . . The opportunities for improvement are severely limited. Stark compromises are inevitable," Blair, D. H., and Pollak, R. A., "Rational Collective Choice," *Scientific American,* August 1983, 88–95.

   The Romanian fuzzy theorist Gheorghe Paun has shown that the negative structure of Arrow's impossibility theorem still holds in a framework of fuzzy sets. Paun, G., "An Impossibility Theorem for Indicators Aggregation," *Fuzzy Sets and Systems,* 1983, 205–10. For a more detailed treatment see Billot, A., *Economic Theory of Fuzzy Equilibria: An Axiomatic Analysis,* 2d edition, Springer-Verlag, 1995.

4. The U.S. Constitution sets up the electoral process in Section 1 of Article 2: "Each state shall appoint, in such Manner as the Legislature thereof may direct, a Number of Electors, equal to the whole Number of Senators and Representatives to which the State may be entitled in the Congress." Voters once voted for electors as well as for presidential candidates. This changed with the Electoral Count Act of 1887. Congress still counts the electoral votes on the January 6th that follows the November election.

5. "**left:** in politics, the more radically progressive wing in any legislative body or party," *The Columbia Encyclopedia,* 5th edition, Houghton Mifflin, 1993, p. 1553.

6. Former House Speaker Newt Gingrich proposed a wide range of market reforms in *To Renew America,* HarperCollins, 1995, even as he called for the Singapore model to deal with drugs: "I strongly favor a mandatory death penalty for entering our territory with a commercial quantity of illegal drugs" (p. 181).

7. *Western Liberalism: A History in Documents from Locke to Croce,* ed. E. K., Bramsted and K. J. Melhuish, Longman, 1978. See also the classic on "conservative" liberalism: Burke, E., *Reflections on the Revolution in France, and on the Proceedings in Certain Societies in London Relative to That Event,* Penguin Books, 1968 (first published 1790).

8. "Roughly *half* of the arrests and court cases in the United States each year involve consensual crimes—actions that are against the law but directly harm no one's person or property except perhaps the 'criminal's.' More than 350,000 people are in jail *right now* because of something they did, something that did *not* physically harm the person or property of another. In addition, more than 1,500,000 people are on parole or probation for consensual crimes. Further, more than 4,000,000 people are arrested *each year* for doing something that hurts no one but potentially themselves. . . . Consensual crimes are sometimes known as *victimless crimes* because it's hard to find a clear-cut victim. The term *victimless crime,* however, has been so thoroughly misused in recent years that as a description it has become almost useless." McWilliams, P., *Ain't Nobody's Business If You Want To: The Absurdity of Consensual Crimes in a Free Society,* Prelude Press, 1993.

9. "Tackling Drugs Together: A Strategy for England 1995–1998," British Government White Paper, May 1995; "National Household Survey on Drug Abuse," U.S. Substance Abuse and Mental Health Administration, 1993; "What American Users Spend on Illegal Drugs, 1988–1993, Office of National Drug Control Policy, Spring 1995; "Tobacco Industry Profile, 1994," Tobacco Institute, Washington, D.C.

10. Johnson, P., *Modern Times: The World from the Twenties to the Eighties,* Harper & Row, 1983. Johnson explores the roots of some of these ideologies in his *The Birth of the Modern: World Society 1815–1830,* HarperCollins, 1991.

11. Many have noted the circularity of the left-right terms: "It is one of the oddities of our era that people of every political persuasion seem to spend precisely one half of their time announcing that the terms 'liberal' and 'conservative' are no longer useful and the other half flinging the same two words around with abandon, like confetti. . . . The final insult of all this [use of the left-right spectrum] is the profligate use of the modifier 'moderate' as a noun, standing there all by itself modifying nothing. I mean: a moderate *what?*" Greenfield, M., "The Tyranny of 'the Spectrum,' " *Newsweek,* 25 September 1995.

12. Political scientists Donald P. Green and Ian Shapiro question whether political scientists have the tools and concepts "to come up with universal theories of politics" in their book *Pathologies of Rational Choice: A Critique of Applications in Political Science,* Yale University Press, 1984. The journal *Critical Review* devotes a double issue to this debate on "rational choice theory" in vol. 9, no. 1–2, Winter/Spring 1995.

13. The cube view of fuzzy sets makes sense only for sets that contain a finite (or countably infinite) number of objects. Suppose a set X contains n objects: $X = \{x_1, x_2, \ldots, x_n\}$. Then X has 2^n subsets in its so-called *power set* 2^X. Each subset defines a binary n-vector of 1s and 0s that show presence or absence of an object in the set. Suppose X contains just 3 objects: $X = \{x_1, x_2, x_3\}$. Then X has 2^3 or 8 subsets. The set $A = \{x_1, x_3\}$ is one of these 8 states. A defines the bit vector (1 0 1). This bit vector states that the first and third elements belong (100%) to the set A and that the second element does not belong to A at all. The binary subsets of a general X map one-to-one to the 2^n corners or vertices of an n-dimensional unit hypercube $[0, 1]^n$ or fuzzy n-cube (and hence the power set 2^X is isomorphic to the cube corners or to the Boolean n-cube).

Fuzzy sets fill in these unit hypercubes as points in the cubes. A set X may contain only n objects but it still has as many fuzzy subsets $A = (a_1, \ldots, a_n)$ as there are real numbers. Here a_i is the fuzzy unit or *fit* value that states the degree to which the ith object x_i belongs to fuzzy set A: $a_i = a(x_i) = Degree(x_i \in A) \in [0, 1]$. The set $X = \{x_1, x_2, x_3\}$ contains the fuzzy subset $A = (a_1, a_2, a_3) = (.3 \ .5 \ .9)$. The fit vector A states that the first object x_1 belongs to A only 30% ($a_1 = .3$) while the third object x_3 belongs 90% ($a_3 = .9$). The second object x_2 belongs to A 50% ($a_2 = .5$) and does not belong to A 50% ($a_2^c = 1 - a_2 = .5$). The second object x_2 lies as much in A as not. (The *fuzzy* power set of X contains all the fuzzy subsets of X and is isomorphic to the unit cube $[0, 1]^n$.) This *sets-as-points* view leads to many geometric theorems on fuzzy sets: Kosko, B., "Fuzzy Entropy and Conditioning," *Information Sciences* 40 (1986) 165–74; 1986; Kosko, B., *Neural Networks and Fuzzy Systems,* Prentice Hall, 1991; Kosko, B., *Fuzzy Engineering,* Prentice Hall, 1996.

14. Economists use a formal theory of revealed preferences to estimate a consumer's preference maps. Nobel laureate economist Paul A. Samuelson gave the first complete mathematical treatment of this technique in his *Foundations of Economics Analysis,* Cambridge: Harvard University Press, 1948.

15. "A free man is he that in those things, which by his strength and wit he is able to do, is not hindered to do what he has a will to do. . . . From the use of the word *freewill,* no liberty can be inferred of the will, desire, or inclination, but the liberty of the man; which consisteth in this, that he finds no stop, in doing what he has the will, desire, or inclination to do," Thomas Hobbes, *Leviathan,* Pocket Books, 1976 (first printing, April 1651). Hobbes viewed the term "free will" as a contradiction in terms. The will is not free because we or our organism or our causal past control it.

16. From the essay "Things and Their Place in Theories" in Quine, W. V. O., *Theories and Things,* Harvard University Press, 1983, p. 11.

17. Barlow, J. P., "Decrypting the Puzzle Palace," *Communications of the Association for Computing Machinery (ACM)* 35, no. 7 (July 1992): 25–31. NSA has barred Microsoft and Novell and other firms from using the RSA encryption algorithm in their export software. Bruce Schneier discusses the practical aspects of the Data Encryption Standard in his article "Differential and Linear Cryptanalysis," *Dr. Dobb's Journal* 21, no. 1 (January 1996): 42–48, and in his book *Applied Cryptography,* 2d edition, Wiley, 1996.

   Much of modern encryption rests on the Fundamental Theorem of Algebra which states that each positive integer factors uniquely into a product of prime numbers. Most computers cannot find these prime factors and that helps keep many encryption schemes secure. Quantum computers could in principle factor numbers well enough to undermine modern encryption schemes: "Such computers could be a threat to what is presently one of the most common methods of encryption." Chuang, I. L., Laflamme, R., Shor, P. W., and Zurek, W. H., "Quantum Computers, Factoring, and Decoherence," *Science* 270 (8 December 1995): 1633–36. See also Lloyd, S., "Quantum-Mechanical Computers," *Scientific American,* October 1995, 140–45.

   MIT physicist Seth Lloyd has shown that the potential computational power of quantum computers lies in their simulation ability to bypass exponential computational explosions. This holds because a quantum bit or *qubit* is the parallel superposition of states such as 0 and 1: "[Richard] Feynman was correct that quantum computers could provide efficient simulation of other quantum systems. A quantum computer with a few tens of quantum bits could perform in a few tens of steps simulations that would require Avogadro's number of memory sites and operations on a classical computer. A mere 30 or 40 quantum bits would suffice to perform quantum simulations of multidimensional fermionic systems such as the Hubbard model that have proved resistant to conventional computational techniques. Hundreds to thousands of bits may be required to simulate accurately systems with continuous variables such as lattice gauge theories or models of quantum gravity," *Science* 273 (23 August 1996): 1073–78.

   Mathematician Timothy Spiller gives an engineering review of quantum computing with applications in cryptography and shows how quantum uncertainty favors security: "The impossibility of measuring quantum systems without disturbing them guarantees the detection of eavesdropping and hence secure information transfer is possible," Spiller, T. S., "Quantum Information Processing: Cryptography, Computation, and Teleportation," *Proceedings of the IEEE* 84, no. 12 (December 1996): 1719–46. Physicist Paul Townsend has carried out the first steps of this research program working with fiber optics: "When quantum-mechanical processes are used to establish the key, any eavesdropping during transmission leads to an unavoidable and detectable disturbance in the received key information. Quantum cryptography has been demonstrated using standard telecommunication fibres linking single pairs of users, but practical implementations will require communication networks with many users." Townsend, P. D., "Quantum Cryptography on Multi-User Optical Fibre Networks," *Nature* 385 (2 January 1997): 47–49. See also Rivest, R. L., "The Case Against Regulating Encryption Technology," *Scientific American,* October 1998, 116–17.

18. "The Myth of the Powerless State," *The Economist,* 7 October 1995, 15–16.

19. Rush Limbaugh presents his mainstream conservative view in *The Way Things Ought to Be,* Pocket Books, 1992. Limbaugh favors a large degree of economic freedom: "Then we have the spaced-out Hollywood left eating beans and rice to focus attention on the evils of capitalism. We are told that there is an inequitable distribution of food and other products under capitalism. Wrong. That's not the problem. The world's biggest problem is the unequal distribution of capitalism. If there were capitalism everywhere, you wouldn't have food shortages. If capitalism is the problem, how is it that the US is the only nation in the world that can feed itself and still feed much of the rest of the planet?" (p. 47).

But Limbaugh would limit political freedom by outlawing abortion and most victimless crimes. He claims that calls to end drug prohibition logically imply a call to end all laws: "A lot of people say we should legalize cocaine and heroin. It would take the crime out of it and hopefully, as a result, some of the violence associated with it. Besides, the people who are inclined to use drugs are going to do so regardless of its illegality. But the fact that people are going to ignore and break laws is not a valid argument for decriminalization. The state would then be sanctioning, even promoting, conduct that is harmful to society. That is similar to today's trendy policies for the free distribution of condoms in public schools and prisons. We are sending a message that such activity is okay, based on the specious argument that we know the activity is going to occur anyway. We might as well just carry this argument to its logical conclusion: The reason laws are broken is that there are laws. People are going to break the law. They are going to steal, rob, and kill. So what we need to do is eliminate laws and get rid of the police. That way there will be no crime" (p. 53). Hence he is an economic libertarian but not a civil libertarian and is thus a conservative.

20. Nobel laureate economist Friedrich A. Hayek may well have been the premiere classical liberal of the twentieth century: "What in Europe was called 'liberalism' was here [in the United States] the common tradition on which the American polity had been built: thus the defender of the American tradition was a liberal in the European sense. This already existing confusion was made worse by the recent attempt to transplant to America the European type of conservatism, which, being alien to the American tradition, has acquired a somewhat odd character. And some time before this, American radicals and socialists began calling themselves 'liberals.' I will nevertheless continue for the moment to describe as liberal the position which I hold and which I believe differs as much from true conservatism as from socialism." From the chapter "Why I Am Not a Conservative" in Hayek, F. A., *The Constitution of Liberty*, University of Chicago Press, 1960, pp. 397–98.

21. "Today people in the United States and around the world who believe in the principles of the American Revolution—individual liberty, limited government, the free market, and the rule of law—call themselves by a variety of terms, including conservative, libertarian, classical liberal, and liberal. We see problems with all those terms. 'Conservative' smacks of an unwillingness to change, a desire to preserve the status quo. . . . 'Libertarian' is an awkward and misinterpreted neologism that has become too closely tied to a particular group of activists. 'Classical liberal' is closer to the mark, but the word 'classical' connotes a backward-looking philosophy all the tenets of which have been carved in stone. Finally, 'liberal' may well be the perfect word in most of the world—the liberals in societies from China to Iran to South Africa to Argentina are supporters of human rights and free markets—but its meaning has clearly been corrupted by contemporary American liberals. 'Market liberal,' by modifying liberal with an endorsement of the free market, thus strikes us as a solid description of a philosophy that is rapidly gaining adherents throughout the world." *Market Liberalism*, ed. D. Boaz and E. H. Crane, Cato Institute, 1993, pp. 8–9.

22. John Stuart Mill describes the early and heady days of the Liberal party in the early 1800s: "At this period, when Liberalism seemed to be becoming the tone of the time, when improvement of institutions was preached from the highest places, and a complete change of the constitution of Parliament was loudly demanded in the lowest, it is not strange that attention should have been roused by the regular appearance in controversy of what seemed a new school of writers, claiming to be the legislators and theorists of this new tendency. The air of strong conviction with which they wrote, when scarcely anyone else seemed to have an equally strong faith in as definite a creed; the boldness with which they tilted against the very front of both the existing political parties [the Whigs and the Tories] . . . made the so-called Bentham [utilitarian Liberal] school in philosophy and politics fill a greater place in the public mind than it had held before, or has ever again held since equally earnest schools of thought have arisen in England." Mill, J. S., *Autobiography*, Penguin, 1989 (first published 1873), pp. 90–91. In time the Liberals came to power and Mill even served as a member of Parliament from 1865 until he lost his bid for reelection in 1868.

23. "Because of their growing disdain for government, more and more Americans appear to be drifting—often unwittingly—toward a libertarian philosophy. That seems particularly true among baby boomers returning to the 'do your own thing' ethos of their youth and among young people involved in the intensely independent computer industry. Indeed, when the

Gallup polling organization last year asked questions about government's role that were designed to distill American's political philosophies, it categorized 22% of the public as 'libertarian.' " Seib, G. F., "Less Is More: Libertarian Impulses Show Growing Appeal Among the Disaffected," *Wall Street Journal,* 20 January 1995, p. A-1.

24. *The Portable Thomas Jefferson,* ed. M. D. Peterson, Viking Press, 1975. Jefferson also had a simple explanation for the growth of large states in the 20th century: "The natural progress of things is for liberty to yield and government to gain ground."

25. "The simplest and most probable explanations for our molecular findings are that Thomas Jefferson, rather than one of the Carr brothers, was the father of [Sally Hemings's last son] Eston Hemings Jefferson, and that [Hemings's first son] Thomas Woodson was not Thomas Jefferson's son. The frequency of the Jefferson haplotype is less than 0.1 per cent, a result that is at least 100 times more likely if the president was the father of Eston Hemings Jefferson than if someone unrelated was the father." Foster, A. E., et al., "Jefferson Fathered Slave's Last Child," *Nature* 396 (5 November 1998) 27–28. This study was not conclusive: "The data establish only that Thomas Jefferson was one of several candidates for the paternity of Eston Hemings, Sally's fifth child," Marshall, E., "Which Jefferson Was the Father?" *Science* 283 (8 January 1999): 153–54.

26. Wilkinson, F., "Sidewalks for Sale: Libertarians are Flourishing on Capitol Hill, on Campus, and On Line," *Rolling Stone,* 6 April 1995. "What liberalism was to the Sixties and conservatism was to the Eighties, libertarianism may be to the youth of the 1990s. . . . Many of the 41 million members of Generation X are turning to . . . libertarianism, a mixture of liberal views on social issues and a conservative bent on pocketbook concerns." From "The GenX Philosophy: Many Reject Politics, Lean Libertarian," *USA Today,* 26 July 1995. Many observers have noted the disproportionate number of libertarians on the Internet and involved in digital disputes over cryptography. This has led to the term "cyber-libertarian" or "crypto-libertarian": "Move over democrats and republicans, liberals, socialists and fascists. Here come the cryptolibertarians, the wirehead heirs of Thomas Jefferson and Henry David Thoreau." McHugh, J., "Politics for the Really Cool," *Forbes,* 8 September 1997, 172–79. Social theorist Charles Murray has made a popular case for libertarianism or classical liberalism in *What It Means to Be a Libertarian: A Personal Interpretation,* Broadway Books, 1997.

27. Cowan, J., and Nelson, R., "Age Discrimination—Against the Young," *Los Angeles Times,* 7 August 1994, Metrosec. Cowan and Nelson direct the youth advocacy group Lead or Leave.

28. Ross Perot has pursued his populism with a folksy vigor in TV ads and popular books in both the 1992 and 1996 presidential races: "It's time to pick up a shovel and clean out the barn! Let's go to work." Perot argues for a new tobacco tax this way: "The federal government might as well increase the cigarette tax because it looks like the states will do it anyway." Perot, R., *Not for Sale at Any Price,* Hyperion, 1993, p. 105. He called for like measures to raise taxes and "spend federal dollars to spread programs that work" in Perot, R., *United We Stand,* Hyperion, 1992, p. 78.

Perot founded the political action group United We Stand America and issued its Mission Statement: "We the people of United We Stand America, recognizing that our republic was founded as a government of the people, by the people and for the people, unite to restore the integrity of our economic and political systems. We commit ourselves to organize, to educate, to participate in the political process, and to hold our public servants accountable. We shall rebuild our country, renew its economic, moral and social strength, and return the sovereignty of America to her people." Perot then founded the Reform Party and ran for president as its first candidate in 1996.

29. *The Columbia Encyclopedia,* 5th edition, Columbia University Press, 1993, p. 926.

30. Quoted in Boaz, D., *Libertarianism: A Primer,* The Free Press, 1997. Boaz edited a companion anthology of classic writings in classical liberalism or libertarianism: *The Libertarian Reader: Classic and Contemporary Writings from Lao-Tzu to Milton Friedman,* The Free Press, 1997.

31. Byock, J. L., *Medieval Iceland: Society, Sagas, and Power,* University of California Press, 1988.

32. Political theorist David F. Nolan seems to be the first to have published the political 2-square in Nolan, D. F., "Classifying and Analyzing Politico-Economic Systems," *The Individualist,* January 1971. Many scholars and textbooks refer to this square or map as the "Nolan chart."

33. Samuelson, P. A., *Economics,* 10th edition, McGraw-Hill, 1976, pp. 884–86.

34. Maddox, W. S., and Lilie, S. A., *Beyond Liberal and Conservative,* Cato Institute, 1984, Table 3.
35. Logical paradoxes of self-reference have a common structure: If A then *not-A*. If *not-A* then A. So A if and only if *not-A*. Hence A and *not-A* are logically equivalent: $A = $ *not-A*. So A and *not-A* have the same truth values: $t(A) = t(not-A)$. This indeed leads to a binary contradiction or "paradox" if we insist that A has the binary truth value of 0 or 1. For if $t(A) = 1$ then the logical equivalence gives $1 = t(A) = t(not-A) = 1 - t(A) = 1 - 1 = 0$. The binary assertion $t(A) = 0$ also implies that $1 = 0$. That should have been sufficient logical cause throughout the ages not to insist on binary truth values. The structure of the self-referential paradoxes tells us only that $t(A) = 1 - t(A)$. A child can solve this simple algebraic expression to find the self-referential truth value: $t(A) = \frac{1}{2}$. Hence the truth lies at the midpoint of the 1-D fuzzy cube $[0, 1]$. There is no good way to round off this midpoint value to either binary extreme of 0 or 1. This may be why midpoint phenomena have taken on the cultural status as "paradoxical."

   Midpoint phenomena can occur in fuzzy cubes of any dimension. A fuzzy two cube houses the medieval dualist paradox:

Socrates:	"What Plato is about to say is true."
Plato:	"Socrates lies."

   We could add a "paradoxical" quote from Aristotle to convert this into a paradox in a 3-D fuzzy cube.

   Philosopher Patrick Grim has shown that we can view paradoxes as discrete dynamical systems in fuzzy cubes. Grim, P., "Self-Reference and Chaos in Fuzzy Logic," *IEEE Transactions on Fuzzy Systems* 1, no. 4 (November 1993): 237–53. A self-referential dynamical system has the simple negation form:

   $$t(S_{n+1}) = 1 - t(S_n)$$

   Then an initial binary truth value of $t(S_0) = 0$ or $t(S_0) = 1$ leads to the infinite binary oscillation of 0, 1, 0, 1, 0, 1, 0,. . . . But non-binary values lead elsewhere. The midpoint truth value $t(A) = \frac{1}{2}$ gives the only fixed-point attractor of the dynamical system. Grim suggests one of infinitely many ways to cast the dualist paradox as a dynamical system in the fuzzy square:

   $$t(S_{n+1}) = 1 - |t(S_n) - t(P_n)|$$
   $$t(P_{n+1}) = 1 - |t(P_n) - (1 - t(S_n))^2|$$

36. Skinner, B. F., *Science and Human Behavior,* Macmillan, 1953, p. 411.
37. "Today 118 of the world's 193 countries are democratic, encompassing a majority of its people (54.8% percent to be exact). . . . Freedom House's 1996–97 survey *Freedom in the World* has separate rankings for political liberties and civil liberties, which correspond roughly with democracy and constitutional liberalism, respectively. Of the countries that lie between confirmed dictatorship and consolidated democracy, 50 percent do better on political liberties than on civil liberties. In other words, half of the 'democratizing' countries in the world today are illiberal democracies. . . . Constitutional liberalism is about the limitation of power, democracy about its accumulation and use." Zakaria, F., "The Rise of Illiberal Democracy," *Foreign Affairs* 76, no. 6 (December 1997): 22–43.
38. "In his 22-page report [from the National Center for Policy Analysis], [University of Texas economist] Professor Gerald Scully notes that democide, a Latin word meaning indiscriminate state killing, and genocide, the murder of minorities, have claimed the lives of 170 million people this century—four times as many as civil and international wars. All in all, 7.3% of the world population in the 20th century succumbed to state-sponsored killing, a steep rise from the 19th century, when only 3.7% of the population met such a demise. . . . He found that the wealthiest nations had the lowest levels of democide, and vice versa. 'A 1% increase in real GDP "buys" about a 1.4% decline in democide. . . . Rising income made killing people who produce goods more costly to kill.' " Bleakley, F. R., "It's No

Wonder Some People Call Economics 'The Dismal Science,' " *Wall Street Journal*, 4 November 1997.

39. "Did somebody say the age of big government was dead? At the beginning of the this century government spending in today's industrial countries accounted for less than one-tenth of national income. Last year, in the same countries, the government's share of output was roughly half. Decade by decade, the change in the government's share of the economy moved in one direction only: up. During war it went up. During peace it went up. . . . The state grows in bad times, it seems, because it has to. It grows in good times as well, only faster, because governments feel more ambitious." "Spend, Spend, Spend," *The Economist* special issue on "The Future of the State," 20 September 1997, 7–11.

40. We can measure the degree of state control $S(t)$ in a 2-D fuzzy cube as the additive inverse of the average of the state's degree $p(t)$ of political liberty at time t and its degree $e(t)$ of economic liberty: $S(t) = 1 - (e(t) + p(t))/2$. We can extend this to an n-D fuzzy cube where each coordinate $x_i(t)$ has its own positive weight w_i:

$$S(t) = 1 - \frac{\sum_{i=1}^{n} w_i x_i}{\sum_{i=1}^{n} w_i}$$

We can further extend the measure to average world government if we add up the 193 state measures $S_1, \ldots, S_{193}$ and divide by 193. In any case $S(t)$ takes values in the unit interval $[0, 1]$. The extreme case $S = 0$ defines political anarchy or a Lockean "state of nature." The government collapses if S falls to zero in time. The other extreme $S = 1$ defines total government control or pure totalitarianism. Now let g be the pro-state sentiment or the degree in $[0, 1]$ to which society likes the government. Then $1 - g$ measures the anti-state sentiment.

A minimal logistic model has the form $\dot{S} = gS(1 - S) - (1 - g)S$ where the overdot term $\dot{S}$ denotes the velocity or time derivative of S. The first term on the right-hand side of the model shows how the state grows quickly at first and then tapers off to an upper asymptote. The second term shows how the anti-government sentiment $1 - g$ causes S to decay in a linear way. The exact solution at any time is

$$S(t) = \frac{(2g - 1)S_0}{gS_0 + [2g - 1 - gS_0]e^{-(2g - 1)t}}$$

for initial condition $S_0 = S(0)$. This solution shows that S converges exponentially quickly to the global equilibrium S_e:

$$S_e = 1 - \frac{1 - g}{g}$$

for any S_0. Too little government will quickly rise to the level S_e just as too much government will quickly fall to the level S_e. It depends only on the ratio of anti-state sentiment $1 - g$ to pro-state sentiment g.

There are many ways to extend this model. The first thing to do is to let the sentiments g and $1 - g$ slowly vary with time and thus become system functionals. More accurate models might let g depend on S or rather $1 - S$. This would say that people favor more government when it is small and favor less when it is large. A still more accurate model would add a spatial element and show how state control "diffuses" in time over an n-D fuzzy cube as in the simple reaction-diffusion model

$$\frac{\partial S}{\partial t} = gS(1 - S) - (1 - g)S + c\sum_{i=1}^{n} \frac{\partial^2 S}{\partial x_i^2}$$

For more details see Kosko, B., "Equations of State," *Liberty* 11, no. 4 (March 1998): 46–47.

4. The Fuzzy Tax Form

1. Mead, C. A., and Conway, L., *Introduction to VLSI Systems,* Addison-Wesley, 1980. Carver Mead also wrote the first text on building neural chips: Mead, C. A., *Analog VLSI and Neural Systems,* Addison-Wesley, 1989. The IEEE awarded Mead its John von Neumann Medal in 1996 "for leadership and innovative contributions to VLSI [very large scale integrated circuitry] and creative microelectronic structures," *IEEE Spectrum* 33, no. 6 (June 1996): 58–59.

2. Rhodes, R., *The Making of the Atomic Bomb,* Simon and Schuster, 1986. Rhodes shows that the hydrogen bomb came in fits and starts without the tighter goals and time pressures of the Manhattan Project: R. Rhodes, *Dark Sun: The Making of the Hydrogen Bomb* (Simon and Schuster, 1995).

3. "Others raise the specter of an industrial policy with the Pentagon picking winners and losers. But all federal research and development programs pick winners and losers. . . . Dual-use technology amounted to $1.8 billion in the 1995 defense budget—about 5% of the total research and development spending." Deutch, J. M., "When the Bullets Meet the Bytes," *Los Angeles Times,* 30 March 1995.

4. "Dual-Use: Fool's Gold or Mother Lode?" briefing, EIA 22nd Annual Spring Technology & Budget Conference, 30–31 March 1993, Washington, D.C. Other researchers see a growing trend to more efficient dual-use production: "Our analyses of 1991 survey data from a large sample of establishments in the machining-intensive durable goods sector show that there are few technical and competitive conditions separating the defense and commercial industrial spheres. Commercial-military integration of production is now the normal practice among the majority of defense contractors in this sector. . . . Defense spending reaches a broad segment of manufacturing in the MDG (machining-intensive durable goods) sector, affecting nearly one-half of all establishments. Contrary to conventional wisdom, commercial-military integration is not only feasible but is largely the normal practice at the end of the Cold War. The vast majority of defense contractors in the MDG sector manufacture military products in the same plants with the same workers and equipment employed in producing items for commercial customers. In fact, commercial customers dominate the sales of most defense contractors in this sector. . . , If our findings for the MDG sector hold true for manufacturing as a whole, we see few technical or organizational barriers to converting most defense plants to further serve commercial markets," Kelley, M. R., and Watkins, T. A., "In From the Cold: Prospects for Conversion of the Defense Industrial Base," *Science* 268 (28 April 1995): 525–32.

5. "Local police in Texas, where the two [dual-use night-vision] products are being tested, have used the system to track fleeing suspects and, by sensing warmth emitted by grow lamps, detect marijuana plants being cultivated behind closed doors. It can also reveal to officers whether a suspect is carrying a concealed weapon: Guns appear on the screen as cold spots on a person's body. . . . Using ATR (automatic target recognition) computer technology police will match gun-shell casings in a fraction of the time it normally requires. Flashy neural-net software is used to automatically identify used gun-shell casings. . . . The company [Booz, Allen & Hamilton] expects the neural-network technology to pay off with dual-use applications, or in this case multi-use—defense, law enforcement, and the medical community. Cornell University Medical Center radiologists are working with the ATR method via X-rays to detect malignant nodules in lungs." Taylor, S. T., "Beating Swords into Police Wares," *Technology Transfer Business,* Winter 1995, 31–36.

    Raytheon [formerly Hughes Electronics] has been among the most successful defense firms in shifting to satellite and other forms of wireless communications: Former chairman Michael Armstrong "slashed jobs by 25% to 47,000. . . . Consolidated defense businesses. Divested 12 noncore units and strengthened key missile segment with an acquisition. . . . Launched DirecTV satellite service, attracting 1.5 million subscribers in first 18 months." Shine, E., and Armstrong, L., "Liftoff: Michael Armstrong Has Made Hughes an Electronics and Telecom Contender," *Business Week,* 22 April 1996, 137.

6. "Astute lawmakers recognize the need for caution. An estimated 70% of active US scientists and engineers holding masters and Ph.D. degrees are in defense-related positions. A sweeping closure of federal laboratories could easily debilitate the nation's 'brain trust.'" Scott, W.

B., "U.S. Labs Embrace Technology Transfer," *Aviation Week & Space Technology*, 23 August 1993, 64–66.

7. The California Air Resources Board (CARB) ruled in 1990 that 2% of all new vehicles sold in California in 1998 must emit no emissions. It claimed this would produce nearly 36,000 electric cars in 1998. The CARB also ruled that 10% of new vehicles must emit no emissions starting in the year 2003.

The CARB rulings have since changed. The CARB first moved to delay the 2% mandate in late December of 1995. RAND analyst David Rubenson suggests that the delay stemmed more from common sense than from the politics of special interests: "The problem is that replacing newer cars with electric vehicles has virtually no effect. Internal combustion vehicles that meet today's new-car clean air standards contribute little to air pollution. Unfortunately, they don't stay that way. So it is only when electric vehicles start to replace 10-, 15-, and 20-year-old cars that we will see substantial benefits. Electric cars are important because of the technological shift they may represent, not because of near-term emission reductions." Rubenson, D., "Forget ZEVs, LEVs, SIPs, and FIPs—Just Fix the Clunkers," *Los Angeles Times*, 29 December 1995, p. B-9.

Then the CARB reversed itself in March of 1996 and repealed most of the short-term mandate: "Friday's unanimous vote eliminated quotas that would have forced auto companies to put 160,000 exhaust-free cars in California showrooms through 2002, starting with 20,000 in the 1998 model year. The only quotas left in effect by the board start in model year 2003, when 10% of vehicles offered for sale, or about 100,000 annually must be zero-emission." Cone, M., *Los Angeles Times*, 30 March 1996, p. A-1.

RAND economists Lloyd Dixon and Steven Garber argued that the CARB's reversal had more to do with common sense than with politics: "If the original mandate had remained, each electric vehicle sold before 2003 could have cost from $6,000 to $28,000 more to produce and operate over its lifetime than gasoline-powered vehicles. If current policies to control emissions from gasoline vehicles turn out to be effective, the electric vehicles would have displaced fairly clean vehicles. Because the mandate linked electric vehicle sales requirements to gasoline vehicle sales, it could have boosted by several hundred dollars the price of new, cleaner gasoline-powered vehicles, slowed their sales and even caused overall vehicle emissions to increase in the short term." Dixon, L., and Garber, S., "Air Board's Revision Was Not a Dirty Deal," *Los Angeles Times*, 26 April 1996. The U.S. Second Circuit Court of Appeals struck down a like plan in New York. Riezenman, M. J., and Jones, W. D., "EV Watch: Court Strikes Down New York EV Plan," *IEEE Spectrum* 13, no. 9 (September 1998): 19–20.

8. Conditioning reflexes suggest Ivan Pavlov but I mean instead B. F. Skinner. Pavlovian or *respondent* conditioning reinforces the stimulus. Skinnerian or *operant* conditioning reinforces the response as in "No work then no pay." Skinner, B. F., *Science and Human Behavior*, The Free Press, 1953. Pavlovian conditioning tries to force a response. It couples a bell or a light to the fresh dog food to produce dog saliva. Operant conditioning waits for good or bad behavior to occur and then rewards or punishes it. It waits for the dog to jump through the hoop before it gives the dog a Milk-Bone.

9. Epstein, R. A., *Takings: Private Property and the Power of Eminent Domain*, Harvard University Press, 1985, pp. 297–300. Epstein uses the takings clause of the Fifth Amendment to argue for a constitutional flat tax: "There is no real cost to reading the eminent domain clause as requiring the flat tax. The flat tax does not place any total revenue constraint upon the government, nor does it hamstring the legislature and the executive from carrying out their constitutional functions, such as appointing judges or in waging war."

The takings clause does not require a fuzzy tax form but it permits one. The tax form would give some compensation to the taxpayer for the money the state takes from him. The compensation is "just" in the sense that it is proportional to how the taxpayer votes his funds. The more money he directs to one social option then the more of that social option he tends to get—even if he gets only a minuscule share of the result.

10. Suppose there are n social options. Then each fuzzy social choice is a point in the n-dimensional unit hypercube $[0, 1]^n$. The social choice defines an n-vector $x = (x_1, \ldots, x_n)$ such that the n fit values x_i sum to unity:

$$\sum_{i=1}^{n} x_i = 1$$

The choice vector (.1, .4, .3, .2) defines a point in the unit cube of 4 dimensions. This means that the space of all fuzzy social choices defines the n-dimensional *simplex* within the unit cube. So each choice vector x has the same form that a probability vector has. But a probability vector would require that all the tax money go to one of the n social options. Then the probability weight x_i would reflect the chance that all the tax money went to the ith social option. The fuzzy-set vector x contains no option uncertainty. It directs x_i-percent of the tax money to the ith option. So a 20% entry for debt reduction means that 20% of that person's tax bill goes to debt reduction. It does not mean that 100% of the tax bill goes there but only with a 20% probability.

11. Presented and published in many places that include Kosko, B., "It's a Perfect Day to Consider Other Ways of Collecting Taxes," *Los Angeles Daily News*, 17 April 1995.

12. Nobel laureate economist Friedrich Hayek called out the role of the frustrated specialist in the growth of the modern state in his most popular book: "The movement for [central economic] planning owes its present strength largely to the fact that, while planning is in the main still an ambition, it unites almost all the single-minded idealists, all the men and women who have devoted their lives to a single task. . . . From the saintly and single-minded idealist to the fanatic is often but a step. Though it is the resentment of the frustrated specialist which gives the demand for planning its strongest impetus, there could hardly be a more unbearable—and more irrational—world than one in which the most eminent specialists in each field were allowed to proceed unchecked with the realization of their ideals." Hayek, F. A., *The Road to Serfdom*, University of Chicago Press, 1944, p. 55

13. Harvard philosopher John Rawls has championed the view that justice is fairness in "Justice as Fairness," *Philosophical Review*, 1955, 164–94. He gave his most complete and forceful statement of this theory in his now seminal *A Theory of Justice*: "The guiding idea is that the principles of justice for the basic structure of society are the object of the original ['state of nature'] agreement. They are the principles that free and rational persons concerned to further their own interests would accept in an initial position of equality as defining the fundamental terms of their association. These principles are to regulate all further agreements; they specify the kinds of social cooperation that can be entered into and the forms of government that can be established. This way of regarding the principles of justice I shall call justice as fairness," Rawls, J., *A Theory of Justice*, Harvard University Press, 1971, p. 11.

14. Cowan, J., and Nelson, R., "Age Discrimination—Against the Young," *Los Angeles Times*, 7 August 1994, Metrosec.

15. Anarcho-capitalist David Friedman has assessed free-market schemes to privatize national defense: "So one solution to the problem of national defense might be the development, for some related purpose, of local defense organizations. These must be organizations permanently endowed for the purpose of providing defense; they cannot be simply local firms with an interest in the protection of their territory, since such firms, having agreed to pay part of the cost of national defense, would be driven out of business by new competitors who had not. This is the problem with Morris and Linda Tannahill's idea of financing national defense through an insurance company or companies which would insure customers against injury by foreign states and finance national defense out of the money saved by defending the customers. Such an insurance company, in order to pay the cost of defense, would have to charge rates substantially higher than the real risk justified, given the existence of its defense system. Since people living in the geographical area defended would be protected whether or not they were insured by that particular company, it would be in their interest either not to be insured or to be insured by a company that did not have to bear the burden of paying for defense and could therefore charge lower rates. The national defense insurance company would lose all its customers and go bankrupt, just as it would if it were simply selling national defense directly to individual customers who would be defended whether or not they paid.

"The same difficulty occurs with Ayn Rand's suggestion of financing national defense by having the government charge for the use of its courts. In order to raise money for defense, such a government must either charge more than competing private court systems or provide a worse product. Such private courts, if permitted, would therefore drive the government out of the business, depriving it of its source of income." Friedman, D., *The Machinery of Freedom: Guide to a Radical Capitalism,* 2d edition, Open Court Press, 1989, p. 140.

16. The back of the U.S. 1040 federal income tax form shows pie charts of federal income and outlays in 1998: 46% of income came from the personal income tax while the state paid 15% of its revenues to service the interest on the national debt. The state spent in contrast only 2% on all "law enforcement and general government."

17. The public-goods view of basic research is old in the economic literature: Nelson, R., "The Simple Economics of Basic Scientific Research," *Journal of Political Economy* 67 (1959): 297–306. Arrow, K., "Economic Welfare and the Allocation of Resources for Invention," in *The Rate and Direction of Inventive Activity,* Princeton University Press, 1962.

The chief scientific advisor to the U.K. government (Robert May) gives a more recent statement as he discusses why modern countries on average spend about 2.2% of their GDP on R&D: "Governments are the principal funders of basic research because the results are unforeseeable and unownable. They also fund applied research to inform public policy and operations. Additionally, governments seek to encourage business R&D to sharpen the competitive edge of their industries and increase national wealth." May, R. M., "The Scientific Investments of Nations," *Science* 281 (3 July 1998): 49–51.

18. "From an economist's point of view, the most appealing attribute of a reward system that is rooted in priority is that it offers non-market-based incentives for producing the public good 'knowledge.' . . . A reward system based on reputation also provides a mechanism for capturing the externalities associated with discovery. The more a scientist's work is used, the larger is the scientist's reputation and the larger are the financial rewards. It is not only that the reward structure of science provides a means for capturing externalities. The public nature of knowledge *encourages* use by others, which in turn enhances the reputation of the researcher." Stephan, P. E., "The Economics of Science," *Journal of Economic Literature* 34 (September 1996): 1199–1235.

19. "A recent policy statement published by the U.S. Committee for Economic Development (CED) reports that two important trends are helping to reshape university research. One is the ability of universities and academic researchers to reap financial rewards from inventions licensed by industry. The other is the growing number of industry-university collaborations. . . . Around 73% of research publications referenced in industrial patents were derived from government-funded research. A gauge of the impact of [the] Bayh-Dole [Act of 1980] is the number of patents issued to academic institutions, which increased from 249 in 1974 to 1,761 in 1994. Much of this improvement in the technology-transfer process . . . has occurred because of the clarification of a university's right to own the patents arising from federally funded research." Horton, B., "Taking Knowledge from Bench to Bank," *Nature* 395 (24 September 1998): 409–10.

20. "The biggest myth in science funding is that published science is freely available. It is not. Access to it is extremely expensive. Consider an analogy with law. No one assumes that legal knowledge is freely available. Anyone could, if they wished, consult the law books and journals to defend themselves in court, but it would take years to master the law and the courtroom lore that is never published, so people employ lawyers and pay them for their accumulated experience. . . . So important is the collection and integration of scientific data, that successful companies see it as the prime role of their scientists. . . . Companies are primarily interested in second-mover research. They need to know what everyone else is doing so that they can exploit advances from all over the world. The only people who can monitor other people's research are the scientists. Companies, therefore, have to employ scientists. But scientists themselves are only really interested in first-mover research, and the best scientists are obsessed by their own first-mover research. Yet even scientists need a salary, so companies and scientists agree a *modus vivendi*. Companies pay scientists to do the first-mover research that they, the scientists enjoy; while in return the scientists, through their reading and attendance at conferences, keep the company informed of developments worldwide. . . . The financial consequences are that companies have to invest very heavily indeed in their researchers' first-mover science to retain them as second-mover consultants." Kealey, T., *The Economic Laws of Scientific Research,* St. Martin's Press, 1996, pp. 228–29.

21. A. J. Lotka first published his empirical "law" in "The Frequency Distribution of Scientific Productivity," *Journal of the Washington Academy of Science* 16, no. 12 (19 June 1926): 317–23. For discussion and more recent studies see De Solla Price, D. J., *Little Science, Big Science . . . And Beyond,* Columbia University Press, 1963. Paula Stephan states Lotka's law this way in her exhaustive survey of the economic analysis of scientists: "Lotka's law states that if k is the number of scientists who publish one paper, then the number publishing n papers is k/n^2. In many disciplines this works out to some five or six percent of the scientists who *publish at all* producing about half of all papers in their discipline. . . . Lotka's law has held up well over time and across disciplines." Stephan, P. E., "The Economics of Science," *Journal of Economic Literature* 34 (September 1996): 1204.

22. Harmon, L. R., "The High School Backgrounds of Science Doctorates," *Science* 133 (10 March 1961): 679–88.

23. "Never before have bettors blown so much money—a whopping $50.9 billion last year [1987]—five times the amount lost in 1980. That's more than the public spent on movies, theme parks, recorded music and sporting events combined." Gold, M., and Ferrell, D., "Going for Broke," *Los Angeles Times,* 13 December 1998, p. A-1.

24. Brennan, R. P., *Dictionary of Scientific Literacy,* Wiley, 1992, p. 58.

25. "The Federal Government alone shells out $125 billion a year in corporate welfare, this in the midst of one of the more robust economic periods in the nation's history. Indeed, thus far in the 1990s, corporate profits have totaled $4.5 trillion—a sum equal to the cumulative paychecks of 50 million working Americans who earned less than $25,000 a year for those eight years." Bartlett, D. L., and Steel, J. B., "Corporate Welfare: Part I," *Time,* 9 November 1998, 36–54. The authors do not explain why more of those workers do not buy part of this corporate profit stream by buying stock in those corporations.

26. Federal Reserve economist Narayana Kocherlakota finds a real "equity premium" of nearly 8% over the very long haul of 100 years: "Over the last hundred years, the average real return to stocks in the United States has been about 6% higher than that on Treasury bills [a so-called risk-free investment]. At the same time, the average real return on Treasury bills has been about one percent per year." Kocherlakota, N. R., "The Equity Premium: It's Still a Puzzle," *Journal of Economic Literature* 34 (March 1996): 42–71.

27. Foreign enterprise funds have had mixed success: "Administrative expenses of the fund [Czech and Slovak American Enterprise Fund or CSAEF] ran to as much as 10% of invested capital or more, depending on how realistically one valued the portfolio. At a comparable private sector fund, the First Hungary Fund, the overhead is closer to 2%. While regulations capped our fund's salaries at $150,000 and prevented anyone from having a performance-based interest in any investment—a normal way to motivate a venture capitalist—the CSAEF, in the best traditions of big government, had no problem running its annual expenses up." Fichera, J. S., and Rubin, R. M., "Uncle Sam, Venture Capitalist," *Wall Street Journal,* 2 May 1996. Nobel laureate economist Milton Friedman thinks socialism might result if the state invests in stock index funds: Friedman, M., "Social Security Socialism," *Wall Street Journal,* 26 January 1999.

28. Gray, G. W., "The Nobel Prizes," *Scientific American* 181, no. 6 (December 1949).

29. "The high-stakes industrial contest [for the $30 million prize] was sponsored by a consortium formed by 24 electric utilities, including Southern California Edison Co., the Los Angeles Department of Water and Power, and Pacific Gas & Electric Co. The competition, which drew 500 responses from around the world, was intended to accelerate development of a refrigerator that would be at least 25% more energy-efficient than today's models and use no ozone-depleting chloro-fluorocarbons, or CFCs." Joshi, P. A., and Parrish, M., "A Cool $30 Million: Whirlpool Wins Prize for Designing Environmentally Safe Refrigerator," *Los Angeles Times,* 30 June 1993, sec. D.

30. Zuckerman, H. A., "The Proliferation of Prizes: Nobel Complements and Nobel Surrogates in the Reward System of Science," *Theoretical Medicine* 13 (1992): 217–31.

31. "*Research! America* commissioned Louis Harris and Associates to conduct a survey of the US public during June 1995. Out of 1,004 adults surveyed, with a margin of error estimated at 3.1%, the survey found that (i) 94% of respondents believed that it is important for the United States to maintain its role as a world leader in medical research; (ii) 65% opposed cuts in federal support for universities and hospitals, and those under the age of

30 opposed such measures by nearly 75%; (iii) 73% would pay more taxes to support medical research, which duplicates the results when the same question was asked in a 1993 Harris poll; (iv) 61% wanted their senators and congresspeople to support legislation that would give tax credits to private industries to conduct more medical research; and (v) 69% agreed with the statement 'Even if it brings no immediate benefits, basic science research which advances the frontiers of knowledge is necessary and should be supported by the Federal Government.' " Wooley, M., "From Rhetoric to Reality," *Science* 269 (15 September 1995): 1495.

5. THE RIGHTS OF GENOMES

1. "Nationally, 86% of counties had no abortion provider in 1996, a figure that has been rising for two decades. . . . Altogether, nearly one-third of women live in counties without providers—defined by the [Alan] Guttmacher Institute as hospitals, clinics and doctors' offices. About 70% of all abortions in 1996 were performed at the nation's 452 abortion clinics. Only 16% of all short-term, general, nonfederal hospitals performed abortions that year." Rubin, A. J., "Abortion Providers at Lowest Mark Since '73," *Los Angeles Times,* 11 December 1998.
2. "More than 100,000 unwanted pregnancies each day, or about 36–53 million each year, end in induced (surgical) abortion. Estimates suggest that more than half of these abortions are performed under unsafe conditions and result in more than 70,000 deaths per year, almost all in developing countries." Ewart, W. R., and Winikoff, B., "Toward Safe and Effective Medical Abortion," *Science* 281 (24 July 1998): 520–21.
3. "Often the question [of abortion] is presented as a women's right to *free choice* regarding the life already existing inside her, that she carries in her womb: the woman should have the right to choose between giving life or taking it away from the unborn child. Anyone can see that *the alternative here is only apparent. It is not possible to speak of the right to choose when a clear moral evil is involved,* when what is at stake is the commandment *Do not kill!*" John Paul II, "The Defense of Every Life," in *Crossing the Threshold of Hope,* Knopf, 1994, p. 205.
4. "In China, abortion has become a chilling tool in the hands of the Communist Party leadership to enforce its one-child per couple demographic policy. In the last decade, an estimated 100 million pregnancies have ended in abortion as the rate climbed to one termination for every two live births. In the world's most populous country, abortion now is often not a matter of choice, but a duty imposed by a salaried state functionary." Dahlburg, J.-T., "Faith and Practice: A Changing World Puts Abortion in the Spotlight," *Los Angeles Times,* 24 January 1995, World Report.
5. The fuzzy vs. binary view of life lines appears in chapter 13 of Kosko, B., *Fuzzy Thinking: The New Science of Fuzzy Logic,* Hyperion, 1993.
6. "The Supreme Court, Mr. Justice Blackmun, held that the Texas criminal abortion statutes prohibiting abortions at any state of pregnancy except to save the life of the mother are unconstitutional; that prior to approximately the first trimester the abortion decision and its effectuation must be left to the medical judgement of the pregnant woman's attending physician, subsequent to approximately the end of the first trimester the state may regulate abortion procedure in ways reasonably related to maternal health, and at the stage subsequent to viability the state may regulate and even proscribe abortion except where necessary in appropriate medical judgement for preservation of life or health of another." *Roe* v. *Wade,* 93 Supreme Court 705 (1973), *Supreme Court Reporter,* vol. 93, p. 705, 1973.
7. "*U. S. News* reporters called 31 of 79 clinics in America that provide abortions after 20 weeks of pregnancy. . . . Eighteen of the clinics contacted, or more than one fifth of those performing abortions after 20 weeks, agreed to in-depth interviews. . . . Only about 5,500 of the approximately 79,000 abortions performed by the clinics that responded (or 6.9%) were on post-20-week fetuses. (This is higher than the 1% found in national studies because late procedures are concentrated at these few clinics.)" Lavelle, M., "When Abortions Come Late in a Pregnancy," *U. S. News & World Report,* 19 January 1998, 31–32.
8. A binary life curve is a step function that rises abruptly from 0% alive to 100% alive. The binary jump can occur anywhere from conception to birth. *Roe* v. *Wade* would in theory draw a life line that jumps from not-life to life at the start of the second trimester:

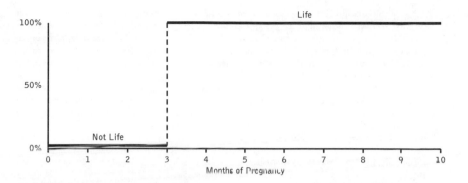

A fuzzy life curve rises up by degrees from 0% to 100%. A person's or group's fuzzy life curve may have an **S**-shape:

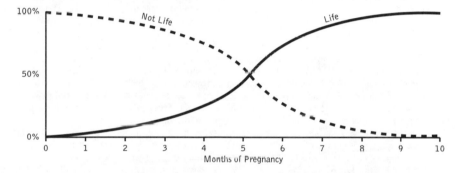

Polls can estimate a life curve in many ways. One way is to call 1,000 persons at random and ask them to pick the month at which month life starts. Those who say it starts at conception pick the 1st month. Those who say it starts at birth pick the 10th month. Each person picks one of these 10 months. Suppose he picks the 7th month. Then we write this choice as a length-10 bit vector with 1s in the 7th and later slots and 0s in the other 6 slots: (0 0 0 0 0 0 1 1 1 1). The poll gives 1,000 of these bit vectors. Add them pointwise and divide each term by 1,000. (The two bit vectors [0 0 0 0 0 0 1 1 1 1] and [0 0 0 1 1 1 1 1 1 1] add to give the bit vector [0 0 0 1 1 1 2 2 2 2] and give the 2-person averaged fuzzy unit or *fit* vector [0 0 0 1/2 1/2 1/2 1 1 1 1].) This average gives a type of histogram or discrete estimate of the group's fuzzy life curve. It has the form of a staircase:

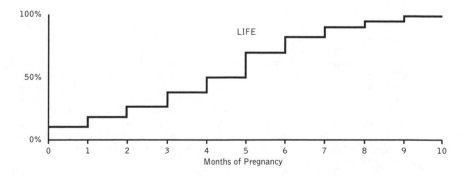

9. "One possible approach is to borrow from John Rawls [in his book *A Theory of Justice*] (whose own writing is silent on the question of abortion) and ask what choices we would make about abortion rights if forced to do so from behind a veil of ignorance. If a person does not know whether he or she will be in the position of the woman seeking abortion or the fetus that will be aborted, what rule of decision would that person choose to maximize the welfare of both?

"Here it seems that the loss of one's own potential life is a far greater loss than the gains on the other side, so that a disembodied view of the subject would incline toward protection of a fetus against the aggression of its mother, even if one does not think that the fetus is a person at the time of an abortion. . . . I think that it is highly likely that each of us would be prepared to surrender any future right to abort in order to escape today the finality of an abortion." Epstein, R. A., "The Killing Grounds: Book Review of Ronald Dworkin's *Life's Dominion: An Argument About Abortion, Euthanasia, and Individual Freedom*," *Reason* 25, no. 6 (November 1993): 58–61.

10. "To a remarkable degree, the Alberts Committee [of the National Resource Council in 1987] read the historical and technical trends correctly. It now appears that even its estimates of time scale and cost—15 years at $200 million per year—were about right." Olson, M. V., "A Time to Sequence," *Science* 270 (20 October 1995): 394–96. "While huge, the central task of the Human Genome Project is bounded by one of the most remarkable facts in all of science: The development of a human being is guided by just 750 megabytes of digital information. In vivo, this information is stored as DNA molecules in an egg or sperm cell. In a biologist's personal computer, it could be stored on a single CD ROM" (p. 396).

The Human Genome Project has evolved in the 1990s. For an earlier perspective see Stephens, J. C., et al., "Mapping the Human Genome: Current Status," *Science* 250 (12 October 1990): 237–44. The Human Genome Project remains on track and on budget: "No one really doubts that the Human Genome Project, started in 1991, will be completed in 2006, on time and under its startling $3 billion budget. The goal of this huge international effort, funded by the U.S. National Institutes of Health, the U.S. Department of Energy, and United Kingdom's Welcome Trust, is to unravel our 100,000 genes. More specifically, the results will be sequences of three billion nucleotide base pairs, each pair taken from the four components (adenine, cytosine, guanine, and thymine) of the famous double helix: dioxyribonucleic acid (DNA). . . . The price to identify a base pair, at a current cost of about $0.35, continues to decrease." Garner, H. R., "The Human Genome Project: On Target for 2006," *IEEE Spectrum* 34, no. 1 (January 1997): 100–101. The Human Genome Project supports a detailed Web site at http://www.ornl.gov/TechResources/Human_Genome/home.html.

The U.S. Patent and Trademark Office (PTO) now routinely approves patents on genes and this is likely to continue apace with discoveries in the Human Genome Project: "John Doll, the PTO's biotech section chief, explains: While 'nobody "owns" the gene in your body, inventors can own the right to exploit it commercially. . . . You can't turn over a rock and find a gene.' Since 1980, Doll says the PTO has received more than 5,000 patent applications based on genes. And it has granted more than 1,500 patents on them. This estimate generally tracks the results of a study published in *Nature* last year by a science policy group at the University of Sussex in Britain, led by S. M. Thomas. Between 1981 and 1995, the Thomas group found, the patent offices of the United States, Europe, and Japan issued 1,175 patents on human DNA sequences. The genes covered by these claims range from DNA coding for human interleukin and interferon—immune system regulating proteins—to genes for bone and brain tissue. Most inventions are aimed at treating medical problems, and in the United States and Europe, more than half of the patents are held by public sector institutions. The single most valuable human DNA patent, however, may be on covering the human erythroprotein gene, which is used to produce a hormone needed by kidney disease patients. In 1991, the U.S. Supreme Court affirmed the validity of this 1987 patent, which now earns its owner, Amgen Inc. of Thousand Oaks, California, more than $1 billion a year." Marshall, E., *Science* 275 (7 February 1997): 780–81. Research progress has brought the end in sight for the project: Waterston, R., and Sulston, J. E., "The Human Genome Project: Reaching the Finish Line," *Science* 282 (6 October 1998): 53–54; Collins, F. S., et al., "New Goals for the U.S. Human Genome Project: 1998–2003," *Science* 282 (23 October 1998): 682–89.

11. The first physical map of the human genome came at the end of 1995 and contained 15,086 sequence-tagged sites (STSs) with an average spacing of 199 kilobases: "A physical map affording ready access to all chromosomal regions is an essential prerequisite for the international effort to sequence the entire human genome. In the shorter term, it is also a key tool for positional cloning of disease genes and for studies of genome organization. The US Human Genome Project has set a target of a physical map consisting of 30,000 STSs spaced at intervals of about 100 kb." Hudson, T. J., et al., "An STS-Based Map of the Human Genome," *Science* 270 (22 December 1995): 1945–54.

The STS map offers a "scaffold" with which to speed up and complete the mapping goal of the Genome Project: "The use of STS-based maps as a scaffold for large-scale sequencing has several advantages: It can be initiated now with the existing STS-based map; it automatically anchors sequences in the genome; it does not require chromosome-specific libraries which involve specialized preparation procedures and often have cryptic biases; it allows improved libraries to be substituted as they become available; and it promotes decentralization by allowing sequencing efforts to focus on regions of any given size in contrast to entire chromosomes. In summary, the physical map must still be refined but it is already adequate to allow initiation of the international project to sequence the entire human genome—a landmark effort that will set the stage for the biology of the next century" (p. 1954).

Science published in 1996 a partial gene map of the human genome and its 100,000 or so genes: Schuler, G. D., et al., "A Gene Map of the Human Genome," *Science* 274 (25 October 1996): 540–46.

12. *Science* reported the first complete gene mapping of a free living creature: Fleischmann et al., *Science* 269 (1995): 496. Meanwhile much of the search for genes that cause or cure diseases involves research at commercial firms: Haseltine, W. A., "Discovering Genes for New Medicines," *Scientific American*, March 1997, 92–97; Gruber, M., "Map the Genome, Hack the Genome," *Wired*, October 1997, 153–56.

13. Fraser, C. M., et al., "The Minimal Gene Complement of *Mycoplasma Genitalium*," *Science* 270 (20 October 1995): 397–403. A worldwide team of scientists sequenced the yeast *saccharomyces cerevisiae* in 1996. This yeast (the first eukaryote sequenced) has about 5,885 genes and 16 chromosomes: Goffeau, A., et al., "Life with 6000 Genes," *Science* 274 (25 October 1996): 546–67. Within a year scientists had sequenced five more bacterium genomes. This includes the 1,668-kbp genome of *Helicobacter pylori*. This bacterium causes peptic ulcers and may live in one out of two human stomachs: Tomb, J.-F., et al., "The Complete Genome Sequence of the Gastric Pathogen *Helicobacter Pylori*," *Nature* 388 (7 August 1997): 539–47.

14. *Nature* published in 1997 a yeast genome directory and soon published (along with *Science*) other breakthroughs in gene mapping: Clayton, R. A., White, O., Ketchum, K. A., and Venter, J. C., "The First Genome From the Third Domain of Life," *Nature* 387 (29 May 1997) 459–62; Blattner, F. R., et al., "The Complete Genome Sequence of *Escherichia coli* K-12," *Science* 277 (5 September 1997): 1453–62; Kunst, F., et al., "The Complete Genome Sequence of the Gram-Positive Bacterium *Bacillus Subtilis*," *Nature* 390 (20 November 1997): 249–56; Klenk, H.-P., et al., "The Complete Genome Sequence of the Hyperthermophilic, Sulphate-Reducing Archaeon *Archaeoglobus Fulgidus*," *Science*, 390 (27 November 1997): 364–70; Fraser, C. M., et al., "Genomic Sequence of a Lyme Disease Spirochaete, *Borrelia Burgdorferi*," *Nature* 390 (11 December 1997): 580–86.

15. Deckert, G., et al., "The Complete Genome of the Hyperthermophilic Bacterium *Aquifex Aeolicus*," *Nature* 392 (26 March 1998): 353–57; Cole, S. T., et al., "Deciphering the Biology of *Mycobaterium tuberculosis* from the Complete Genome Sequence," *Nature* 393 (11 June 1998): 537–44; Fraser, C. M., et al., "Complete Genome Sequence of *Treponema palladium*, the Syphilis Spirochete," *Science* 281 (17 July 1998): 375–88; Stephens, R. S., et al., "Genome Sequence of an Obligate Intracellular Pathogen of Humans: *Chlamydia trachomatis*," *Science* 282 (23 October 1998): 754–60; Gardner, M. J., et al., "Chromosome 2 Sequence of the Human Malaria Parasite *Plasmodium falciparum*," *Science* 282 (6 November 1998): 1126–32; Meinke, D. W., et al., "*Arabidopsis thaliana*: A Model Plant for Genome Analysis," *Science* 282 (23 October 1998): 662–81; Andersson, S. G. E., et al., "The Genome Sequence of *Rickettsia prowazekii* and the Origin of Mitochondria," *Nature* 396 (12 November 1998): 133–40; Wilmut, I., "Cloning for Medicine," *Scientific American*, December 1998, 58–63; The *C. ele-*

gans Sequencing Consortium, "Genome Sequence of the Nemotode *C. elegans:* A Platform for Investigating Biology." *Science* 282 (11 December 1998): 2012–18.

16. Harvard biologist Edward O. Wilson uses a simple argument to give a low estimate of the base-pair count of the world's biodiversity: "To the nearest order of magnitude, or powers of ten, 10^8 (100 million) species, multiplied by 10^9 (1 billion) nucleotide pairs on average per species; hence a total of 10^{17} (100 quadrillion) nucleotide pairs specifying the full genetic diversity among species. Nucleotide diversity, it should be noted in passing, is limited to a maximum of four kinds of nucleotide per site and hence does not add as much as an order of magnitude. That figure, 10^{17}, is in one sense the entire diversity of life." Wilson, E. O., *The Diversity of Life,* Harvard University Press, 1992, p. 161.

17. " 'So the vision is that very powerful computational tools, coupled with genomic information, will revolutionize medicine,' Professor Leroy Hood says. 'We'll see emerge in the next 10 to 15 years a whole new field of biomedicine that will be manned by mathematicians. They'll be in a position to do DNA fingerprints and compute your future health for you. And then we will have moved into an era of preventive medicine.' " Garrett, L., "The Dots are Almost Connected . . . Then What?," *Los Angeles Times Magazine,* 3 March 1996, p. 49. Dr. Hood's epigraph at the start of the chapter comes from the same source.

18. Kosko, B., *Nanotime,* Avon Books, 1997.

19. The following volume gives a good sample of applied simulated annealing and other types of random hill climbing: Davis, L, editor, *Genetic Algorithms and Simulated Annealing,* Morgan Kaufmann, 1987. The notes to chapter 10 list many primary sources for genetic algorithms.

20. "Molecular computers also have the potential for extraordinary efficiency. In principle, one joule is sufficient for approximately 2×10^{19} ligation operations. This is remarkable considering that the second law of thermodynamics dictates a theoretical maximum of 34×10^{19} (irreversible) operations per joule at room temperature. Existing supercomputers are far less efficient executing at most 10^9 operations per joule." Adleman, L. A., "Computing with DNA," *Scientific American,* August 1998, 54–61. The first article on DNA computing was Adleman, L. A., "Molecular Computation of Solutions to Combinatorial Problems," *Science* 266 (11 November 1994): 1021–24.

21. Service, R. F., "Microchip Arrays Put DNA on the Spot," *Science* 282 (16 October 1998): 396–99.

6. THE RIGHTS OF WHALES

1. Former Czech President Vaclav Havel noted the fuzzy and relative borders of the vague object we call Europe: "Europe represents a common destiny, a common complex history, common values, and a common culture and a way of life. More than that, it is also in a sense a region characterized by particular forms of behavior, a particular quality of will, a particular understanding of responsibility. As a consequence, the borders of Europe may at times seem fuzzy or variable: it cannot be defined by looking at a school atlas or studying a list of member states of the European Union or of countries that could join should they wish, such as Norway, Switzerland, or Iceland." Havel, V., "The Hope for Europe," *The New York Review of Books,* 20 June 1996, 38.

2. The complete quote shows how Locke tries to draw a line between complete or exclusive self-ownership and inclusive ownership of the "common":

> "Though the earth and all inferior creatures be common to all men, yet every man has a property in his own person. This nobody has any right to but himself. The labor of his body, and the work of his hands, we may say, are properly his. Whatsoever, then, he removes out of the state of nature that nature hath provided and left it in, he hath mixed his labor with, and joined to it something that is his own, and thereby makes it his property. It being by him removed from the common state nature placed it in it hath by this labor something annexed to it that excludes the common right of other men. For this labor being the unquestionable property of the laborer, no man but he can have a right to what that is once joined to, at least where there is enough and as good left in common for others." Locke, J., *The Second*

Treatise of Government: An Essay Concerning the True Original, Extent, and End of Civil Government, in *Political Writings of John Locke,* ed. David Wootton, Mentor, 1993 (first published 1681), p. 274.

3. Harvard philosopher Robert Nozick suggests that labor mixings may dilute one's ownership rather than extend it: "But why isn't mixing what I own with what I don't own a way of losing what I own rather than a way of gaining what I don't? If I own a can of tomato juice and spill it in the sea so that its molecules (made radioactive, so I can check this) mingle evenly throughout the sea, do I thereby come to own the sea, or have I foolishly dissipated my tomato juice?" Nozick, R., *Anarchy, State, and Utopia,* Basic Books, 1974, pp. 174–75.

   Nozick's tomato juice poses a problem for a binary view of Lockean property rights but perhaps not for a fuzzy view. His fuzzy rights degrade in proportion to how his tomato-juice molecules spread through the ocean volume. They also increase with the amount of owned tomato juice that one pours.

   Consider the extreme case where most of the ocean boils away. Then suppose some rich person or space alien pours a huge can of tomato juice into the volume to fill it up. It would make sense to say that the owner of the can owned the ocean fluid to a high degree. The degree would reflect the large proportion of his tomato juice in the ocean mix. A thirsty populace might contract with such an agent to fill the dry chasm with tomato juice or to fill it with some other fluid and so grant the agent a large fuzzy property right in it.

4. "About 50 miles of postwar border has remained unmarked because of disagreement over where the line should be. And that line is what the recent fighting has been about." Long, W. R., "Pride Pushed Peru, Ecuador into Their Lethal Border Clash," *Los Angeles Times,* 5 February 1995, p. A-7.

5. Jean-Jacques Rousseau's noble savage has often lost more than his dignity in pursuit of the noble metals: "Men who had once painted their faces and hunted naked in the jungle were living in town [Redencao, Brazil] sporting designer jeans and sunglasses, driving new pickups, hiring pilots for their private planes. Amazon Indians are often portrayed as wise conservationists whose native customs blend harmoniously with their natural habitat. Brazil's Kayapo are an example of how a good relationship with Mother Nature can turn bad. Over the past 15 years, logging and mining have done serious environmental damage. And although most Kayapo received relatively little benefit from the mahogany and gold, it was just enough to change their way of life. The easy money has undermined the values and customs of a once proud and self-sufficient native society." Long, W. R., "How Gold Led Tribe Astray," *Los Angeles Times,* 29 August 1995, p. A-1.

6. Coase, R. H., "The Problem of Social Cost," *The Journal of Law & Economics* 3, no. 1 (October 1960): 1–44. Coase expands these concepts of exchange and the legal framework that houses them in *The Firm, The Market, and The Law,* University of Chicago Press, 1988.

7. Coase discussed the lack of math in his work when he gave his Nobel Prize acceptance speech to the Swedish Academy on 9 December 1991: "My remarks have sometimes been interpreted as implying that I am hostile to the mathematization of economic theory. This is untrue. Indeed, once we begin to uncover the real factors affecting the performance of the economic system, the complicated interrelationships between them will clearly necessitate a mathematical treatment, as in the natural sciences, and economists like myself, who write in prose, will take their bow." Coase, R. H., "The Institutional Structure of Production," *American Economic Review* 82, no. 4 (September 1992): 713–19.

   Nobel laureate Gerard Debreu presented an equally eloquent case for the use of advanced math in economics in his presidential address to the American Economic Association on 29 December 1990: "Mathematics provides him [the mathematical economist] with a language and a method that permit an effective study of economic systems of forbidding complexity. But it is a demanding master. It ceaselessly asks for weaker assumptions, for stronger conclusions, for greater generality." Debreu, G., "The Mathematization of Economic Theory," *American Economic Review* 81, no. 1 (March 1991): 1–7. See also Debreu, G., *Mathematical Economics: Twenty Papers of Gerard Debreu,* Cambridge University Press, 1983.

8. Most economists point out that the Coase Theorem does not depend on who owns what just as long as the traders own all things. This means that the result is robust and does not depend on perfect competition or on "initial conditions":

"*Coase's Theorem:* Regardless of the specific initial assignment of property rights, in market equilibrium the final outcome will be efficient—provided that the initial legal assignment is well-defined and that the transactions involving exchange of rights are costless." Hirschleifer, J., *Price Theory and Applications,* 2d edition, Prentice Hall, 1980, p. 536.

Nobel laureate economist George Stigler also focused on how the assignment of legal rights does not matter for a Coase-based Pareto optimum: "Coase asserted an almost incredible proposition: if transaction costs were zero, the theorem on maximum satisfaction would always hold, no matter how the rights and duties of parties were assigned by the law. Whether the factory owner or the housewives were responsible for the damage done to laundry by chimney soot, exactly the socially optimum amount of soot would be produced. Whether the automobile driver or the pedestrian was liable for injury to the latter, each would take the socially optimum amount of care to avoid accidents. In this regime of zero transaction costs, no monopoly would restrict output below the optimum level because consumers would pay the monopolist not to do so. Such miraculous corollaries of the Coase theorem have greatly enlivened an early week in courses in economic theory." Stigler, G. J., "The Economists' Traditional Theory of the Economic Functions of the State," in *The Citizen and the State: Essays on Regulation,* University of Chicago Press, 1975, p. 107.

Makowski and Ostroy have observed that the Coase Theorem has much the same form as the so-called First Theorem of welfare economics that says that ideal competitive markets are Pareto efficient: "The message of the Coase Theorem is similar to the First Theorem. Once property rights are fully articulated, efficiency will be achieved. Besides complete property rights (the replacement for complete markets), the other key assumption of the Coase Theorem is *zero transactions costs* (the replacement for price-taking behavior). In addition to eliminating the typical frictions ignored in much of economic theory (e.g., the need for a title search in property transactions), this assumption is used to eliminate the 'transactions costs' which are due to imperfect competition—as if zero transactions costs make individuals with monopoly power behave as efficiently as price-takers. The conclusion of the Coase Theorem is that all appropriability problems stem from ownership problems. This is more or less supported by conventional interpretations of the First Theorem which trace departures from efficiency to incompleteness of markets." Makowski, L, and Ostroy, J. M., "Appropriation and Efficiency: A Revision of the First Theorem of Welfare Economics," *American Economic Review* 85, no. 4 (September 1995): 808–27. Nobel laureate Kenneth Arrow first proved this First Welfare Theorem that perfectly competitive markets are Pareto optimal: Arrow, K. J., "An Extension of the Basic Theorems of Classical Welfare Economics," in J. Neyman, ed., *Proceedings of the Second Berkeley Symposium on Mathematical Statistics and Probability,* University of California Press, 1951, pp. 507–32.

9. "An allocation is *Pareto optimal* or *Pareto efficient* if production and distribution cannot be reorganized to increase the utility of one or more individuals without decreasing the utility of others. Conversely, an allocation is *Pareto nonoptimal* if someone's utility can be increased without harming anyone else." Henderson, J. M., and Quandt, R. E., *Microeconomic Theory: A Mathematical Approach,* 3d edition, McGraw-Hill, 1980, p. 286.

Nobel laureate economist Paul Samuelson casts a Pareto optimum in terms of socialist planning: "If *A* is the competitive equilibrium [and thus a Pareto optimum from the first welfare theorem of economics], there is no point *B* the planner could devise which could be approved over *A* by *unanimous* vote. Some would be hurt in going to *B,* some might gain. But the gainers could never find it worthwhile to give big enough bribes to win the losers over to approving the move to *B.*" Samuelson, P. A., *Economics,* 10th edition, McGraw-Hill, 1976, p. 634.

More formal definitions of Pareto optimality depend on the math of agent endowments and preference maps: Mas-Colell, A., *The Theory of General Economic Equilibrium: A Differentiable Approach,* Cambridge University Press, 1985; Border, K. C., *Fixed Point Theorems with Applications to Economics and Game Theory,* Cambridge University Press, 1985; Hildenbrand, W., and Kirman, A. P., *Introduction to Equilibrium Analysis,* North Holland, 1976.

10. Cheung, S. N. S., "The Fable of the Bees: An Economic Investigation," *Journal of Law & Economics* 16 (April 1973).

11. "The Coase Theorem dispenses with the heavy assumptions of perfect competition [of the First Welfare Theorem], but replaces them with the strong assumption that no mutually beneficial agreement is missed. So while it economizes on institutions, it demands a lot of coordination and negotiation. . . . So we cannot assume that all mutually beneficial contracts are signed, unless we assume that everyone knows everything about everyone, which they do not. The strong form of the Coase Theorem—the claim that voluntary negotiation will lead to fully efficient outcomes—is implausible unless people know one another exceptionally well." Farrell, J., *Journal of Economic Perspectives* 1, no. 2 (Fall 1987): 113–29.

Paul Samuelson presents a harsher criticism of the Coase Theorem and its binary framework of rights and costs: "Allocation of property rights—and how they are to be defined—matters mightily. They are the chips in the game of dickering, threatening, and litigating. These processes are unavoidable. And it is only in my MIT seminar room that they go on all morning long in dry runs that converge to (an infinity of) Pareto-optimal solutions; and that at high noon an omnipotent go-between finalizes at one of these calculated optima. In the real world rivalry is incessant. Only in certain Santa Claus situations—constant returns to scale, infinite divisibility, free entry, dispersed ownership of each grade of factor, shared knowledge, complete markets—only then will [Adam] Smithian self-interest be compelled to achieve Pareto optimality.

"To try to capture all that which contributes to deadweight loss under the verbal rubric of 'transactions costs' weakens a useful concept without gaining understanding of incompleteness of markets, asymmetries of information, and insusceptibilities of various technologies to decentralized pricing algorithms." Samuelson, P. A., *Japan and the World Economy*, vol. 7, 1995, pp. 1–7.

12. Economist Leland Yeager has observed that an equality or identity in economics like the velocity-of-money equation $MV = PQ$ (the quantity of money times its turnover or velocity in circulation equals the average price of a unit of output times the quantity of that output) has the same form as Newton's second law $F = m\,a$ (force equals mass times acceleration) in mechanics. Yeager, L. B., "Tautologies in Economics and the Natural Sciences," *Eastern Economic Journal* 20, no. 2 (Spring 1994): 157–69. The Coase Theorem has the same "trivial" status just because it is a theorem.

Philosopher Carl Hempel has explored how we map these abstract symbols into measurable quantities and thus how we convert logical tautologies into testable hypotheses in Hempel, C. G., "The Theoretician's Dilemma: A Study in the Logic of Theory Construction," in Hempel's anthology *Aspects of Scientific Explanation,* The Free Press, 1965, pp. 173–226. Hempel wrestles with William Craig's famous theorem that we can dispense with the throughput terms and equations of logically consistent theories in favor of simpler maps from input to output. Craig, W., "On Axiomatizability Within a System," *Journal of Symbolic Logic* 18 (1953): 30–32.

13. A theorem is a sentence in a formal system. The formal system contains two types of sentences. The first type are axioms. The second type are theorems or sentences that follow from the axioms when one applies a rule of inference such as *modus ponens* (if '*A*' is a sentence in the system and if 'If *A* then *B*' is a sentence in the system then '*B*' is a sentence in the system). Hunter, G., *Metalogic: An Introduction to the Metatheory of Standard First Order Logic,* University of California Press, 1973.

14. The fuzzy *modus ponens* theorem holds for the popular multivalued if-then or implication operator of Jan Lukasiewicz: $t_L(A \to B) = \min(1, 1 - t(A) + t(B))$ where $t(S) \in [0,1]$ gives the truth value or degree of truth of statement S. Other implication operators give a syllogism with the same form but not in general with the same numerical results. The Lukasiewicz-based extension of *modus ponens* gives a syllogistic schema:

$$
\begin{array}{ll}
\textit{If} & t_L(A \to B) = c \\
\textit{and} & t(A) \geq a \\
\textit{then} & t(B) \geq \max(0, c + a - 1)
\end{array}
$$

This holds since $c \leq 1 - t(A) + t(B)$ holds for the Lukasiewicz implication operator. So $t(B) \geq c + t(A) - 1 \geq c + a - 1$ holds by hypothesis. The result follows because no truth value can be less than zero. Q.E.D.

Note that Aristotle's binary *modus ponens* holds in the special case when $c = a = 1$. Then $t(B) \geq \max(0, 1 + 1 - 1) = \max(0, 1) = 1$. So $t(B) = 1$ or the consequent or then-part is certain (100% true) if both the implication and antecedent or if-part are certain.

Now equate symbols to statements as follows:

A	$=$	The apple is ripe
B	$=$	I will eat the whole apple
$A \rightarrow B =$		If the apple is ripe then I will eat the whole apple

Then assign the truth values $c = .9$ and $a = .7$. So the formal statement $t(A) \geq .7$ stands for the claim that the apple is at least 70% ripe. Then the theorem says that the degree to which it is true that I will eat the whole apple is $t(B) \geq \max(0, .9 + .7 - 1) = \max(0, .6) = .6$. So the conclusion is at least 60% true.

15. The fuzzy *modus tollens* theorem also holds for the Lukasiewicz implication operator: $t_L(A \rightarrow B) = \min(1, 1 - t(A) + t(B))$. The theorem gives the syllogistic schema

If	$t_L(A \rightarrow B) = c$
and	$t(B) \leq b$
then	$t(A) \leq \min(1, 1 - c + b).$

This holds since $c \leq 1 - t(A) + t(B)$ holds for the Lukasiewicz implication operator. So $t(A) \leq 1 - c + t(B) \leq 1 - c + b$ holds by hypothesis. The result follows because no truth value can exceed unity. Q.E.D.

Note that Aristotle's binary *modus tollens* holds in the special case when $c = 1$ and $b = 0$. Then $t(A) \leq \min(1, 1 - 1 + 0) = \min(1,0) = 0$. So $t(A) = 0$ or the antecedent or if-part is 100% false if the implication is 100% true and if the consequent or then-part is 100% false.

Again equate symbols to statements as follows:

A	$=$	The apple is ripe
B	$=$	I will eat the whole apple
$A \rightarrow B =$		If the apple is ripe then I will eat the whole apple.

Then assign the truth values $c = .9$ and $b = .6$. So the formal statement $t(B) \leq .6$ stands for the claim that it is at most 60% true (or at least 40% false) that I will (or did) eat the apple. Then the theorem says that the degree to which it is true that the apple is ripe is $t(A) \geq \min(1, 1 - .9 + .6) = \min(1, .7) = .7$.

Now assign the truth values $c = .9$ and $b = .2$. So it is only at most 20% true that I will eat the apple. Then "The apple is ripe" is true at least to degree .3 or $t(A) \geq \min(1, 1 - .9 + .2) = \min(1, .3) = .3$. For details and other multivalued inference schemas see chapter 1 of Kosko, B., *Fuzzy Engineering*, Prentice Hall, 1996.

16. Dunham, W., *Journey Through Genius: The Great Theorems of Mathematics,* Wiley, 1990, p. 138.

17. Bivalent math defines a probability measure $P: \mathfrak{I} \rightarrow [0,1]$ as a mapping from certain bivalent sets or "events" to numbers. The probability measure P assigns a number in $[0, 1]$ to the bivalent sets $A \in \mathfrak{I}$ in some sigma-algebra $\mathfrak{I} \subset 2^X$ of subsets of some sample space X. These sets A obey the Aristotelian laws of noncontradiction and excluded middle: $A \cap A^c = \varnothing$ and $A \cup A^c = X$. So set contradictions have measure zero and define "impossible" events: $P(A \cap A^c) = P(\varnothing) = 0$. Vagueness or fuzziness holds for a set A if and only if it breaks the Aristotle's laws: $A \cap A^c \neq \varnothing$ and $A \cup A^c \neq X$. So the probability framework excludes fuzziness by definition.

Lotfi Zadeh showed how to extend the probability framework to define probability measures on fuzzy sets and so to put chance numbers on vague statements like "The stock market will rise slightly tomorrow." A fuzzy set $A \subset X$ has a multivalued set function $a: X \rightarrow [0,1]$ such that $a(x) = Degree(x \in A)$ for an object $x \in X$. A bivalent set A has set membership values of only 0 and 1 and so defines the binary map $a: X \rightarrow \{0,1\}$. Then the probability $P(A)$ of event A is just the expected value $E[a]$ of the (measurable) binary set function a:

$$P(A) = E[a] = \int_X a(x) \, dP(x)$$

for a suitable abstract integral. Then $P(A)$ gives the probability of a fuzzy event A by using the multivalued set function in the integral: Zadeh, L. A., "Probability Measures of Fuzzy Events," *Journal of Mathematical Analysis and Applications* 10 (1968): 421–27. Fred Watkins uses the fact that the sigma-algebra $\mathfrak{S} \subset 2^X$ of measurable fuzzy set functions (the fuzzy events) is a subset of the unit ball in L^∞ (X, P) to show that the above integral defines a measure on the sigma-algebra that properly extends the original measure P. Watkins, F. A., "Fuzzy Engineering," Ph.D. dissertation, Department of Electrical Engineering, University of California at Irvine, 1994 (University Microfilms International, 300 North Zeeb Road, Ann Arbor, Michigan 48106).

18. A fuzzy Coase Theorem with binary if-then strength has the *modus ponens* form

$$
\begin{array}{ll}
\textit{If} & t_L(A \rightarrow B) = 1 \\
\textit{and} & t(A) \geq a \\
\\
\textit{then} & t(B) \geq a
\end{array}
$$

because $c = 1$ implies that $t(B) \geq \max(0, c + a - 1) = \max(0, a) = a$. Here the statement A stands for the if-part conjunction "Property rights are well-defined (binary) and transactions costs are zero." The min operator can factor this statement into a truth function of its two conjuncts: $t(A) = \min(t(P), t(T))$ where P stands for the statement "Property rights are well-defined (binary)" and T stands for "Transaction costs are zero." Product could also define the conjunction operator.

19. The Federal Communications Commission put up for auction the small part of the spectrum between 1,850 megahertz to 1,990 megahertz. Ronald Coase had already called for such an auction as early as 1959. Coase, R.H., "The Federal Communications Commission," *Journal of Law & Economics* 2 (October 1959): 1–40. The FCC auction involved the latest thinking in auction theory and statistical game theory. McMillan, J., "Selling Spectrum Rights," *Journal of Economic Perspectives* 8, no. 3 (Summer 1994): 145–62.

20. "Minerals essential to industrial economies are not now in short supply, nor are they likely to be for the next several generations. Accordingly, given the lack of pending crises in raw materials availability, mining can no longer presume *de facto* acclaim as the best of all possible uses for land; it must compete with compelling demands for alternative uses." Hodges, C. A., "Mineral Resources, Environmental Issues, and Land Use," *Science* 268 (2 June 1995): 1305–12.

This abundance of metals has also led to a new abundance of metal pollution in the atmosphere: "The Industrial Revolution brought about unprecedented demand for metals and an exponential increase in the intensity of metal emissions, both in absolute masses and in the number and type of toxic metal compounds released. The emissions may be compared with mine productions of Cd, Cu, Pb, Ni, and Zn. About 90% of the mine outputs were consumed in this century. These inventories show that the cumulative industrial releases of heavy metals into our environment are indeed massive and pervasive and have overwhelmed the natural biogeochemical cycles of the metals in many ecosystems." Nriagu, J. O., "A History of Global Metal Pollution," *Science* 272 (12 April 1996): 223–24.

21. The Greeks were among the first peoples to mine and smelt copper and silver. They may have been the first to privatize a social asset: "Because silver-bearing ore usually contains other metals, most commonly lead, the quest for silver involved not only removing the metals from the ore, but separating the metals from one another. How the ancients learned to do it is unknown, but by the sixth century B.C. the Greeks were recovering large quantities of silver by smelting ore mined at Laurium on the southern tip of the Greek peninsula. The mines were controlled by the city-state of Athens, which leased them for operation by private citizens in return for a 4% royalty on the silver produced." St. John, J., *Noble Metals,* Time-Life Books, 1984, p. 87.

22. The cowboy socialism of the real American West has a lot more to do with politics than the rugged individualism of legend and film: "The American West in all its magnificence was built on federal dollars. The interstate highway system that connects one patch of desert to another happened because federal dollars flowed easily. Employment skyrocketed because waterless shipyards were built on the arid plains of Colorado, because air force bases, scien-

tific labs, and federal centers mushroomed overnight, because a slew of federal bureaus provided enough jobs to keep rural outposts from becoming ghost towns, because 10 times as many defense dollars were spent in the West as in the rest of the nation, and because 10 western states representing 9% of the nation's population controlled a fifth of the votes in the US Senate." Hess, K., Jr., "The West at War With Itself," *Reason*, 27, no. 2 (June 1995): 18–25.

Such nonmarket pricing has lead to bizarre and costly capital outlays: "Federally funded trappers have worked for decades to purge the western range of wolves, bears, cougars, and coyotes that prey on domestic livestock. Not only must taxpayers shoulder the costs (about $30 million a year), but they must also pay a bill that is sometimes far greater than the value of the livestock that is lost. Thanks to government efficiency and the USDA animal damage control program, it's possible for you and me to pay $1,000 or more in places like New Mexico to shoot or poison a single sheep-eating coyote. . . . In New Mexico, for example, rainfall has been above average for five of the past six years, yet so has federal drought relief, averaging 15% of net ranching income. Nevada ranchers, the most vocal of sagebrush rebels and the most intent on kicking Uncle Sam out of the West, receive on average $18,000 per year for every man and woman in the program" (pp. 22–23).

23. The smog auctions went well but the state gave so many extra smog credits that it depressed the trading price. The state predicted a trading price near $1,000 per smog credit but often received less than half of that. Zorpette, G., "A Slow Start for Emissions Trading," *IEEE Spectrum*, July 1994, 49–52.

The political framework of trading in smog credits and the like has not led to as free of a market as many economists had hoped: "Whatever the future of emissions trading, pollution-rights markets can't work exactly like real markets. They *are* delivering at least some of the economists' dreams of cost-saving efficiencies. But the nature of air pollution makes it hard for each of us to choose in a market how much clean air is worth to us. And the nature of the political conflict over pollution—environmentalists married at least rhetorically to the notion that any pollution is intolerable versus polluting industries and their customers concerned about costs—guarantees that managed markets in air pollution credits won't take the politics out of an inherently political system." Doherty, B., "Selling Air Pollution," *Reason* 28, no. 1 (May 1996): 32–37.

Yet the smog-credit experiment seems to have been a qualified success: "Little is known empirically about the impact of trading on technological change. Also, much more empirical research is needed on how the pre-existing regulatory environment affects the operation of permit trading programs. Moreover, all the successes with tradeable permits have involved air pollution: acid rain, leaded gasoline, and chlorofluorocarbons. Our experience (and success rate) with water pollution is much more limited, and in other areas, we have no experience at all. . . . Despite these and other uncertainties, market-based instruments for environmental protection—and, in particular, tradeable permit systems—now enjoy proven successes in reducing pollution at low cost." Stavins, R. N., "What Can We Learn from the Grand Policy Experiment? Lessons from SO_2 Allowance Trading," *Journal of Economic Perspectives* 12, no. 3 (Summer 1998): 69–88.

24. "It's not clear that a trading system will work with a half-dozen greenhouse gases, where trades among different industries and across the world would be required. And a key factor in the greenhouse case is the stringency of the emission cap—the final figure of allowable emissions. If it's too low, flexibility is reduced along with the price competition it encourages. . . . But for reining in pollution without choking industry [a trading system in greenhouse gases] looks like a good place to start." Kerr, R. A., "Acid Rain Control: Success on the Cheap," *Science* 282 (6 November 1998): 1024–27.

25. "Every workday, between 5 A.M. and 10 P.M., certain cars were banished from the city's streets, according to their license numbers. . . . The law bans cars one day a week according to the last digit of each license plate. If a plate ends with a 1 or 2, for example, the car may not be driven on Monday; a 3 or 4 bans the car every Tuesday, and so on through 9 and 0 on Friday. By buying additional cars, drivers could acquire a selection of plates that allowed them to drive every day, undermining the system. . . . But now, a new study by Mexico City's Metropolitan Commission for Pollution Control and Prevention has confirmed what most of the city's estimated 20 million residents have suspected for years: The law does not work,

and it has not worked since soon after it came into force." Fineman, M., "Future Is Cloudy for Mexico City Smog Law," *Los Angeles Times,* 25 November 1995, p. A-2.

26. "Low diversities of mitochondrial DNA (mtDNA) have recently been found in four species of matrilineal whale. No satisfactory explanation for this apparent anomaly has been previously suggested. Culture seems to be an important part of the lives of matrilineal whales. The selection of matrilineally transmitted cultural traits, upon which neutral mtDNA alleles 'hitchhike,' has the potential to strongly reduce genetic variation. Thus, in contrast to other nonhuman mammals, culture may be an important evolutionary force for the matrilineal whales." Whitehead, H., *Science* 282 (27 November 1998): 1708–11.

27. " 'By the way,' interjected Franklin, 'is the fence purely electrical?'
   " 'Oh no. Electric fields control fish pretty well but don't work satisfactorily on mammals like whales. The fence is largely ultrasonic—a curtain of sound from a chain of generators half a mile below the surface. We can get fine control at the gates by broadcasting specific orders; you can set a whole herd stampeding in any direction you wish by playing back a recording of a whale in distress.' " Clarke, A. C., *The Deep Range,* Harcourt Brace Jovanovich, 1968 (first published 1957).
   Such high-tech property fences remain in the domain of science fiction. Governments in the meantime are trying to use some mix of taxes and quotas and outright bans to control fishing of the commons. North Sea cod is the latest fish species to suffer from the tragedy of the ocean commons: "In common with many fish stocks in the North Sea, cod are heavily exploited with as much as 60% of the fishable stock being removed annually. . . . Cod can live many years and only reach maturity in significant numbers by the age of four. However, fish are caught as early as age one and by age two young fish are fully exploited by the fishery. This means fish suffer substantial fishing mortality before they have a chance to reproduce. At present exploitation rates, only 4% of fish aged one will survive to the age of four. Without a substantial reduction in the rate of fishing, the North Sea cod stock may well collapse." Cook, R. M., Sinclair, A., and Stefansson, G., "Potential Collapse of North Sea Cod Stocks," *Nature* 385 (6 February 1997): 521–22.

28. "[Aquaculture] is growing rapidly, more than doubling in volume to 28M tons and almost trebling in value to $43 billion between 1986 and 1995, according to a new report by the Food and Agriculture Organization [in Rome]. About half of this takes place in freshwater ponds. Aquaculture is usually pilloried for the pollution it causes. The culprits are shrimp and fin fish such as salmon. A study in 1994 found that 80 kg of nitrogen compounds and 7.5 kg of phosphates were produced for each ton of farmed salmon, and that 70%–80% of the antibiotics that were supposed to protect the fish went to waste. Shrimp cultivation can also pollute, but spectacular outbreaks of disease in Asia farms have been a great incentive for better management." "Survey: The Deep Green Sea," *The Economist,* 23 May 1998, 13; Boyd, C. E., and Clay, J. W., "Shrimp Aquaculture and the Environment," *Scientific American,* June 1998, 58–65.

29. "Of 200 major fish stocks accounting for 77% of world marine landings, 35% are currently classified as overfished. Currently, overfishing is diminishing the production of fish as food, limiting the economic productivity of fisheries, restricting subsistence and recreational uses, and reducing genetic diversity and ecological resilience." Costanza, R., et al., "Principles for Sustainable Governance of the Oceans," *Science* 281 (10 July 1998): 198–99.

30. Harvard zoologist Edward O. Wilson has argued for debt-for-nature swaps with the area-species curve of population ecology (the number of species S in a land area such as an island grows with the cube root or even the sixth root of the land area A or $S = cA^d$ for constant $c > 0$ and exponent d in the range ($\frac{1}{6} \leq d \leq \frac{1}{3}$): "The area-species relation governing biodiversity shows that maintenance of existing parks and reserves will not be enough to save all the species living within them. Only 4.3 percent of the earth's land surface is currently under legal protection, divided among national parks, scientific stations, and other classes of reserves. . . . So we should try to expand reserves from 4.3% to 10% of the land surface. One of the more promising means to attain this goal is by debt-for-nature swaps. As currently practiced, conservation organizations such as Conservation International, the Nature Conservancy, and the World Wildlife Fund raise funds to purchase a portion of a country's commercial debt at a discount, or else they persuade creditor banks to donate some of it. . . .

By early 1992 a total of twenty such agreements totaling $110 million had been arranged in nine countries, including Bolivia, Costa Rica, Dominican Republic, Ecuador, Mexico, Madagascar, Zambia, the Philippines, and Poland." Wilson, E. O., *The Diversity of Life,* W. W. Norton & Company, 1992, pp. 337–38.

Thomas Lovejoy of the World Wildlife Fund first put forth debt-for-nature swaps in 1984 to help protect the forests of Costa Rica. Simons, P., "Costa Rica's Forests are Reborn," *New Scientist,* 22 October 1988, pp. 44–45. Meanwhile the area-species curve remains an empirical fact of our changing ecosystems if not a theoretical one: "Two patterns in the distribution of species have become firmly but independently established in ecology: The species-area curve, which describes how rapidly the number of species increases with area, and the positive relation between species' geographical distribution and average local abundance. There is no generally agreed explanation of either pattern, but for both the two main hypotheses are essentially the same: divergence of species along the ecological specialist-generalist continuum and colonization-extinction dynamics." Hanski, Ikka, and Gyllenberg, M., "Uniting Two General Patterns in the Distribution of Species," *Science* 275 (17 January 1997): 397–402.

31. John Baden and Richard Stroup have gone further and proposed that the state give large tracts of common wilderness to groups like the Sierra Club or Audubon Society or Ducks Unlimited. Congress would approve these "endowment boards." They might permit some oil drilling or cattle grazing or off-road race driving but only if it met their standards. They could use the proceeds from these permits or royalties to support other parts of their legal mission as ecological stewards.

The Audubon Society has allowed oil drilling for decades on its Rainey Wildlife Sanctuary in Louisiana. Consolidated Oil and Gas drills "oil wells in Rainey which are a potential source of pollution, yet Audubon experience in the past few decades indicates that oil can be extracted without measurable damage to the marsh." Baden, J., and Stroup, R., "Saving the Wilderness," *Reason* 13 (July 1981): 28–36; Baden, J., and Stroup, R., "Endowment Areas: A Clearing in the Policy Wilderness," *Cato Journal* 2 (Winter 1982): 691–708. The North Maine Woods is a related experiment. A joint effort of 25 timber and paper firms owns and maintains its almost three million acres of working forests. This private wilderness has a few user-fee stations but no paved roads or park rangers.

Simpler privatization schemes allow hunters in effect to buy binary property rights in wild game: "The 'poachers' villified and even shot in the name of biodiversity are as often as not dirt-poor villagers starving to death. Declaring land off-limits to hungry people is not a strategy for saving nature or feeding people. Worst of all may be the utter futility of trying to save wildlife by preventing its use with absolute bans. The costs of effective antipoaching efforts in African parks have been enormous: more than $500 per square mile per year. Most African countries have been able to afford a 10th that at most. Killing animals to save them sounds paradoxical, even cynical and perverse. Yet economics and history offer a strong brief for 'sustainable use.' In the United States, the market for sport hunting has been a spectacular success in saving targeted species. There were fewer than 5,000 pronghorn antelope in 1910. There are over a million today. Wild turkeys were virtually extinct in 1890. Today there are 4 million. And a growing number of wildlife officials and environmental groups worldwide are taking heart in the practical success that regulated markets for wildlife can achieve in even the poorest countries facing the greatest environmental pressures. Where wildlife have commercial value to people who share the land with them, a local constituency is created with a stake in the animals' continued existence. Preservation by contrast is a never-ending battle of enforcement by punitive costs." Budiansky, S., "Killing with Kindness," *U.S. News & World Report,* 25 November 1996, 47–49. Like arguments may soon lead to a new and legal trade in African elephant ivory. Satchell, M., "Save the Elephants: Start Shooting Them," *U.S. News & World Report,* 25 November 1996, 51–53.

Some Indian reservations in the United States have also privatized their animal "common": "The White Mountain Apache of east-central Arizona, by contrast, have shown what can happen in Indian country if you pay attention to incentives. This tribe is managing its trophy elk population and other wildlife opportunities on a sustainable basis—and making a profit. The Fort Apache Reservation covers 1.6 million acres—diverse habitat ranging from oak chaparral at lower elevations to mixed coniferous forests at the heights. This habitat supports about 12,000 free-ranging elk. To get some idea of the success elk hunters enjoy, consider the reservation's track record. From 1977 to 1995, nontribal hunters have bagged 90 bull elk that made either Boone and Crockett or Safari Club record books. In comparison, this is about the same number of record elk taken from the entire state of Montana since record keeping began in 1932. Since 1980, nontribal hunters have enjoyed a 90 to 95 percent success rate. The average score for antlers has been 366 Boone and Crockett points. Such scores are the equivalent of a foursome averaging three under par for a round of golf." Anderson, T. L., "Dances with Myths," *Reason* 28, no. 9 (February 1997): 45–50.

32. "Both Mexico and Indonesia are losing about 1% of their forest lands every year to logging and to slash-and-burn agriculture. But . . . many woodlands outside the tropics are quietly prospering. . . . Based on an analysis of decades of Forest Service data, these researchers conclude that forest growth in the United States has outpaced forest clearing over the last 50 years, increasing the country's total timber volume by 30%. . . . Logging didn't erase the gains, even though consumption of all timber products grew 70% between 1900 and 1993 because existing woodlands are producing more efficiently, harvest data show. This has been achieved by better spacing of trees in plantations, planting rapidly growing species and harvesting them on a shorter life cycle, and taking advantage of early, faster growth." Moffat, A. S., "Temperate Forests Gain Ground," *Science* 282 (13 November 1998): 1253.

33. "When the whale comes to the surface to breathe, a pressure sensor will tell the system that it is on the surface and thus able to communicate with US Air Force GPS satellites and its Japanese relate satellite. The system will automatically calculate the whale's position by acquiring data from two or three GPS spacecraft and then transmit the location and other data to the Japanese Whale Ecology spacecraft when it flies overhead." Covault, C., "Japan Monitors Whales from Space," *Aviation Week & Space Technology,* 16 October 1995, 25.

34. Palsboll, P. J., et al., "Genetic Tagging of Humpback Whales," *Nature* 388 (21 August 1997): 767–69.

35. Humans have uplifted dolphins and chimpanzees while other galactic races have uplifted other races in physicist David Brin's novels: *Startide Rising* (Bantam, 1983) and *The Uplift War* (Bantam, 1987). Brin defines *uplift* as "the process by which older spacefaring races bring new species into Galactic culture, through breeding and genetic engineering."

36. Kellogg, W. W., and Schneider, S. H., "Climate Stabilization: For Better or Worse?" *Science* 186 (1974): 1163–72. Schneider presents his no-fault insurance scheme in a broader context in Schneider, S. H., and Londer, R., *The Coevolution of Climate and Life,* Sierra Club Books, 1984, p. 438.

Schneider believes the weather changer would offer the insurance to those groups that could use past statistics to argue that the weather change damaged them. The groups could always find scientists who think the weather change caused their damage. The weather changer would expect this and offer insurance on a no-fault basis. Schneider also believes that the world may someday hold countries like the US to blame for the fossil fuels that it burns and that lead to a rise in carbon dioxide and other greenhouse gases.

37. "The Chicago Board of Trade sells 'hurricane futures,' which pay out an amount tied to hurricane-related claims in specified periods and places. After a sluggish beginning, trade in these is now picking up. Also, Wall Street firms are developing 'Act-of-God Bonds,' traditional debt but with a built-in forgiveness clause that releases the firm from some of its obligations in the event of a natural disaster." "A Brief History of Derivatives," *The Economist,* Survey of Corporate Risk Management, 10 February 1996, 10. For formal pricing details see Baxter, M., and Rennie, A., *Financial Calculus: An Introduction to Derivative Pricing,* Cambridge University Press, 1996.

38. Ponte, L, *The Cooling,* Prentice Hall, 1976, pp. 149–51. Science writer Lowell Ponte argued in the 1970s that the earth had entered a period of global cooling: "The United Nations may eventually develop a system of world weather welfare to correct deficient weather and climate as needed and enhancing them where desired. The alternative is a world of increasing drought, as the cooling continues, in which injured nations will rightly or wrongly blame all their weather woes on countries with big weather-modification projects, like the US and USSR, or on neighboring countries that try rainmaking methods. Even if untrue, politicians may find it expedient to accuse other nations of weather modification if the alternative is to accept blame themselves for failed policies" (pp. 153–54). The cooling thesis suggests that greenhouse gases may cause more global warming than we realize.

Earlier authors wrote on weather modification with more enthusiasm: "Instead of merely observing the formation of a hurricane and plotting its course, we and our colleagues have attempted to interfere in a critical area with the delicately balanced forces that sustain a mature hurricane. The basic technique is not new: it involves 'seeding' a certain cluster of clouds near the 'eye,' or center, of the storm with tiny crystals of silver iodide in order to release the latent heat energy of the clouds. In this way we hope to trigger a self-sustaining chain of events that will lead to a reduction of wind speed near the eye." Simpson, R. H., and Malkus, J. S., "Experiments in Hurricane Modification," *Scientific American* 211, no. 6 (1964).

Recent concerns over how global warming can affect climate and ocean water levels may also have passed into exaggeration: "So long as the West Antarctic ice sheet remains reasonably behaved, the real question facing residents of coastal regions may be how greenhouse warming affects local weather extremes and the size of damaging storm surges. Yet for those kind of changes, scientists are especially hard put to offer predictions." Schneider, D., *Scientific American,* March 1997, 112–17.

39. Political scientist Evan McKenzie discusses the mixed experience of private homeowner associations in *Privatopia: Homeowner Associations and the Rise of Residential Private Governments* (Yale University Press, 1994): "To many residents, association boards often seem to operate as though wearing blinders, rigidly enforcing technical rules against people's use of their own homes and ignoring the consequences of such intrusive behavior. Many CID [common interest developments] residents causally accept the rationality of such management-mania, but others become angry and file lawsuits or refuse to obey and find themselves sued by the association. A California study found that 44% of board members had been harassed or threatened with lawsuits during the preceding year. The CID resident must choose between conformity and conflict" (p. 19). Economist Fred Foldvary gives an alternative view of CIDs in his review of *Privatopia* in the *Cato Journal* 15, no. 1 (Spring 1995): 143–45.

40. "If ocean colonization is such a good idea, why aren't the rich industrialists already doing it? They've got the giant ships, the subs, the drilling platforms, floating factory complexes, the whole works—why aren't they colonizing already?" Eichman, B., "Ocean Colonization: A Practical Approach," *Extropy: The Journal of Transhuman Thought* 6, no. 1 (Spring 1994): 5–14. Eichman's article gives many diagrams of city designs to process solar energy and ocean thermal energy.

41. The second or Aquarius step of the proposed Millennial Project would colonize and farm the seas to combat Malthusian growth in human population amid fixed land resources: "Each ton of sea water contains as much energy as two pounds of gasoline. The energy contained in the world's sea water is equivalent to filling the ocean basins twenty feet deep in high-octane fuel. The world's oceans contain 5×10^{21} BTUs of potential energy—an amount equal to a million billion barrels of oil. There is enough latent energy in the oceans to supply the entire world power demand for 25,000 years. And it is renewable. The world's oceans contain 550 billion metric tons of nitrates. This is 36 times more nitrogen than is held in the planet's entire biomass. . . . The pulsing heart of Aquarius is an OTEC (Ocean Thermal Energy Converter). The OTEC produces electrical power by exploiting the temperature differential [about 40 degrees Fahrenheit] between warm surface waters and cold deep waters. Unlike conventional power plants, OTECs are net energy producers. An OTEC consumes only 700 calories of energy for every 1000 it produces." Savage, M. T., *The Millennial Project: Colonizing the Galaxy in Eight Easy Steps,* Little, Brown, 1994, pp. 26–35.

The First Millennial Foundation publishes *First Foundation News!* and supports a wide range of ambitious experiments: "Aquarius Rising is our three-part development to be built on the island of St. Croix in the U.S. Virgin Islands. The project consists of: 1) A 450-room, beachfront hotel that includes a duty-free shopping mall, a world-class casino, and a convention center. 2) An Ocean Thermal Energy Conversion (OTEC) plant producing all of the electric power, fresh water, and air conditioning required by the project. 3) A prototype sea/space colony housing 100 colonists engaged in marine and space research." Savage, M., and Spangle, K., "Aquarius Rising: A Preview of the New Millennium," *First Foundation News!* 1, no. 5 (April 1995): 4–5. The First Millennial Foundation is a nonprofit corporation: First Millennial Foundation, P. O. Box 347, Rifle, Colorado, 81650.

42. "If extraterrestrial intelligent beings exist and have reached a high level of technical development, one by-product of their energy metabolism is likely to be the large-scale conversion

of starlight into far-infrared radiation." Dyson, F. J., "Search for Artificial Stellar Sources of Infra-red Radiation," *Science* 131 (3 June 1960): 1667–68. Dyson discusses how what we now call Dyson spheres could lead to the "Greening of the Galaxy" in Dyson, F. J., *Disturbing the Universe,* Basic Books, 1979.

7. SMART WARS

1. "Logistics, transport, roads, and maps; timekeepers, standards, trumpets, and the ability to write—in their totality, these essentially nonmilitary technologies probably did as much to shape warfare, particularly the strategic dimension of warfare, as did any number of weapons and arms. Their combined effect was that there were no real lines of communication to be cut, hardly any 'bases' to be guarded or occupied. Also, the technical problems of orientation, of timekeeping, of communication, did not allow forces operating far apart to be easily coordinated and directed against a common objective." Van Creed, M., *Technology and War: From 2000 B.C. to the Present,* The Free Press, 1989, pp. 48–49.
2. MIT Media Lab's Nicholas Negroponte looks at the shift from atoms to bit in the overall information economy in *Being Digital,* Knopf, 1995. Other works discuss the shift in terms of smart weapons of the present and future: Kosko, B., "Smarter Weapons, Harder Fights," *Liberty* 5, no. 1 (September 1991): 37–38; Kosko, B., "Invest in Higher Machine IQ," *Reason,* October 1992, pp. 42–44.
3. "About 70% of the world's 60,000,000 square miles of dry land is either too high, too cold or too waterless for the conduct of military operations. The poles, North and South, demonstrate the effect of such conditions with starkness. The Antarctic continent's inaccessibility and the extreme climatic conditions that prevail there secluded it from warmaking for millennia, though several states laid claim to territory. The icecap, moreover, is known to cover valuable mineral deposits." Keegan, J., *A History of Warfare,* Knopf, 1993, pp. 68–69.
4. "From a strictly technical standpoint, accurately delivered modern conventional munitions are competitive with nuclear munitions for some applications and superior for others. This means that some operations which in the past were thought of as requiring nuclear munitions can, with modern technology, be conducted with nonnuclear munitions." Hudson, C. I., and Hass, P. H., "New Technologies: The Prospects" in *Beyond Nuclear Deterrence,* ed. J. J. Holst and U. Nerlich, Crane, Russak & Company, 1977, p. 108. Carl Builder extends this thesis of the "precisely placed" smart weapon: "Most manmade objects are susceptible to destruction by a properly placed projectile or explosive charge. It is only when the destructive device cannot be properly placed that massing weapons effects in rapid-fire guns or in large bombs, including atomic bombs, becomes attractive." Builder, C. H., *Strategic Conflict Without Nuclear Weapons,* RAND Corporation Report R-2980-FF/RC, April 1983, p. 22.
5. CEPs have shrunk with the use of global positioning satellites. GPS can still lead to target errors of 30 feet. Synthetic aperture radar can reduce this error: "Block 20 capabilities would allow demonstration of relative targeting and a 20-foot CEP by mid-1996," "Precision Bomb Programs May Merge," *Aviation Week & Space Technology,* 27 September 1993, 45.
6. Nuclear weaponeers use an exponential formula to compute the odds that a strategic target survives a nuclear missile that has made it through the defense system:

$$p_{survive} = \left(\frac{1}{2}\right)^{a}$$

where the exponent has the form

$$a = \frac{c\, Y^{2/3}}{(PSI)^{b}\, (CEP)^{2}}$$

with constant coefficients c and b that depend on the nature of the nuclear yield and the target. "The latter probability is a function of the hardness of the target (PSI) and the warhead characteristics of accuracy (CEP) and yield (Y)." Perkins, F. M., "Optimum Weapon Deployment for Nuclear Attack," *Operations Research,* 1961, pp. 77–94. Then the kill probability is

$$p_{kill} = 1 - \left(\frac{1}{2}\right)^a$$

This shows that the kill probability grows as the CEP shrinks. The a term grows larger as the CEP shrinks and grows smaller as the CEP expands. The survival probability falls to zero (and the kill probability grows to unity) as the CEP shrinks to zero and thus as the a term grows to infinity. This relation still holds in theory even if the yield Y is small and perhaps non-nuclear so long as the CEP is sufficiently small.

7. "Ten men, battalions, or divisions fighting fifteen men, battalions, or divisions, conquer— that is, kill or take captive—all of them, while losing only four themselves, so that on one side four and on the other fifteen are lost. Consequently the four are equal to the fifteen and $4x = 15y$. Therefore $x:y::15:4$. This equation does not give us the value of the unknown factors but gives us the ratio between the two unknowns. And by bringing into such equations variously selected historical units (battles, campaigns, periods of war) a series of numbers is obtained in which certain laws should exist and could be discovered." Tolstoy, L., *War and Peace,* trans. A. Dunnigan, Penguin, 1968 (first published in Russian in 1869), p. 1235.

Tolstoy states the above force identity after he puts forth his own version of Newton's second law of motion ($F = m$ a or the force that acts on object equals the mass of the object times the acceleration of the object): "The spirit of an army is the factor which, multiplied by the mass, gives the resulting force [$F = m \times spirit$]. To define and explain the significance of this unknown factor, the spirit of an army, is a problem of science" (p. 1235). This looks like Napoleon's strength or momentum of an army ($S = m \times velocity$) but replaces the speed or velocity of the troop with its acceleration. The two men may have differed in their meaning or one may have jumped too quickly to a physics analogy. I suspect Tolstoy erred and meant momentum rather than force and thus sought a troop "spirit" that acted more like its physical velocity and thus he agreed with Napoleon.

8. Tolstoy begins Part Three of Book Three of *War and Peace* with a review of the "new branch of mathematics" we now call the integral calculus and how it gives historical movement as a sum of wills: "This new branch of mathematics, unknown to the ancients [the Greeks], by admitting infinitesimal magnitudes when examining the problem of motion, that is, predicating the chief condition of motion (absolute continuity), thereby corrects the inevitable error which the human mind cannot avoid when it examines discontinuous units of motion instead of continuous motion. In seeking laws of historical movement, exactly the same thing happens" (p. 986).

9. Tolstoy vents much of his wrath in *War and Peace* on the "great man" theory of history. He argues that the great-man view begs the question of what causes historical events: "The theory of the transference of the will of the masses to historical persons is merely a paraphrase— a restatement of the question in different words. What is the cause of historical events? Power. What is power? Power is the collective will vested in one person. On what condition is the people's will vested in one person? On the condition that that person expresses the will of the whole people. That is, power is power. In other words, power is a word the meaning of which we do not know" (p. 1429).

Tolstoy also deals with this earlier in the book in an attempt to dismiss the Russian terrors Napoleon and Genghis Khan in favor of blind and distributed causal determinism: "Accordingly all of them—myriads of causes—coincided to bring about what occurred. And so there was no single cause for the war [against Napoleon], but it happened simply because it had to happen. Millions of men, renouncing human feelings and reason, had to move from west to east to slay their fellows, just as some centuries' earlier hordes of [Mongol] men had moved from east to west slaying their fellows. . . . Consciously man lives for himself, but unconsciously he serves as an instrument for the accomplishment of the historical, social ends of mankind. The higher a man stands in the social scale, the more connections he has with people and the more power he has over them, the more manifest is the predetermination and inevitability of his every act. 'The hearts of kings are in the hand of God.' A king is the slave of history. History, that is, the unconscious, common, swarm life of mankind uses every moment of the life of kings as an instrument for its own ends" (pp. 733–34).

10. Lanchester combat models describe how the variables x and y evolve in time: $x(t)$ describes the red troop strength and $y(t)$ describes the blue troop strength at time t. The Lanchester "linear" model of ancient warfare says that both troops decay in proportion to how they interact:

$$\frac{dx}{dt} = -axy$$

$$\frac{dy}{dt} = -bxy$$

where $a > 0$ is the combat effectiveness of the y or blue troops and $b > 0$ is the combat effectiveness of the red or x troops. This simple model assumes that neither side receives new recruits. Both sides just pile onto each other. The orbits of the system solve the equation:

$$\frac{dy}{dx} = \frac{bxy}{axy} = \frac{b}{a}$$

Then integration gives the linear law $ay - bx = C_1$ for some constant C_1. So x or red wins if $C_1 < 0$ and blue wins if $C_1 > 0$. Some models view the ratios $a/(a+b)$ and $b/(a+b)$ as the default odds or probability that blue wins or that red wins. Brown, R. H., "Theory of Combat: The Probability of Winning," *Operations Research* 11 (1963): 418–25.

Lanchester's model of red conventional forces versus blue conventional forces has the forced form

$$\frac{dx}{dt} = -ay + r_x(t)$$

$$\frac{dy}{dt} = -by + r_y(t)$$

The forcing terms $r_x(t)$ and $r_y(t)$ measure the number of new red and blue recruits that arrive at time t. Suppose there are no new recruits. Then $r_x = r_y = 0$ and the orbits of the unforced combat system solve the equation

$$\frac{dy}{dx} = \frac{bx}{ay}.$$

Integration gives the solutions as a family of hyperbolas in the famous "square law":

$$ay^2 - bx^2 = C_2$$

for some constant C_2. Red wins if $C_2 < 0$ and blue wins if $C_2 > 0$.

Much research has focused on how to let the kill rates a and b vary with time. They do after all stand for all aspects of a battle except the raw number of troops. See Taylor, J. G., " 'Simple-Approximate' Battle-Outcome-Prediction Conditions for Variable-Coefficient Lanchester-Type Equations of Modern Warfare," *Naval Research Logistics Quarterly* 30 (1983): 113–31; Taylor, J. G., and Brown, G. G., "Annihilation Prediction for Lanchester-Type Models of Modern Warfare," *Operations Research* 31, no. 4 (July 1983): 752–71.

11. Analyst Franklin C. Brooks introduced the term *stochastic determinism* into mathematical warfare in 1965 after studying the effects of large computer battle simulations that modeled tens of thousands of aircraft and weapons attacking thousands of ranked targets: "Such [random or stochastic] battle models are said to be *stochastically determined* if the gross results—for

example, the casualties to each side—show low variance in repeated plays in which only the random numbers are changed. Here, low variance means a standard deviation small compared to the initial number of weapons engaged. When stochastic determinism exists in a large battle model, only a few replications of the model are needed to get good estimates of the gross results." Brooks, F. C., "The Stochastic Properties of Large Battle Models," *Operations Research,* January 1965, pp. 1–17. Brooks showed that Lanchester linear-law and square-law battles are stochastically determined.

Brooks also saw how his theorems touched on the Tolstoyan debate over the "great man" theory of history. Brooks's group theorems support Tolstoy's view that history changes through the combination of wills. But unlike Tolstoy he believes this weighs in favor of the "great man" theory: "Although the familiar quotation [of 17th-century English poet George Herbert] 'For want of a nail the shoe is lost, for want of a shoe the horse is lost, for want of a horse the rider is lost,' may be true, it is not likely that the missing nail will lead to loss of the whole war. On the other hand the highest echelons who make the strategic decisions, direct selection and training, support research and development, and the like, can win or lose the war by virtue of their actions. If this logic is applied to history as a whole, it tends to support the 'great man theory of history'—that single individuals with widespread control or influence *can* change the course of history" (p. 14).

12. "The artillery concentrations of World War I, however, did exhibit Lanchester's [square] law at work, especially when the Germans, obstinately refusing to risk yielding any ground, remained under fire of much stronger enemy concentrations of artillery." Jones, A., *The Art of War in the Western World,* University of Illinois Press, 1987, p. 640.

Archer Jones cites data to show that the square law held when the French fought the British in Spain at the Battle of Albuera during the Napoleonic Wars and in Nelson's 1805 Battle of Trafalgar at sea: "Nelson pitted all 27 of his ships against 23 [ships] of the enemy. Since all ships could fire at all others, the British advantage, by Lanchester's theory, compare as 27^2 to 23^2, or 729 to 529. The dominance given the seasoned British by their better seamanship and gunnery made the odds in their favor greater than those indicated by Lanchester's augmentations" (p. 379). But the British "seasoning" would just decrease their own attrition rate or increase that of the French. Jones also cites data to support that the square law held in Rommel's Panzer fight with French tanks at Sedan in World War II and in some of the air battles in the German-English Battle of Britain.

Herbert Weiss found the square law at work in some Civil War battles in "Combat Models and Historical Data: The US Civil War," *Operations Research* 14: 759–90. Weiss observed that losses grew with the size of a force. He followed others and put forth a logarithmic model to describe it: "It is suggested that the vulnerability of a force as a target increases directly with its number, but that its effectiveness in delivering firepower increases less rapidly than the number of units comprising it" (p. 788). This new logarithmic model has the form

$$\frac{dx}{dt} = -ax \ln y.$$

13. MIT researcher J. H. Engel first showed that Lanchester's square law held in the battle of Iwo Jima. He ended his famous paper with a grand call for further research that still rings throughout the aerospace industry: "It may be possible after sufficient research to ascertain the influence of such factors as terrain, time to prepare a defensive position, length and nature of artillery preparation and air support, ratio of combatant troops to troops in support, amount of troop experience, morale, etc., on the values of such parameters as [the attrition coefficients] *a* and *b,* the casualty-producing rates of the forces about to enter into a specific battle." Engel, J. H., "A Verification of Lanchester's Law," *Operations Research* 2 (1954): 163–71. Engel estimated *a* and *b* to be $a = 0.0544$ and $b = 0.0106$. Mathematician Martin Braun presents the Lanchester theory in a more general framework and reworks Engel's example in *Differential Equations and Their Applications,* Springer-Verlag, 1978, pp. 291–99.

14. The Lanchester model allows one to capture guerrilla conflict as a type of ancient warfare of red-blue interaction:

$$\frac{dx}{dt} = -axy + r_x(t)$$

$$\frac{dy}{dt} = -by + r_y(t).$$

Here the red or x troop is a guerrilla troop and the y or blue is a conventional troop. A pure guerrilla-versus-guerilla battle would replace the term $-by$ with $-bxy$ in the second equation and thus give back the ancient or "linear" model of combat.

The guerrilla-conventional dynamical system leads to a family of parabolas as its solution: $ay^2 - 2bx = C_3$ for some constant C_3. This mixed law says that red wins if $C_3 < 0$ and that blue wins if $C_3 > 0$. See Schaffer, M. B. "Lanchester Models of Guerrilla Engagements," *Operations Research* (May 1968). 457–88, Kisi, T., and Hirose, T., "Winning Probability in an Ambush Engagement," *Operations Research* (1967): 1137–38.

15. "This research raises the question of whether the conclusions and applications of past modeling might have been tainted by the unsuspected presence of mathematical chaos and it casts a shadow over the claimed reliability of combat (and other) models even when they do not (or have not been proved to) exhibit blatantly non-monotonic behavior." Dewar, J. A., Gillogly, J. J., and Juncosa, M. L., *Non-Monotonicity, Chaos, and Combat Models,* RAND Report R-3995-RC, ISBN: 0-8330-1140-5, RAND Corporation, 1991.

16. The journal *Phalanx* ran short articles and letters on the chaos and "non-monotonicity" controversy in combat starting in late 1992. Much of this grew out of the 1991 RAND Report *Non-Monotonicity, Chaos, and Combat Models* and like work from RAND researchers. The 60th Military Operations Research Society held a symposium on the subject in June of 1992.

Analysts from MITRE Corporation and RAND first raised the issue for Lanchester models in Tsai, H. K., Ellenbogen, J. C., "Bounding Potentially Pathological Nonlinear Behavior in Combat Models and Simulations," *Phalanx*, December 1992; Palmore, J., "Dynamical Instability in Combat Models," *Phalanx*, December 1992; Davis, P. K., "Dynamic Instability . . . ," *Phalanx*, December 1992; Louer, P. E., "More on Nonlinear Effects," *Phalanx*, March 1993; Helmbold, R. L., "Combat Analysis," *Phalanx*, March 1993.

The next volley in the exchange used 84 million computer runs to make its point: "In this paper we present a counter-example to the claim that stochastic thresholds will eliminate non-monotonic effects. . . . We demonstrate that stochastic thresholds not only can fail to eliminate non-monotonic behavior, but can actually make the non-monotonic behavior worse." Allen, P., Gillogly, J., and Dewar, J., "Non-Monotonic Effects in Models with Stochastic Thresholds," *Phalanx*, December 1993, 15–20. Other responses followed including Cooper, G., "Non-Monotonicity and Other Combat Modeling Ailments," *Phalanx*, June 1994.

17. "A new [classified] report on missile proliferation currently being circulated among senior US Defense Department officials indicates that low-flying cruise missiles are becoming the No. 1 proliferation threat. The report has led some senior US officials to question whether the Strategic Defense Initiative Organization's [Star Wars] research should be focused primarily on ballistic missile defenses. The report concludes that Syria, Iran, and China will have cruise missiles with some low observable or stealth capabilities between 2000 and 2010. All three countries are expected to have both chemical and biological warheads for their cruise missiles." *Aviation Week & Space Technology,* 1 February 1993, 26–27.

18. "Even a small business jet could be turned into a cruise missile with five tons of high explosives, a television camera for takeoff, autopilot, GPS, and a cheap navigation system." Fulghum, D. A., "US Developing Plan to Down Cruise Missiles," *Aviation Week & Space Technology,* 22 March 1993, 46–47.

19. "In 1964, six years after the integrated circuit was invented, Gordon Moore observed that the number of transistors that semiconductor makers could put on a chip was doubling every year. Moore, who cofounded Intel Corporation in 1968 and is now an industry sage, correctly predicted that this pace would continue into at least the near future. The phenomenon became known as Moore's Law, and it has had far-reaching implications." Hutcheson, G. D., and Hutcheson, J. D., "Technology and Economics in the Semiconductor Industry," *Scientific American,* January 1996, 54–62.

20. "Every 18 months microprocessors double in speed. Within 25 years, one computer will be as powerful as all those in Silicon Valley today." Patterson, D. A., "Microprocessors in 2020," *Scientific American,* September 1995, 62–67.

21. "The pop-up/terminal-dive maneuver was facilitated solely by a software change, and this is characteristic of what has been done throughout the history of the Tomahawk [cruise missile] program. The aim has been to make the hardware as universally capable as possible, then add specific performance features through software changes. Software refinements are obviously much easier and more cost-effective to implement than hardware modifications. Once the software for a new flight program has been developed it is only necessary to create as many copies as there are launch platforms. When a launch platform loads a revised program into a Tomahawk's onboard computer, that missile immediately has the new capability, without any other work being performed. In short, the missile has been instantaneously modified." Macknight, N., *Tomahawk Cruise Missile,* Mil-Tech Series, Motorbooks International, 1995, p. 75.

22. "The GPS planners chose a compromise solution [between low-earth and high-earth orbits for the satellites], launching the satellites into orbits that were neither particularly low nor high; the satellites were set to orbit at an altitude of about 20,000 kilometers. At that altitude, 17 satellites would be sufficient to ensure that at least four of them—the minimum number needed to fix a position—would always be available from any location on the earth's surface. The final configuration adopted for the GPS has 21 primary satellites and 3 spares in orbit." Herring, T. A., "The Global Positioning System," *Scientific American,* February 1996, 48. For more details on the GPS system see the January 1999 special issue on GPS in the *IEEE Proceedings.*

   GPS plays an increasing role in the guidance of cruise missiles: "The most important technological innovation in air-to-surface munitions is GPS. Should GPS integration prove successful in guided bombs such as the US Joint Direct Attack Munition that is being competitively developed by teams headed by McDonnell Douglas and Lockheed Martin, this technology could supplant semiactive laser guidance. France has embarked on its own equivalent GPS-guided bomb program, the advanced air-to-surface missile, and other countries are likely to follow. The other key area for GPS is in stand-off munitions, particularly dispensers. Development of these glide bombs has been inhibited by the expense of inertial guidance packages or alternatives such as the Tercom system used on the BGM-109 Tomahawk. GPS promises high accuracy, almost regardless of range, while unit cost should be reasonable. The big if is whether GPS will prove vulnerable to electronic jamming." Zaloga, S., "Missiles Ride Gulf Air Wave," *Aviation Week & Space Technology,* 8 January 1996, 129. Europeans are testing fiber-optic guidance systems. Morrocco, J. D., "New Roles Envisioned for Fiber-Optic Guided Missiles," *Aviation Week & Space Technology,* 9 November 1998, 86–87.

23. The journal *Neural Networks* devoted a double issue to automatic target recognition (vol. 9, no. 7/8, 1995). The articles' subjects range from how eye-like systems process synthetic aperture radar to how bats sense moths with sonar. Neural articles that also use fuzzy techniques include Bradski, G., and Grossberg, S., "Fast-Learning VIEWNET Architectures for Recognizing Three-Dimensional Objects from Multiple Two-Dimensional Views," pp. 1053–80, and Rubin, M. A., "Application of Fuzzy ARTMAP and ART-EMAP to Automatic Target Recognition Using Radar Range Profiles," pp. 1109–16.

   The *IEEE Transactions on Image Processing* has also devoted a special issue to automatic target recognition. Many of these ATR articles use neural or neural-like math schemes while others use more standard techniques from image processing and pattern recognition. See Bhanu, B., Dudgeon, D. E., Zelnio, E. G., Rosenfeld, A., Casasent, D., and Reed, I. S., "Introduction to the Special Issue on Automatic Target Detection and Recognition," *IEEE Transactions on Image Processing* 6, no. 1 (January 1997): 1–6.

24. "ARPA [Advanced Research Projects Agency] also has contractors working on advanced algorithms and techniques that will pick targets from massive amounts of data acquired by dozens of sources. These advanced processing tools rely on 'fuzzy logic' and other artificial intelligence techniques.

   " 'We need to develop algorithms that can search any terrain. They can't be optimized for just grass or trees or desert,' Larry B. Stotts [assistant for sensors and processing in ARPA's Advanced Systems Technology Office] said. 'They have to handle changing conditions, such as tree lines next to open ground. We want algorithms that can adapt. Using advanced algorithms, fuzzy logic, neural networks and combinations of these, we think we can get close to optimum performance.' " Scott, W. B., *Aviation Week & Space Technology,* 31 May 1993, 37–38. The article further stresses the need for "smart" techniques to help in automatic target recognition: "These [ATR] must be able to identify targets when viewed from any angle. 'Template'-type systems that must store 10,000–50,000 views for comparison to a detected object are probably too unwieldy."

25. "A cruise missile based on an unmanned aerial vehicle could cost less than $100,000 each. . . . The low end of the cruise missile market is expected to be dominated by the Chinese, who have already purchased components and technical advice from the Russians." Fulghum, D. A., "Cheap Cruise Missiles A Potent New Threat," *Aviation Week & Space Technology,* 6 September 1993, 54.

26. "Defensive weapons are much cheaper than offensive weapons, and some can be held in the hands of a single defender. One such is the Stinger missile, which Afghan rebels used effectively to knock down helicopters during the Soviet Union's intervention of Afghanistan in the 1980s. The Patriot missile, which destroyed Iraqi Scud missiles in the Persian Gulf War of 1991 and can destroy attacking aircraft, costs only a fraction of a Scud's price and about 1 percent of a fighter bomber's." Alexander, B., *How Great Generals Win,* W. W. Norton, 1993, p. 22.

27. Richard Rhodes describes when US Air Force General Curtis LeMay took advantage of the Soviet's lack of antiballistic missile defense in the 1950s: "Soviet defense forces had no way of knowing if LeMay's crisscrossing reconnaissance aircraft carried nuclear weapons or not. If Soviet aircraft had crisscrossed US cities under similar circumstances, SAC would certainly have preempted. The Soviets hunkered down because they had no adequate response, but their lack of defenses predictably emboldened LeMay." Rhodes, R., *Dark Sun: The Making of the Hydrogen Bomb,* Simon and Schuster, 1995, p. 565.

28. "Of the several ways of making VX, Iraq chose to synthesize it from EMPTA, which stands for O-ethyl methylphosphonothioic acid. According to a US intelligence source in *Chemical & Engineering News,* Iraq and Sudan are the only countries to have taken the EMPTA route to VX. Three US Army chemists developed the approach, which was the subject of a secret patent application in 1958. After being declassified in 1975, the patent became publicly available (it is now on the Internet)." Zorpette, G., and Frank, S. J., "Patent Blunder," *Scientific American,* November 1998, 42.

29. "Perhaps 110 million mines lurk in 64 nations around the world, and each year they kill or maim about 30,000 people, usually civilians. The heaviest concentrations of mines are in poor countries like Cambodia, Somalia, Bosnia, Mozambique, Afghanistan, and Angola that have survived for years or even decades of civil war. Five million new mines are laid each year, and only 100,000 are cleared. A new mine costs $3. Uprooting one costs between $200 and $1,000." Fedarko, K., Kroon, R., Purvis, A., and Thompson, M., "Land Mines: Cheap, Deadly, and Cruel," *Time,* 13 May 1996, 54–55. See also Fialka, J. J., "Killing Fields: Land Mines Prove to Be Even Harder to Detect than They Are to Ban," *Wall Street Journal,* 17 May 1996, p. A-1; Strado, G., "The Horror of Land Mines," *Scientific American,* May 1996, 40–45; Nadis, S., "Political Will and Cash 'Needed to Speed up Removal of Landmines,' " *Nature* 385 (9 January 1997): 101; Mecham, M., "Kaman Mine Detector Lidar System Debuts in Singapore," *Aviation Week & Space Technology,* 9 March 1998, 71.

30. Neural networks and fuzzy systems have entered the search for land mines: Plett, G. L., Doi, T., and Torrieri, D., "Mine Detection Using Scattering Parameters and an Artificial Neural Network," *IEEE Transactions on Neural Networks* 8, no. 6 (November 1997): 1456–67; Gader, P., Keller, J. M., Frigui, H., Liu, H., and Wang, D., "Landmine Detection Using Fuzzy Sets with GPR Images," *Proceedings of the 1998 IEEE World Congress on Computational Intelligence* 1 (May 1998): 232–36.

   Scientists have also proposed fluorescent chemosensors to detect the TNT in landmines: Czarnik, A. W., "A Sense of Landmines," *Nature* 394 (30 July 1998): 417–18. Other neural-fuzzy approaches use infrared sensing: Filippidis, A., Jain, L. C., and Martin, N. M., "Using Genetic Algorithms and Neural Networks for Surface Land Mine Detection," *IEEE Transactions on Neural Networks,* vol. 47, no. 1, January 1999, 176–88.

31. Edwards describes the Igloo White Program's strengths and failings in Edwards, P. N., *The Closed World: Computers and the Politics of Discourse,* MIT Press, 1994. See also General

Kennedy's perspective in Kennedy, W. V., *Intelligence Warfare: Penetrating the Secret Worlds of Today's Advanced Technology Conflict,* Crescent Books, 1987.

32. "Battleground circa 2020 may replace massed troops and armor with networks of intelligent mines and unpiloted drones that can perform reconnaissance and launch or plant weapons. Highly dispersed special forces may scout for targets and evaluate battle damage. Remotely fired missiles may become the main instrument for destroying enemy targets." Stix, G., *Scientific American,* December 1995, 92–98.

"The rapid commercial advancements of powerful microprocessors, optical computing, infrared transceivers that link small computers with intelligent, high-bandwidth reliable communication networks and new high-security techniques will continue to outrun military development in the same areas. Many of these "off-the-shelf" commercial technologies and systems could be used in space [information] warfare." Scott, W. B., " 'Information Warfare' Demands New Approach," *Aviation Week & Space Technology,* 13 March 1995, 86.

Control of information in space and at high altitude will play an increasing role in future arms races: "Future warfare may be more a gigantic artillery duel fought with exceptionally sophisticated munitions than a chesslike game of maneuver and positioning. As all countries gain access to the new forms of air power (space-based reconnaissance and unmanned aerial vehicles), hiding large-scale armored movements or building up safe rear areas chock-a-block with ammunition dumps and truck convoys will gradually become impossible." Cohen, E. A., "A Revolution in Warfare," *Foreign Affairs,* March 1996, 37–54.

See also Schwartau, W., *Information Warfare: Chaos on the Electronic Superhighway,* New York: Thunder's Mouth Press, 1994; Urban, E. C., "The Information Warrior," *IEEE Spectrum Magazine,* November 1995, 66–81; "Cyber Wars: Logic Bombs May Soon Replace More Conventional Munitions," *The Economist,* 13 January 1996, 77–78; Schwartz, P., "Warrior in the Age of Intelligent Machines," *Wired,* April 1995, 137–38; Munro, N., "Infowar Disputes Stall Defense Policy," *Washington Technology,* 25 May 1995; Fialka, J. J., "Pentagon Studies Art of 'Information Warfare' to Reduce Its Systems' Vulnerability to Hackers," *Wall Street Journal,* 3 July 1995, p. A-10; Waller, D., "Onward Cyber Soldiers," *Time,* 21 August 1995, 38–46; Fulghum, D. A., "Duplication Enemy Voices Becoming a Combat Skill," *Aviation Week & Space Technology,* 8 July 1996, 48–49; Mann, P., "Cyber Threat Expands with Unchecked Speed," *Aviation Week & Space Technology,* 8 July 1996, 63–64; Venzke, B., "Information Warrior: Winn Schwartau," *Wired,* August 1996, 136–37; McKenna, J. T., "Rome Lab Targets Info Warfare Defenses," *Aviation Week & Space Technology,* 12 August 1996, 65–67; Rapaport, R., "World War 3.1: The Shape of Things to Come?" *Forbes ASAP,* 7 October 1996, 125–32; Regis, E., "BioWar," *Wired,* November 1996, 142–53; McKenna, J. T., "Tighter Security Urged for Defense Computers," *Aviation Week & Space Technology,* 20 January 1997, 60–61; Mann, P., "Government/Industry Alliance Urged Cyber Threats," *Aviation Week & Space Technology,* 13 July 1998, 65–68.

33. "*Nanotechnology:* The term *nanotechnology* is here used to refer to an anticipated technology giving thorough control of the structure of matter at the molecular level. This involves *molecular manufacturing,* in which materials and products are fabricated by the precise positioning of molecules in accord with explicit engineering design." *Foresight Update,* no. 19 (1995): 7.

The quarterly British journal *Nanotechnology* publishes "hard" articles on nanotechnology as does *Science* and a few others. A typical article is Ralph Merkle's article on the dynamic stability of diamondoid molecular gears: "A Proof About Molecular Bearings," Merkle, R. C., *Nanotechnology* 4 (1993): 86–90.

Eric Drexler named and popularized the field of nanotechnology in his book *Engines of Creation:* "Tough, omnivorous 'bacteria' could outcompete real bacteria: they could spread like blowing pollen, replicate swiftly, and reduce the biosphere to dust in a matter of days." Drexler, K. E., *Engines of Creation: The Coming Era of Nanotechnology,* Anchor Press, 1986. Drexler gives a rigorous treatment of nanotech mechanics in terms of energy minimization and physical chemistry in his text *Nanosystems: Molecular Machinery, Manufacturing, and Computation,* Wiley, 1992.

Scientific American ran a critical but nontechnical update article on nanotechnology: "On the dark side, assemblers would streamline the production of superweapons, allowing rapid fabrication of a tank or a surface-to-air missile. And then there is the 'gray goo' problem— the possibility that nanodevices might be designed to replicate uncontrollably, like malignant tumor cells, and reduce everything to dust within days." Stix, G., "Trends in Nanotechnology: Waiting for Breakthroughs," *Scientific American*, April 1996, 94–99. The update article ends with a short-term demand for product proof that would have laid waste to the first years of research on Maxwell's equations or the atomic model or information theory: "Until the nanoists can make an assembler and find something useful to do with it, molecular nanotechnology will remain just a latter-day cargo cult." For a rebuttal see http://www.foresight.org/SciAmDebate/. Meanwhile research in nanotechnology continues: Heath, J. R., Kuekes, P. J., Snider, G. S., and Williams, R. S., "A Defect-Tolerant Computer Architecture: Opportunities for Nanotechnology," *Science* 280 (12 June 1998): 1716–21.

See also the nanotech web sites at http://nano.xerox.com/nano and http://www.foresight.org and the Internet bulletin board at sci.nanotech.

9. PATCH THE BUMPS

1. Mathematician Richard Bellman introduced the phrase "curse of dimensionality" into the technical literature in his pioneering work on multi-stage optimization in *Dynamic Programming*, Princeton University Press, 1957.

2. The sun is a nonlinear dynamical system that consists of a huge number of variables. Simple models that use only a few variables can never expect to explain much of its behavior. Just the rough wave structure of the sun's atmosphere can tax most formal models: "Helioseismology is probing the interior structure and dynamics of the sun with ever-increasing precision, providing a well-calibrated laboratory in which physical processes can be studied under conditions that are unattainable on Earth. Nearly 10 million resonant modes of oscillation are observable in the solar atmosphere, and their frequencies need to be known with great accuracy in order to gauge the sun's interior." Gough, D. O., Leibacher, J. W., Scherrer, P. H., and Toomre, J., "Perspectives on Helioseismology," *Science* 272 (31 May 1996): 1281–83. It is still an open research problem which subset of the many solar variables best explains the sun's 11- and 22-year solar cycles: Kuhn, J. R., Bush, R. I., Schleick, X., and Scherrer, P., "The Sun's Shape and Brightness," *Nature* 392 (12 March 1998): 155–57.

3. The fuzzy approximation theorem (FAT) says that an additive fuzzy system $F:C \subset R^n \to R^p$ with a finite number m of rules can uniformly approximate any continuous or bounded measurable function $f:C \subset R^n \to R^p$ if the domain C is a compact (closed and bounded) set. In this sense fuzzy systems are dense in the space of continuous (or bounded measurable) functions much as the rational numbers are dense in the real numbers. The if-part fuzzy sets $A_j \subset R^n$ can have any multivalued set function $a_j:R^n \to [0,1]$ and so can have any shape. The then-part sets $B_j \subset R^p$ can also be arbitrary so long as the then-part set functions $b_j:R^p \to [0,1]$ are integrable.

The proof of the FAT theorem is constructive and uses a fuzzy cover of rule patches $A_j \times B_j \subset R^n \times R^p$ with a product Cartesian set function $R_{A_j \to B_j}(x,y) = a_j(x)b_j(y)$ for the jth rule of verbal form "If X is A_j then Y is B_j" or $A_j \to B_j$. (The FAT theorem also holds for many other types of combination operators as well and subsumes many approximation results as special cases.) This constructive graph cover in theory allows clustering techniques and other learning schemes to learn the rules given enough time and enough accurate input-output samples from the approximand function f. For details see Kosko, B., "Fuzzy Systems as Universal Approximators," *IEEE Transactions on Computers* 43, no. 11 (November 1996): 1329–33. An earlier version appears in the *Proceedings of the First IEEE International Conference on Fuzzy Systems (FUZZ-92)*, March 1992, 1153–62; Dickerson, J. A. and Kosko, B., "Fuzzy Function Approximation with Ellipsoidal Rules," *IEEE Transactions on Systems, Man, and Cybernetics* 26, no. 4 (August 1996): 542–60.

4. Ebrahim Mamdani published many papers on fuzzy systems in the 1970s and these launched the modern field of fuzzy engineering: Mamdani, E. H., and Assilian, S., "An Experiment in Linguistic Synthesis with a Fuzzy Logic Controller," *International Journal of Man-Machine Studies* 7 (1977): 1–13; Mamdani, E. H., "Application of Fuzzy Logic to Approximate Reasoning Using Linguistic Synthesis," *IEEE Transactions on Computers* 26, no. 12 (December 1977): 1182–91. Mamdani worked out his own math structure for a scalar fuzzy system

$F:R^n \rightarrow R$ but based much of his language and terminology on the philosophical ideas in an earlier paper of Lotfi Zadeh: Zadeh, L. A., "Outline of a New Approach to the Analysis of Complex Systems and Decision Processes," *IEEE Transactions on Systems, Man, and Cybernetics* 3 (1973): 28–44. He reviewed much of this work in fuzzy control in Mamdani, E., "Twenty Years of Fuzzy Control: Experience Gained and Lessons Learned," *Proceedings of the 2nd IEEE International Conference on Fuzzy Systems (FUZZ-93)*, March 1993, 339–44. Didier Dubois and Ronald Yager reprint many of these early papers in the volume *Readings in Fuzzy Sets for Intelligent Systems*, ed. Dubois, D., Prade, H., and Yager, R. R., Morgan Kaufmann, 1993.

Today the Mamdani model has evolved to the simplest fuzzy system that we call a COG or center of gravity system. Michio Sugeno and others extended this model to allow the then-part sets to have a simple linear structure in Sugeno, M., "An Introductory Survey of Fuzzy Control," *Information Sciences* 36 (1985): 59–83; Terano, T., Asai, K., and Sugeno, M., *Fuzzy Systems Theory and Its Applications*, Academic Press, 1992. This additive model sometimes goes by the name of a TS or TSK model in homage to the paper of Takagi, T., and Sugeno, M., "Fuzzy Identification of Systems and its Applications to Modeling and Control," *IEEE Transactions on Systems, Man, and Cybernetics* 15 (1985): 116–32. These fuzzy systems are all special cases of the general additive fuzzy system model as discussed in Kosko, B., *Fuzzy Engineering*, Prentice Hall, 1996.

Many other fuzzy texts discuss these fuzzy systems from a variety of perspectives in control and the decision sciences and reflect the rapid growth of the field: Zimmerman, H. J., *Fuzzy Set Theory and its Application*, Kluwer, 1985; Kandel, A., *Fuzzy Mathematical Techniques with Application*, Addison-Wesley, 1986; Klir, G. J., and Folger, T. A., *Fuzzy Sets, Uncertainty, and Information*, Prentice Hall, 1988; Kosko, B., *Neural Networks and Fuzzy Systems: A Dynamical Systems Approach to Machine Intelligence*, Prentice Hall, 1991; Yager, R. R., and Filev, D. P., *Essentials of Fuzzy Modeling and Control*, Wiley, 1994; Klir, G. J., and Yuan, B., *Fuzzy Sets and Fuzzy Logic: Theory and Applications*, Prentice Hall, 1995; Pedrycz, W., *Fuzzy Sets Engineering*, CRC Press, 1995; Ross, T. J., *Fuzzy Logic with Engineering Applications*, McGraw-Hill, 1995; Yen, J., and Langari, R., *Fuzzy Logic: Intelligence, Control, and Information*, Prentice Hall, 1999.

5. The classic glucose-hormone math model has a form much like the Lanchester combat models discussed in chapter 7:

$$\frac{dg}{dt} = f_1(g,h) + r(t)$$

$$\frac{dh}{dt} = f_2(g,h).$$

Here $g(t)$ is the concentration of glucose in the bloodstream and $h(t)$ is the net hormonal concentration at time t. The exogenous variable $r(t)$ measures the external rate at which the blood glucose rises or falls. The functions f_1 and f_2 are the crux of the matter. Most scientists guess at these two functions or linearize them by replacing them with the first linear terms from a Taylor series expansion. For details see Ackerman, E., Gatewood, L., Rosevear, J., and Molnar, G., "Blood Glucose Regulation and Diabetes," in *Concepts and Models of Biomathematics*, ed. F. Heinmets, Marcel Dekker, 1969, pp. 131–56.

6. "The man who proposes a theory makes a choice—an imaginative choice which outstrips the facts. The creative activity of science lies here, in the process of induction. For induction imagines more than there is ground for and creates relations which at bottom can never be verified. Every induction is a speculation and it guesses at a unity which the facts present but do not strictly imply." Bronowski, J., "The Creative Process," *Scientific American*, September 1958.

7. "I shall not require of a scientific system that it shall be capable of being singled out, once and for all, in a positive sense. But I shall require that its logical form shall be such that it can be singled out, by means of empirical tests, in a negative sense: *it must be possible for an empirical scientific system to be refuted by experience.* . . . It is possible by means of purely deductive inferences (with the help of the *modus tollens* of classical logic) to argue from the truth of singular statements to the falsity of universal statements. Such an argument to the falsity of uni-

versal statements is the only strictly deductive kind of inference that proceeds, as it were, in the 'inductive direction'; that is, from singular to universal statements." Popper, K. R., *The Logic of Scientific Discovery,* Harper & Row, 1959, p. 41.

8. A fuzzy system $F:C \subset R^n \to R^p$ needs on the order of k^{n+p-1} rules to approximate a continuous (or bounded measurable) function $f:C \subset R^n \to R^p$ on a compact set C. So the popular scalar fuzzy systems $F:R^n \to R$ need on the order of k^n to approximate a scalar approximand function f. This holds for "blind" approximation where we do not know or use the mathematical form of f. Then it takes on the order of k scalar if-part fuzzy sets to cover each of the n input axes.

9. Most texts on quantum mechanics apply the Schroedinger wave equation to the "hydrogen-*like*" atom: "Instead of treating just the hydrogen atom, we consider a slightly more general problem: the *hydrogenlike* atom. By this we mean a system consisting of one electron and a nucleus of charge Ze. For $Z = 1$, we have the hydrogen atom. For $Z = 2$, the He^+ (helium) ion. For $Z = 3$, the Li^{++} (lithium) ion, etc. The hydrogenlike atom is the single most important system in quantum chemistry. An exact solution of the Schroedinger equation for atoms with more than one electron cannot be obtained due to the interelectronic repulsions." Levine, I. N., *Quantum Chemistry,* 2d edition, Allyn and Bacon, 1974, p. 98. The Schroedinger equation itself has the deceptively simple form of a first-order differential equation:

$$i\hbar \frac{\partial \psi}{\partial t} = H\psi$$

where $i = \sqrt{-1}$, h is Planck's constant, $\hbar = h/2\pi$, ψ is the quantum matter wave function, and H is the Hamiltonian operator. The Hamiltonian includes both the system's local potential energy and the Laplacian operation $\nabla^2\psi$ or the sum of second partial spatial derivatives. The latter term implies that the Schroedinger equation has the form of a reaction-diffusion equation: How the matter wave function changes in time depends on how it concentrates in space.

10. Brememermann's limit 10^{93} follows from the quantum uncertainty between time and energy $\Delta E\, \Delta t \geq \hbar$, Einstein's energy approximation $E = mc^2$, and Claude Shannon's channel capacity theorem. Brememermann, H. J., "Optimization Through Evolution and Recombination," in *Self-Organizing Systems,* ed. M. C. Yovits, Spartan Books, 1962, pp. 93–106.

11. "Chess aside, what does the chess match [between Deep Blue and Garry Kasparov] mean for the future of computers for complex, knowledge-rich tasks and for the role of AI methods in performing these tasks? First, speed alone cannot solve complex problems and must be supplemented with knowledge. Moreover, if there is enough knowledge, the ability to recognize cues and thereby access knowledge associated with particular kinds of situations will gradually replace speed and brute-force search as the main tool for building high-performance systems." Simon, H. A., and Munakata, T., "The Implications of Kasparov vs. Deep Blue: AI Lessons," *Communications of the ACM* 40, no. 8 (August 1997): 23–25.

12. Michie, D., "Problems of Computer-aided Concept Formation," in *Applications of Expert Systems,* vol. 2, ed. J. R. Quinlan, Addison-Wesley, 1987.

13. Leech, W. J., "A Rule-based Process Control Method with Feedback," *Advances in Instrumentation* 41 (1987): 169–75.

14. Fayyad, U. M., Smyth, P., Weir, N., and Djorgovski, S., "Automated Analysis and Exploration of Image Databases: Results, Progress, and Challenges," *Journal of Intelligent Information Systems* 4 (1995): 1–19.

15. AI pioneers Pat Langley and Herbert Simon review some of the most successful application of AI search trees in Langley, P., and Simon, H. A., "Applications of Machine Learning and Rule Induction," *Communications of the ACM* 38, no. 11 (November 1995): 55–64.

16. Most modern research in Bayesian networks stems from the work of UCLA computer scientist Judea Pearl: Pearl, J., "On the Evidential Reasoning in a Hierarchy of Hypotheses," *Artificial Intelligence* 28 (1986): 9–15; Pearl, J., *Probabilistic Reasoning in Intelligent Systems: Networks of Plausible Inference,* Morgan Kaufmann, 1988; Shafer, G. and Pearl, J., editors, *Readings in Uncertain Reasoning,* Morgan Kaufmann, 1990. See also Heckerman, D., Mamdani, A., and Wellman, M., "Real-World Applications of Bayesian Networks," *Communications of the ACM* 38, no. 3 (1995): 24–26.

Fuzzy petri nets also propagate uncertainty in networks: Pedrycz, W., and Gomide, F., "A Generalized Fuzzy Petri Net Model," *IEEE Transactions on Fuzzy Systems* 2, no. 4 (November 1994): 295–301; Scarpelli, H., Gomide, F., and Yager, R. R., "A Reasoning Algorithm for High-Level Fuzzy Petri Nets," *IEEE Transactions on Fuzzy Systems* 4, no. 3 (August 1996): 282–94.

17. For a survey of Japanese fuzzy applications see Nakamura, K., "Applications of Fuzzy Logical Thinking in Japan: Current and Future," *Proceedings of the IEEE International Conference on Fuzzy Systems (FUZZ-95),* March 1995, 1077–82; Hirota, K., and Sugeno, M., editors, *Industrial Applications of Fuzzy Technology in the World,* World Scientific, 1995. For a European perspective on fuzzy applications see von Altrock, C., *Fuzzy Logic and NeuroFuzzy Applications Explained,* Prentice Hall, 1995; Baldwin, J., editor, *Fuzzy Logic,* Wiley, 1996. See also Patyra, M. J., and Mylnek, D. M., *Fuzzy Logic: Implementation and Applications,* Wiley, 1996. Two key consumer applications are fuzzy systems that stabilize the jitter in a camcorder and that brake a car: Egusa, Y., Akahori, H., Morimura, A., and Wakami, N., "An Application of Fuzzy Set Theory for an Electronic Video Camera Image Stabilizer," *IEEE Transactions on Fuzzy Systems* 3, no. 3 (August 1996): 351–56; Mauer, G. F., "A Fuzzy Logic Controller for an ABS Braking System," *IEEE Transactions on Fuzzy Systems* 3, no. 4 (November 1995): 381–88. Fuzzy systems can also control communication networks: Beauchamp, J. N., and Kandel, A., "A Linguistic Approach for the Control of Information Flow in a Battlefield Environment," *IEEE Transactions on Fuzzy Systems* 6, no. 4 (November 1998): 588–95.

18. Engineers use a mix of smart techniques in the Brazilian oil fields that produce more than 75,000 cubic meters of oil each day: "Fuzzy logic and neural networks are being used in one of these developments [of expert decision aids]. Because the language of fuzzy logic control is closer to that of engineers and technicians than that of programmable logic control, fuzzy logic control is beginning to replace programmable logic in Brazil's oil industry and elsewhere." Da Rocha, A. F., Morooka, C. K., and Alegre, L., "Smart Oil Recovery," *IEEE Spectrum* 33, no. 7 (July 1996): 48–51. A fuzzy system also assists control of the Carajas train system in Brazil: Vieira, P. and Gomide, F., "Computer-Aided Train Dispatch," *IEEE Spectrum* 33, no. 7 (July 1996): 50–51.

19. An *additive* fuzzy system $F{:}R^n \rightarrow R^p$ stores m rules of the patch form $A_j \times B_j \subset R^n \times R^p$ or of the word form "If $X = A_j$ then $Y = B_j$" and adds the "fired" then-parts $B'_j(x)$ to give the output set $B(x)$:

$$B(x) = \sum_{j=1}^{m} w_j\, B'_j(x) = \sum_{j=1}^{m} w_j\, a_j(x)\, B_j(x)$$

for scalar rule weights $w_j > 0$. The factored form $B'_j(x) = a_j(x)\, B_j$ makes the additive system a *standard* additive model or a SAM system. The fuzzy system F computes its output $F(x)$ by taking the centroid of the output set $B(x)$: $F(x) = \text{Centroid}(B(x))$. The SAM Theorem then gives the centroid as a simple ratio:

$$F(x) = \frac{\displaystyle\sum_{j=1}^{m} w_j\, a_j(x)\, V_j c_j}{\displaystyle\sum_{j=1}^{m} w\, a_j(x)\, V_j}$$

$$= \sum_{j=1}^{m} p_j(x) c_j.$$

The convex coefficients or discrete probability weights $p_1(x), \ldots, p_m(x)$ depend on the input x through the ratios

$$p_j(x) = \frac{w_j\, a_j(x)\, V_j}{\displaystyle\sum_{k=1}^{m} w_k\, a_k(x)\, V_k}.$$

V_j is the finite positive volume (or area if $p = 1$ in the range space R^p) and c_j is the centroid of then-part set B_j:

$$V_j = \int_{R^p} b_j(y_1, \ldots, y_p) \, dy_1 \ldots dy_p > 0$$

$$c_j = \frac{\displaystyle\int_{R^p} y \, b_j(y_1, \ldots, y_p) \, dy_1 \ldots dy_p}{\displaystyle\int_{R^p} b_j(y_1, \ldots, y_p) \, dy_1 \ldots dy_p}.$$

The popular scalar case of $p = 1$ reduces the volume and centroid terms to

$$V_j = \int_{-\infty}^{\infty} b_j(y) \, dy$$

$$c_j = \frac{\displaystyle\int_{-\infty}^{\infty} y \, b_j(y) \, dy}{\displaystyle\int_{-\infty}^{\infty} b_j(y) \, dy}.$$

The general additive model first appears in Kosko, B., *Neural Networks and Fuzzy Systems: A Dynamical Systems Approach to Machine Intelligence,* Prentice Hall, 1991. For a detailed analysis of SAM systems and their applications to many fields in information science see Kosko, B., *Fuzzy Engineering,* Prentice Hall, 1996.

Many fuzzy applications ignore the rule weights and the then-part set volumes. This means they assume the weights and volumes have the same value: $w_1 = \cdots = w_m > 0$ and $V_1 = \cdots = V_m > 0$. This reduces the SAM ratio above to the simpler and less powerful *center of gravity* or COG model:

$$F(x) = \frac{\displaystyle\sum_{j=1}^{m} a_j(x) c_j}{\displaystyle\sum_{j=1}^{m} a_j(x)}.$$

The first fuzzy system models were *ad hoc* COG models. A few researchers combined the fired rule then-parts $B'_j(x)$ with pairwise maximum rather than with the pairwise sum but this does not lead to a tractable input-output model. The COG model allows one to ignore the fuzziness of the then-part set B_j and replace it with a unit rectangle centered at the centroid value c_j or with a delta-pulse "spike" at c_j. This gives a fuzzy system F with the same output values $F(x)$ as in the COG ratio. But the variance or uncertainty of the output $F(x)$ changes when one changes the shape of B_j.

COG models further reduce to neural *radial basis function* networks of neural network theory if the m if-part joint set functions each factor into n scaled Gaussian probability densities:

$$a_i^j(x_i) = s_i^j \exp\left[-\frac{1}{2}\left(\frac{x_i - \bar{x}_i^j}{\sigma_i^j}\right)^2\right].$$

for scaling constant $0 < s_i^j \leq 1$. Then the joint set function $a_j : R^n \to [0,1]$ takes the factored form

$$a_j(x) = \prod_{i=1}^{n} a_i^j(x_i) = \prod_{i=1}^{n} s_i^j \exp\left[-\frac{1}{2}\sum_{i=1}^{n}\left(\frac{x_i - \bar{x}_i^j}{\sigma_i^j}\right)^2\right].$$

Some radial basis nets use the rule weights w_j to give a variance-weighted SAM structure: $w_j = 1/\sigma_{B_j}^2$, where $\sigma_{B_j}^2$ is the variance of the jth then-part set B_j. These simple Gaussian COG/SAM systems correspond to the radial-basis nets of Specht, D. F., "A General Regression Neural Network," *IEEE Transactions on Neural Networks* 4, no. 4 (1991): 549–57; Moody, J., and Darken, C., "Fast Learning in Networks of Locally Tuned Processing Units," *Neural Computation* 1 (1989): 281–94.

Detailed simulations have found that the sinc function of signal processing often gives an efficient if-part scalar set function for fuzzy function approximation:

$$a_j^i(x_i) = \frac{\sin\left(\dfrac{x_i - m_j^i}{d_j^i}\right)}{\dfrac{x_i - m_j^i}{d_j^i}}$$

for centering value m_j^i and width value $d_j^i > 0$. Many of these simulation comparisons appear in Mitaim, S., and Kosko, B., "What Is the Best Shape for a Fuzzy Set in Function Approximation?" *Proceedings of the IEEE International Conference on Fuzzy Systems (FUZZ-96)*, vol. 2, September 1996, 1237–43.

20. A fuzzy rule patch is a just a multivalued subset of a product state space: $A_j \times B_j \subset R^n \times R^p$. Its structure depends on the structure of its component fuzzy sets and on the pointwise operator (usually product) that combines the sets. A rule patch has a still simpler matrix form in the discrete case as discussed in Kosko, B., *Neural Networks and Fuzzy Systems*. Lotfi Zadeh views rule patches as fuzzy graphs in his linguistic view of functions and relations and "soft computing": Zadeh, L. A., "The Concept of a Linguistic Variable and Its Application to Approximate Reasoning (Part 3)," *Information Sciences* 9 (1975): 43–80; Zadeh, L. A., "Fuzzy Logic, Neural Networks, and Soft Computing," *Communications of the ACM* 37, no. 3 (1994): 77–84.

21. Kim, H. M., Dickerson, J. A., Kosko, B., "Fuzzy Throttle and Brake Control for Platoons of Smart Cars," *Fuzzy Sets and Systems* 84, no. 3 (1997); Dickerson, J. A., Kim, H. M., and Kosko, B., "Fuzzy Control for Platoons of Smart Cars," *Proceedings of the IEEE International Conference on Fuzzy Systems (FUZZ-94)*, July 1994, 1632–37.

22. A feedback fuzzy cognitive map can use the size and structure of its fuzzy sets and rules to define the virtual reality's concept size or granularity: Dickerson, J. A., and Kosko, B., "Virtual Worlds as Fuzzy Cognitive Maps," *Presence* 3, no. 2 (Spring 1994): 173–89.

23. The mean-value theorem of differential calculus leads to a simple proof that lone optimal fuzzy rules "patch the bumps" or cover the extrema of an approximand function f for any L^p metric. Kosko, B., "Optimal Fuzzy Rules Cover Extrema," *International Journal of Intelligent Systems* 10, no. 2 (February 1995): 249–55; an earlier version appears in the *Proceedings of the 1994 World Congress on Neural Networks (INNS WCNN-94)*, vol. 1, June 1994, 697–98.

24. Fred Watkins has shown that a scalar SAM fuzzy system $F:R \rightarrow R$ can not just approximate but exactly represent a bounded nonconstant function $f:R \rightarrow R$ with just two rules. The catch is that the rules just scale and translate f and so require complete knowledge of the approximand f. Watkins has also shown that in higher dimensions there are many functions that a fuzzy system can never represent even though it can uniformly approximate them. Watkins, F. A., "The Representation Problem for Additive Fuzzy Systems," *Proceedings of the IEEE FUZZ-95*, vol. 1, March 1995, 117–22.

25. Mitaim, S., and Kosko, B., "What Is the Best Shape for a Fuzzy Set in Function Approximation?" *Proceedings of the IEEE International Conference on Fuzzy Systems (FUZZ-96)*, vol. 2, September 1996, 1237–43.

26. Control theorist Anthony Michel gives a formal overview of classical and modern stability theory in Michel, A. N., "Stability: The Common Thread in the Evolution of Feedback Control," *IEEE Control Systems*, June 1996, 50–60.

27. Fuzzy cognitive maps use fuzzy rules to connect fuzzy-set concept nodes much as neural networks use synapses to connect to neurons. But the fuzzy nodes and rule edges change with distinct nonlinear signal and learning laws. Kosko, B., "Hidden Patterns in Combined and Adaptive Knowledge Networks," *International Journal of Approximate Reasoning* 2, no. 4

(1988): 377–93. Fuzzy cognitive maps also apply to social and medical modeling as well as to problems of engineering: Taber, W. R., "Knowledge Processing with Fuzzy Cognitive Maps," *Expert Systems with Applications* 2., no. 1 (February 1991): 82–87; Styblinski, M. A., and Meyer, B. D., "Signal Flow Graphs vs. Fuzzy Cognitive Maps in Application to Qualitative Circuit Analysis," *International Journal of Man-Machine Studies* 35 (1991): 175–86; Taber, W. R., "Matters of Degree," *Helix* 4, no. 3 (1995): 50–55; Pelaez, C. E., and Bowles, J. B., "Using Fuzzy Cognitive Maps as a System Model for Failure Modes and Effects Analysis," *Information Sciences* 88 (1996): 177–99; Schneider, M., Shnaider, E., Kandel, A., and Chew, G., "Automatic Construction of FCMs," *Fuzzy Sets and Systems* 93 (1998): 161–72.

Fuzzy cognitive maps are similar to other tree-based and graph-based knowledge schemes: Lendaris, G. G., "On the Human Aspects in Structural Modeling," *Technological Forecasting and Social Change* 14 (1979): 329–51. *Societal Systems,* Wiley, 1976.

28. Most feedback fuzzy systems are either discrete or continuous generalized standard additive models (SAMs):

$$x(k + 1) = \sum_{j=1}^{m} p_j(x(k))\, B_j x(k) \quad \text{or} \quad \dot{x} = \sum_{j=1}^{m} p_j(x)\, B_j x$$

for the usual SAM convex coefficients in note 19 above. Now the then-part "set" B_j is not a fuzzy set but an n-by-n matrix. Sugeno and Tanaka showed how this matrix can house the coefficients of polynomial then-part sets and how its eigenvalue structure affects the global asymptotic stability of the entire discrete feedback SAM system. Tanaka, K., and Sugeno, M., "Stability Analysis and Design of Fuzzy Control Systems," *Fuzzy Sets and Systems* 45, no. 2 (24 January 1992): 135–56.

The continuous feedback SAM system is globally asymptotically stable if each real then-part matrix B_j is negative definite (and thus has negative eigenvalues). The discrete feedback SAM system is globally asymptotically stable if each then-part matrix B_j is both diagonal and stable (has eigenvalues in the unit circle of the complex plane). For details see Kosko, B., "Global Stability in Feedback Additive Fuzzy Systems," *Proceedings of the IEEE FUZZ-96,* vol. 3, September 1996, 1924–30; Kosko, B., "Global Stability of Generalized Additive Fuzzy Systems," *IEEE Transactions on Systems, Man, and Cybernetics* 28, no. 3 (August 1998): 441–52.

Kazuo Tanaka used an unweighted discrete feedback SAM to back up a truck and five trailers in Tanaka, K., and Yoshioka, K., "Design of Fuzzy Controller for Backer-Upper of a Five-Trailers and Truck," *Proceedings of the IEEE FUZZ-95,* March 1995, 1543–48. An adaptive fuzzy system can also approximate and control the noise parameters of a feedback dynamical system: Mitaim, S., and Kosko, B., "Adaptive Stochastic Resonance," *Proceedings of the IEEE: Special Issue on Intelligent Signal Processing* 86, no. 11 (November 1998): 2152–83.

10. OPTIMAL BRAIN DAMAGE

1. Some commercial systems use neural networks in the process of screening pap smears: Mango, L. J, Tjon, R., and Herriman, J. M., "Computer-Assisted Pap Smear Screening Using Neural Networks," *Proceedings of the INNS World Congress on Neural Networks (WCNN-94),* vol. 1, June 1994, 84–89. The July 1997 issue of the *IEEE Transaction on Neural Networks* is "Special Issue on Everyday Applications of Neural Networks." Topics range from fraud detection in credit-card operations to cork classification.

2. Many texts tell the tale of both feedforward and feedback neural networks: Grossberg, S., *Studies of Mind and Brain,* Reidel, 1982; Rumelhart, D. E., and McClelland, J. L., editors, *Parallel Distributed Processing: Explorations in the Microstructure of Cognition,* vol. 1, MIT Press, 1986; Kohonen, T., *Self-Organization and Associative Memory,* 3d edition, Springer-Verlag, 1988; Minsky, M. L., and Papert, S. A., *Perceptrons: An Introduction to Computational Geometry,* expanded edition, MIT Press, 1988; Anderson, J., and Rosenfeld, E., editors, *Neurocomputing: Foundations of Research,* MIT Press, 1988; Anderson, J. A., Pellionisz, A., and Rosenfeld, E., editors, *Neurocomputing 2: Directions for Research;* Grossberg, S., *Neural Networks and Natural Intelligence,* MIT Press, 1988; Mead, C. A., *Analog VLSI and Neural Systems,* Addison-Wesley, 1989; Wasserman, P. D., *Neural Computing: Theory and Practice,* Van Nostrand Reinhold, 1989; Hecht-Nielsen, R., *Neurocomputing,* Addison-Wesley, 1990; Kosko, B., *Neural Networks*

and Fuzzy Systems: A Dynamical Systems Approach to Machine Intelligence, Prentice Hall, 1991; Kosko, B., *Neural Networks for Signal Processing,* Prentice Hall, 1991; White, H., *Artificial Neural Networks: Approximation and Learning Theory,* Blackwell, 1992; Chester, M., *Neural Networks: A Tutorial,* Prentice Hall, 1993; Anderson, J. A., *Introduction to Practical Neural Modeling,* MIT Press, 1994; Haykin, S., *Neural Networks: A Comprehensive Foundation,* Macmillan, 1994; Bose, N. K., and Liang, P., *Neural Network Fundamentals with Graphs, Algorithms, and Applications,* McGraw-Hill, 1996. For personal views of 17 neural-net researchers who discuss their work and the first 50 years of neural-net research see Anderson, J. A., and Rosenfeld, E., editors, *Talking Nets: An Oral History of Neural Networks,* MIT Press, 1998.

3. "Worldwide neural technology sales are projected to reach $1 billion by 1997. Nearly 70% of Britain's top 100 companies—from British Gas to Boots the chemist—are using neural technologies in some way or planning to do so within two years, according to a study by the Department of Trade and Industry." *The Economist,* 15 April 1996, 76.

4. Identity or replicator networks compress patterns or images in their internal "hidden" layers and also act as information-theoretic approximators: "Replicator neural networks self-organize by using their inputs as desired outputs. They internally form a compressed representation for the input data. A theorem shows that a class of replicator networks can, through minimization of mean-squared reconstruction error (for instance, by training on raw data), carry out optimal data compression for arbitrary data vector sources." Hecht-Nielsen, R., "Replicator Neural Networks for Universal Optimal Source Coding," *Science* 269 (29 September 1995): 1860–63. The first experiment in image compression with such networks appeared in Cottrell, G. W., Munro, P., and Zipser, D., "Image Compression by Back Propagation: An Example of Extensional Programming," in *Advances in Cognitive Science,* vol. 3, ed. N. E. Sharkey, Ablex, 1987.

5. The adaptive resonance theory (ART) model lets its synapses learn only if the neural fields they connect resonate on an incoming pattern: "According to ART, only resonance-producing stimuli become conscious percepts." Grossberg, S., "The Attentive Brain," *American Scientist* 83 (September 1995): 438–49; Grossberg, S., "How Does a Brain Build a Cognitive Code?" *Psychological Review* 87 (1980): 1–51. The related family of adaptive bidirectional associative memories (ABAMs) use a Lyapunov function to prove that both the changing synapses and changing neurons converge exponentially quickly for many types of unsupervised learning (that includes Hebbian, competitive, and differential Hebbian learning). Kosko, B., "Unsupervised Learning in Noise," *IEEE Transactions on Neural Networks* 1, no. 1 (March 1990): 44–57; Kosko, B., "Adaptive Bidirectional Associative Memories," *Applied Optics* 26, no. 23 (December 1987): 4947–60. Adaptive feedback supervised neural networks involve more computation and often learn spatio-temporal patterns: Werbos, P. J. "Generalization of Backpropagation with Application to a Recurrent Gas Market Model," *Neural Networks* 1 (1988): 338–56; Williams, R. J., and Zipser, D., "A Learning Algorithm for Continually Running Fully Recurrent Neural Networks," *Neural Computation* 1, no. 2 (Summer 1989): 270–80.

6. The "optimal brain damage" pruning scheme ranks a synapse by how the synapse affects the Hessian matrix (of second-order changes) of the neural net's average error. This scheme amounts to a diagonal quadratic expansion. It adjusts only the diagonal elements since the size of the Hessian (second-derivative) matrix can be prohibitively large. The scheme is optimal because it assumes supervised learning has driven the synaptic web to a local minimum in the synaptic weight space and hence the error gradient is null: LeCun, Y., Denker, J. S., and Solla, S. A., "Optimal Brain Damage," in *Advances in Neural Information Processing Systems 2,* ed. D. S. Touretzky, Morgan Kaufmann, 1990. This scheme and others use the second partial derivative of the net's average squared error

$$s_k = \frac{1}{2} \frac{\partial^2 E}{\partial w_k^2} \, w_k^2$$

as a saliency or importance measure of the kth synaptic weight w_k.

A related neural-net pruning scheme has the equally provocative name "optimal brain surgeon": Hassibi, B., and Stork, D. G., "Second Order Derivatives for Network Pruning: Optimal Brain Surgeon," in *Advances in Neural Information Processing Systems 5,* ed. S. J. Hanson, J. D. Cowan, and C. L. Giles, Morgan Kaufmann, 1993, pp. 164–71; Reed, R.,

"Pruning Algorithms—A Survey," *IEEE Transactions on Neural Networks* 4., no. 5 (September 1993): 740–47. Young human brains undergo severe synaptic pruning as they age. Some neural models try to capture this by maximizing a neural signal-to-noise ratio: Chechik, G., Meilijson, I., and Ruppin, E., "Synaptic Pruning in Development: A Computational Account," *Neural Computation* 10 (1998): 1759–77.

7. The LMS algorithm estimates the unknown expected squared error with the observed instantaneous squared error: $E[e^2(k)] \approx e^2(k)$. This greatly simplifies the algorithm but increases its random nature since it replaces the deterministic expectation with a random variable. Then the LMS algorithm has the simple weight-vector form

$$w(k + 1) = w(k) + 2\mu e(k)x(k)$$

for a wide-sense stationary random signal vector $x(k) \in R^n$ and instantaneous desired minus-actual error $e(k) = d_k - y(k)$ for some desired signal d_k. This gradient descent algorithm updates the "one neuron" linear combiner $y(k) = \sum_{i=1}^n w_i(k)x_i(k)$. The weight vector w will converge to the optimal weight vector if the constant learning rate μ obeys the bound $0 < \mu < \lambda_{max}$ where λ_{max} is the largest eigenvalue of the signal vector's positive semi-definite correlation matrix. This algorithm lies at the heart of most adaptive signal processing from long-distance telephone calls to radar and antenna theory. Widrow, B., and Stearns, S. D., *Adaptive Signal Processing*, Prentice Hall, 1985. Most supervised learning schemes for nonlinear multilayer neural networks extend the LMS algorithm.

8. The early 1980s saw a new surge of interest in simulated annealing: Kirkpatrick, S., Gelatt, C. D., and Vecchi, M. P., "Optimization by Simulated Annealing," *Science* 220, no. 4598 (13 May 1983): 671–80. Stuart Geman and Chii-Ruey Hwang showed that a Brownian motion diffusion can minimize a cost function $C:[0,1]^n \to R$ if one slowly cools or anneals the stochastic differential diffusion equation

$$dx_t = -\nabla C(x_t) \, dt + \sqrt{2T} dw_t$$

where w is standard Brownian motion: Geman, S., and Hwang, C.-R., "Diffusions for Global Optimization," *SIAM Journal of Control and Optimization* 24, no. 5 (September 1986): 1031–43. The catch is that the annealing schedule $T(t)$ must decrease very slowly according to $T(t) = c/\ln(2 + t)$ for some extremely large positive value c. Then the solution tends to concentrate near the global cost minimum in the form of a Gibbs probability measure. The stochastic convergence is the weakest kind (convergence in distribution as with the central limit theorem): "Almost sure convergence [with probability one] to the set minimizing C is, in general impossible. . . . If $T(t)$ decreases to 0 sufficiently slowly to guarantee escape from local minima, then repeated escapes from global minima are also guaranteed (albeit with increasing rareness)." A like result holds in the discrete case: Geman, S., and Geman, D., "Stochastic Relaxation, Gibbs Distributions, and the Bayesian Restoration of Image," *IEEE Transactions on Pattern Analysis and Machine Intelligence* 6, no. 6 (November 1984): 721–41.

A Cauchy probability density can also pick the next random search site. This often helps the annealing process search distant regions on the cost surface and so helps the search to bounce out of local minima and cool faster: Szu, H., "Fast Simulated Annealing," in *Neural Networks for Computing*, ed. J. Denker, American Institute of Physics, 1986, pp. 420–25. The modern literature on genetic algorithms seems to have overlooked this type of Cauchy search and in general search with impulsive (infinite-variance) noise processes.

9. The field of genetic algorithms has grown steadily since the 1970s: Holland, J. H., *Adaptation in Natural and Artificial Systems*, University of Michigan Press, 1975; Davis, L., editor, *Genetic Algorithms and Simulated Annealing*, Morgan Kaufmann, 1987; Goldberg, D. E., *Genetic Algorithms in Search, Optimization, and Machine Learning*, Addison-Wesley, 1989; Koza, J. R., *Genetic Programming: On the Programming of Computers by Means of Natural Selection*, MIT Press, 1992; Holland, J. H., *Adaptation in Natural and Artificial Systems*, MIT Press, 1992; Kauffman, S. A., *The Origins of Order: Self-Organization and Selection in Evolution*, Oxford University Press, 1993; Koza, J. R., *Genetic Programming II: Automatic Discovery of Reusable Programs*, MIT Press, 1994. Mitchell, M., *An Introduction to Genetic Algorithms*, MIT Press, 1996.

10. Holland, J. H., "Genetic Algorithms," *Scientific American*, July 1992, 66–72.

11. GA pioneer David Goldberg reviews many GA applications in Goldberg, D. E., "Genetic and Evolutionary Algorithms Come of Age," *Communications of the ACM* 37, no. 3 (March 1994): 113–19.

12. Packard, N., "A Genetic Learning Algorithm for the Analysis of Complex Data," *Complex Systems* 4 (1990): 573–86; Caldwell, C., and Johnston, V. S., "Tracking a Criminal Suspect Through Face-Space with a Genetic Algorithm," *Proceedings of the Fourth International Conference on Genetic Algorithms,* 1991, 416–21. GAs have also helped schedule the processing tasks involved in brewing beer and distilling spirits: Lipton, M. J., Rosenof, H. P., and Liston, G., "Genetic Algorithms," *PC AI,* September 1996, 16–27. Engineers have also applied to genetic algorithms to problems of signal processing and the design of adaptive filters: Tang, K. S., et al., "Genetic Algorithms and Their Applications," *IEEE Signal Processing* 13, no. 6 (November 1996): 22–37; Ng, S. C., et al., "The Genetic Search Approach," *IEEE Signal Processing* 13, no. 6 (November 1996): 38–46.

13. The gradient descent of supervised learning can slow to a crawl on flat regions of the error surface or get trapped in local minima: Hush, D. R., and Horne, B. G., "Progress in Supervised Neural Networks," *IEEE Signal Processing Magazine* 10, no. 1 (January 1993): 8–39.

14. "Rolling steel mill control techniques have progressed remarkably since beginning with classical control theory. The advances of optimal control theory with state-space dynamical equations improved control performance dramatically. These developments allow the multivariable synthesis of control systems. But the optimal control theory approach still showed certain limitations. It required a set of linear dynamical equations to describe the objective plants despite their complicated nonlinear properties. Furthermore, current product rigor quality requirements make it almost impossible to describe the whole phenomenon in precise qualitative forms. Various designers have developed fuzzy control techniques to address those problems. But they have been generally unsuccessful until adding neural technology." Funabashi, M., Maeda, A., Morooka, Y., and Mori, K., "Fuzzy and Neural Hybrid Expert Systems: Synergetic AI," *IEEE Expert* 10 (August 1995): 32–40.

15. The fuzzy standard additive model (SAM) of chapter 9 has in the scalar-valued case $F:R^n \to R$ the ratio form

$$F(x) = \frac{\sum_{j=1}^{m} w_j \, a_j(x) \, V_j \, c_j}{\sum_{j=1}^{m} w \, a_j(x) \, V_j} = \sum_{j=1}^{m} p_j(x) c_j$$

for the simplest model. Supervised learning adjusts the SAM parameters to reduce the approximation squared error between F and f:

$$\frac{1}{2}\varepsilon_t^2 = \frac{1}{2}(f(x_t) - F(x_t))^2.$$

The centroids, rule weights, and then-part set volumes have simple learning laws:

$$c_j(t + 1) = c_j(t) + \mu \varepsilon_t p_j(x_t)$$

$$w_j(t + 1) = w_j(t) + \mu \varepsilon_t \frac{p_j(x_t)}{w_j(t)} [c_j - F(x_t)]$$

$$V_j(t + 1) = V_j(t) + \mu \varepsilon_t \frac{p_j(x_t)}{V_j(t)} [c_j - F(x_t)]$$

for learning coefficients μ_t (often linearly decreasing). Complexity arises when supervised learning tunes the if-part set functions. Factorable set functions $a_j(x) = a_j^1(x_1) \cdots a_j^n(x_n)$ require that we tune each scalar set function and include the partial derivative

$$\frac{\partial a_j}{\partial a_j^k} = \frac{a_j(x)}{a_j^k(x_k)}$$

in the general learning law that updates each scalar if-part set function in terms of any one of its own parameters:

$$m_j^k(t + 1) = m_j^k(t) + \mu \varepsilon_t \, \frac{p_j(x_t)}{a_j(x_t)} \, [c_j - F(x_t)] \, \frac{\partial a_j}{\partial a_j^k} \, \frac{\partial a_j^k}{\partial m_j^k}.$$

The sinc set function

$$a_j^k(x_k(t)) = \frac{\sin\left(\dfrac{x_k - m_j^{ki}}{d_j^k}\right)}{\dfrac{x_k - m_j^{ki}}{d_j^k}}$$

is one of the most powerful set-functions for learning but has a trigonometric structure in its partial derivatives:

$$\frac{\partial a_j^k}{\partial m_j^k} = \left(a_j^k(x_k(t)) - \cos\left(\frac{x_k(t) - m_j^k}{d_j^k}\right)\right) \frac{1}{x_k(t) - m_j^k} \quad \text{if } x_k \neq m_j^k$$

$$\frac{\partial a_j^k}{\partial m_j^k} = 0 \quad \text{if } x_k = m_j^k$$

$$\frac{\partial a_j^k}{\partial d_j^k} = \left(a_j^k(x_k(t)) - \cos\left(\frac{x_k(t) - m_j^k}{d_j^k}\right)\right) \frac{1}{d_j^k}.$$

For a comparison of this adaptive SAM set function with many others see Mitaim, S., and Kosko, B., "What is the Best Shape for a Fuzzy Set in Function Approximation?" *Proceedings of the IEEE International Conference on Fuzzy Systems (FUZZ-96)*, vol. 2, September 1996, 1237–43,
   Problems in higher dimensions n require an unfactorable joint set function since the product of factors tends to zero. Transformed metrics offer a simple way to construct (and tune) such joint set functions and preserve the correlation and nonlinear structure among the input variables. They replace a product of scalar set functions with one of the scalar set functions applied to an n-dimensional metric or distance measure (or metric norm on a function space such as L^p). So an unfactorable (Euclidean) metric sinc has the form $a_j[d_j(x)] = (\sin d_j(x))/d_j(x)$ for metric $d_j^2(x) = (x - m_j)^T K_j(x - m_j)$ with tunable mean vector m_j and tunable positive-definite weight matrix K_j. For details and derivations and other joint set functions and learning laws see Kosko, B., *Fuzzy Engineering*, Prentice Hall, 1996; Mitaim, S., and Kosko, B., "Adaptive Joint Fuzzy Sets for Function Approximation," *Proceedings of the IEEE 1997 International Conference on Neural Networks (ICNN-97)* 1, June 1997, 537–42.

16. Herrnstein, R. J., and Murray, C., *The Bell Curve: Intelligence and Class Structure in American Life*, The Free Press, 1994.

17. Fractal pioneer Benoit Mandelbrot was one of the first researchers to apply alpha-stable statistics in engineering and economics: Berger, J. M., and Mandelbrot, B., "A New Model for Error Clustering in Telephone Circuits," *IBM Journal* 7 (July 1963): 224–36; Mandelbrot, B., "The Variation of Certain Speculative Prices," *Journal of Business* 36 (October 1963): 394–419. For more recent applications see Nikias, C. L., and Shao, M., *Signal Processing with Alpha-Stable Distributions and Applications*, Wiley, 1995; Grigoriu, M., *Applied Non-Gaussian Processes*, Prentice Hall, 1995.

18. For learning in the presence of alpha-stable noise see Kim, H. M., and Kosko, B., "Fuzzy Prediction and Filtering in Impulsive Noise," *Fuzzy Sets and Systems* 77, no. 1 (15 January 1996): 15–34; Kosko, B., *Fuzzy Engineering*, Prentice Hall, 1996. Fuzzy rules can watch random number generators to learn "random" rules that securely spread and despread signals in

spread-spectrum wireless communications: Pacini, P. J., and Kosko, B., "Adaptive Fuzzy Frequency Hopper," *IEEE Transactions on Communications* 43, no. 6 (June 1995): 2111–17. Adaptive fuzzy systems can still find the optimal amount of noise to add to a nonlinear dynamical system when impulsive alpha-stable noise drives the dynamical system: Mitaim, S., and Kosko, B., "Adaptive Stochastic Resonance," *Proceedings of the IEEE* 86, no. 11 (November 1998): 2152–83.

19. Many researchers have used genetic algorithms (and sometimes neural networks as well) to tune fuzzy systems: Karr, C. L., "Genetic Algorithms for Fuzzy Controllers," *AI Expert* 6, no. 2 (1991): 26–33; Sanchez, E., "Genetic Algorithms, Neural Networks, and Fuzzy Logic Systems," *Proceedings of the 2nd International Conference on Fuzzy Logic and Neural Networks (Iizuka-92),* vol. 1, July 1992, 17–19; Karr, C. L., and Gentry E. J., "Fuzzy Control of pH Using Genetic Algorithms," *IEEE Transactions on Fuzzy Systems* 1, no. 1 (February 1993): 46–53; Homaifar, A., and McCormick, E., "Simultaneous Design of Membership Functions and Rules Sets for Fuzzy Controllers Using Genetic Algorithms," *IEEE Transactions on Fuzzy Systems* 3, no. 2 (May 1995): 129–39; Perneel, C., Themlin, J.-M., Render, J.-M., and Acheroy, M., "Optimization of Fuzzy Expert Systems Using Genetic Algorithms and Neural Networks," *IEEE Transactions on Fuzzy Systems* 3, no. 3 (August 1995): 300–12; Kim, J., and Zeigler, B. P., "Hierarchical Distributed Genetic Algorithms: A Fuzzy Logic Controller Design Application," *IEEE Expert,* June 1996, 76–84; Jang, J.-S., R., Sun, C.-T., and Mizutani, E., *Neuro-Fuzzy and Soft Computing: A Computational Approach to Learning and Machine Intelligence,* Prentice Hall, 1997; Herrera, F., and Magdalena, L., "Introduction: Fuzzy Genetic Systems," *International Journal of Intelligent Systems* 13, no. 10 (October 1998); 887–90. Wang, C.-H., Hong, T.-P., and Tseng, S.-S., "Integrating Fuzzy Knowledge by Genetic Algorithms." *IEEE Transactions on Evolutionary Computing,* vol. 2, no. 4, November 1998, 134–49.

20. The Sanyo Electric Company produced the first chaotic appliance in 1992. It was a chaotic kerosene fan that uses chaos to slightly perturb the temperature value around a desired set point (based on studies that have shown users prefer such variance to a constant temperature). Katayama, R., Kajitani, Y., Kuwata, K., and Nishida, Y., "Developing Tools and Methods for Applications Incorporating Neuro, Fuzzy, and Chaos Technology," *Computers and Industrial Engineering* 24, no. 4 (October 1993): 579–92. Most research in chaos engineering has come from the Far East: Ditto, W., and Munakata, "Principles and Applications of Chaotic Systems," *Communications of the ACM* 38, no. 11 (November 1995): 96–102; Aihara, K., and Katayama, R., "Chaos Engineering in Japan," *Communications of the ACM* 38, no. 11 (November 1995): 103–7; Osana, Y., Hattori, M., and Hagiwara, M., "Chaotic Bidirectional Associative Memory," *Proceedings of the IEEE International Conference on Neural Networks (ICNN-96),* June 1996, 816–21. Modern control theory has also begun to venture from its linear assumptions and explore the deliberate use of nonlinear chaos: Vincent, T. L., "Control Using Chaos," *IEEE Control Systems Magazine,* December 1997, 65–76.

21. "In a quarter of a century the global financial marketplace has undergone a transformation equivalent to replacing a village shop with a shopping mall. In 1986 the total outstanding value of derivatives markets was just over $1 trillion. In 1994 it was $20 trillion." "Corporate Risk Management," *The Economist,* 10 February 1996, survey, 6–9. *The Economist* suggests that much of this market involves raw speculation as well as prudent hedging: "Using derivatives will not actually increase the value of many companies: opportunities for their effective use are a lot rarer than is widely believed, and even when they arise in principle they can be hard to exploit in practice. . . . Firms may be using derivatives not to hedge their transactions but to speculate on the direction of markets, although few would be willing to admit as much."

Others are more optimistic about derivatives: "Like other technological breakthroughs such as cheap computing power, financial engineering—the use of derivatives to manage risk and create customized financial instruments—has the potential not only to reduce the cost of existing activities but also to make possible the development of new products, services, and markets. . . . Forward-looking managers need to keep abreast of their rivals' successful uses of promising breakthroughs like financial engineering." Tufano, P., "How Financial Engineering Can Advance Corporate Strategy," *Harvard Business Review,* January 1996, 136–46.

Derivatives also reduce the cost of risk hedging a complex portfolio: "Index futures and options increase economic efficiency by providing investors with a previously unavailable means of limiting their exposure to broad market declines—one that substantially reduces the transaction costs of quickly rebalancing a large portfolio." Merton, R. C., "Financial Innovation and Economic Performance," *Journal of Applied Corporate Finance,* Winter 1991, 12–22.

22. The mathematical analysis of options and other derivatives stems from research in the early 1970s: Black, F., and Scholes, M., "The Pricing of Options and Corporate Liabilities," *Journal of Political Economy* 81 (1973): 637–54; Merton, R., "Theory of Rational Options Pricing," *Bell Journal of Economics and Management Science* 4 (1973): 41–183; Cox, J., and Ross, S., "The Valuation of Options for Alternative Stochastic Processes," *Journal of Financial Economics* 3 (1976): 145–66.

The Black-Scholes model assumes no brokerage "friction" fees or taxes and assumes a continuous 24-hour market. The model also assumes a log-normal diffusion structure for the underlying stock of a European call option (which one can exercise only on the date of expiration *T*). Then Ito's Lemma of the stochastic calculus leads to the Black-Scholes stochastic (Brownian) diffusion equation

$$\frac{\partial C}{\partial t} + \frac{1}{2}\frac{\partial^2 C}{\partial S^2}\,\sigma^2 S^2 + rS\,\frac{\partial C}{\partial t} - rC = 0$$

where $C(S, t)$ is the value of the call option at time t of a stock with share price S at time t and that pays no dividend. The competitive rate of return r may refer to a government treasury bond or other "riskless" asset. The volatility σ^2 describes the variance or dispersion of the underlying share price S and there seems no exact way to measure it. There is a chance for arbitrage only if the call-option price C does *not* satisfy this stochastic differential equation since it describes the "no-arbitrage" solution. The call-price solution $C(S, t)$ to the Black-Scholes stochastic partial differential equation depends exponentially on the remaining time $\tau = T - t$. Then simple boundary conditions give the equilibrium solution for the arbitrage-free call price as

$$C(S,t) = S\,N(d_1) - E\,e^{-r\tau}\,N(d_2)$$

for the standard normal or Gaussian cumulative probability distribution N:

$$N(x) = \frac{1}{\sqrt{2\pi}}\int_{-\infty}^{x} e^{-y^2/2}dy$$

Here E is the exercise price of the option at time T. The log-normal arguments depend on the remaining time $\tau = T - t$ to expiration of the option:

$$d_1 = \frac{1n\dfrac{S}{E} + (r + \dfrac{\sigma^2}{2})\tau}{\sigma\sqrt{\tau}}$$

$$d_2 = d_1 - \sigma\sqrt{\tau}.$$

The value of a call or put option on a forward contract has a like exponential structure: Black, F., "The Pricing of Commodity Contracts," *Journal of Financial Economics,* September 1976, 167–79; Malliaris, A. G., "Ito's Calculus in Financial Decision Making," *SIAM Review* 25, no. 4 (October 1983): 481–96. For detailed practical applications of the equilibrium solution see Kolb, R. W., *Understanding Futures Markets,* 3d edition, Simon & Schuster, 1991.

23. "Banks' willingness to lend to LTCM [Long-Term Capital Management] was incredibly generous given a debt-equity ratio of 50:1, though Chase Manhattan, for one, claims that this ratio was nowhere near as scary at the time the loans were made. Maybe so, but at the very least, Wall Street's finest were blinded by the reputations of LTCM's founders, who included John Meriwether, a legendary former head of Salomon Brothers' bond-arbitrage unit, and Robert Merton and Myron Scholes, who last year shared the Nobel Prize for economics for their contributions to the understanding of financial risk." "Long-term Sickness?" *The Economist,* 3 October 1998, 81–83.

24. Kolb, R. W., *Financial Derivatives,* 2d edition, Blackwell Business, 1996, pp. 242–44.
25. There is a vast literature on the use of neural networks in financial markets. Samples include Moody, J., "Principled Architecture Selection for Neural Networks: Application to Corporate Bond Rating Prediction," in *Advances in Neural Information Processing Systems,* vol. 4, ed. J. Moody, S. J. Hanson, and R. P. Lippmann, Morgan Kaufmann, 1991; Malliaris, M., and Salchenberger, L., "A Neural Network Model for Estimating Option Prices," *Journal of Applied Intelligence* 3 (1993): 193–206; Wu, L., and Moody, J., "A Smoothing Regularizer for Feedforward and Recurrent Neural Networks," *Neural Computation* 8, no. 3 (1996): 463–91; Glaria-Bengoechea, A., et al., "Stock Market Indices in Santiago de Chile: Forecasting Using Neural Networks," *Proceedings of the IEEE International Conference on Neural Networks (ICNN-96),* June 1996, 2172–75. The bimonthly *NeuroVe$t Journal* in Virginia publishes articles on neural finance. For a detailed review of neural financial engineering see Refenes, A.-P., N., Burgess, A. N., and Bentz, T., "Neural Networks in Financial Engineering: A Study in Methodology," *IEEE Transactions on Neural Networks* 8, no. 6 (November 1997): 1222–67; Saad, E. W., Prokhorov, D. V., and Wunsch, II, D. C., "Comparative Study of Stock Trend Prediction Using Time Delay, Recurrent, and Probabilistic Neural Networks," *IEEE Transactions on Neural Networks* 9., no. 6 (November 1998): 1456–70.
26. Research in black-box finance techniques often combines neural or fuzzy or genetic techniques: Deboeck, G., editor, *Trading on the Edge: Neural, Genetic, and Fuzzy Systems for Chaotic Financial Markets,* Wiley, 1994; Bauer, R. J., *Genetic Algorithms and Investment Strategies,* Wiley, 1994; Cox, E. D., *Fuzzy Logic for Business and Industry,* Charles River Media, 1995; John, G. H., Miller, P., and Kerber, R., "Stock Selection Using Rule Induction," *IEEE Expert,* October 1996, 52–58; Trippi, R. R., and Lee, J. K., *Artificial Intelligence in Finance & Investing: State-of-the-Art Technologies for Securities Selection and Portfolio Management,* Irwin Professional Publications, 1996.
27. "It seems very clear that under scientific scrutiny chart reading must share a pedestal with alchemy. There has been a remarkable uniformity in the conclusions of all studies done on all forms of technical analysis. Not one has consistently outperformed the placebo of a buy-and-hold strategy. Technical methods cannot be used to make useful investment strategies. This is the fundamental conclusion of the random-walk theory. . . . The technicians do not help produce yachts for customers, but they do help generate the trading that provides yachts for the brokers." Malkiel, B. G., *A Random Walk Down Wall Street,* 1990, pp. 150–51.

    A formal analysis shows that a perfectly competitive market produces a martingale process and hence that stock prices "vibrate randomly" in the language of Nobel laureate economist Paul Samuelson: Samuelson, P. A., "Proof that Properly Discounted Present Values of Assets Vibrate Randomly," *Bell Journal of Economics and Management Science* 4 (1973): 369–74. The analysis starts with the now classic view that the current share price P_t at time t is the discounted future stream of per-share profit or dividends D_{t+n}:

$$P_t = \sum_{n=1}^{\infty} \frac{D_{t+n}}{(1+r)^n}$$

for the competitive risk-free discount rate r. All share information up to time t forms an information set or sigma-algebra I_t (so that a filtration or "learning without forgetting" holds: $I_n \subset I_{n+1}$). The conditional expected share price has the share value

$$p_t = E[P_t \mid I_t] = \sum_{n=1}^{\infty} E\left[\frac{D_{t+n}}{(1+r)^n} \mid I_t \right]$$

in light of the information set I_t. Combining this term with the like future term p_{t+1} shows that if the discount rate r equals the conditional expected return of the stock (if $r = E[D_{t+1} \mid I_t]/p_t$) then the stock value is a random martingale process: $E[p_{t+1} \mid I_t] = p_t$. So if everyone knows everything then the best prediction of tomorrow's share price or of any future share price is just today's share price.
28. Renowned venture capitalist Don Valentine of the Sequoia Fund does not invest in high-tech systems or gadgets for their own sake or in high-tech solutions in search of problems.

He invests in demand rather than in supply: "We're not picking people. We're picking markets. Products are our second concern. People are third. If the market demand is significant, I'll take that any time over anything else. . . . You can have a terrific product, but unless somebody cares—and a lot—I'm not interested. Market demand can minimize risks of not so terrific people." Valentine, D., and Sheff, D., "Don Valentine Interview," *Upside*, May 1990, 63–77.

29. The discounted-earnings formula for the current share price P in note 27 above simplifies if the current per-share profit or dividend D grows each year at a constant rate g in the face of the competitive risk-free discount rate r:

$$P = D \sum_{n=1}^{\infty} \left(\frac{1+g}{1+r} \right)^n$$

The sum simplifies greatly if the growth rate g is less than the discount rate r: $g < r$. Then the term in parentheses is less than unity and the formula for summing a geometric series applies: $P = D (1 + g)/(r - g)$. Economists have argued that a rapid rise in the 30-year Treasury bond and hence in r (and a related fall in g) can explain the 1987 crash of the United States stock market as rational or efficient: "There was a substantial increase in interest rates over the late summer and early fall [of 1987]. Yields on long-term Treasury bonds increased from about 9% to 10½% just before the crash." Malkiel, B. G., "Is the Stock Market Efficient?" *Science* 243, (10 March 1989) 1313–18. Investors often use a two-stage analysis for firms whose profit growth rate $g_1 > r$ exceeds the discount rate for the next N years but whose growth rate falls to $g_2 < r$ from then on. Then putting $a = (1 + g_1)/(1 + r)$ and using the result for summing a finite non-unity geometric series gives the "intrinsic value" of the share price as

$$P = D \frac{a - a^{N+1}}{1 - a} + Da^N \frac{1 + g_2}{r - g_2}.$$

One must in practice divide P by the total number of shares outstanding to get the final estimate of intrinsic value per share. See Fischer, D. E., and Jordan, R. J., *Security Analysis and Portfolio Management*, 6th edition, Prentice Hall, 1995, pp. 264–76.

30. Warren Buffett states that he invests in firms where there is a reasonable chance of estimating their future stream of profits: "We try to stick to businesses we believe we understand. That means they must be relatively simple and stable in character. If a business is complex or subject to constant change, we're not smart enough to predict future cash flows." Buffett, W. E., *1992 Annual Report of Berkshire Hathaway Incorporated*, Spring 1993, p. 11.

31. "Like steel, chips have become just like any other commodity. From the point of view of national economies, it matters little where they are made provided that they are available to be cheaply bought." Editorial, "That Astonishing Microchip," *The Economist*, March 1996, 13, 23.

32. "Two parallel studies using positron emission tomography, one conducted in neurological patients with brain lesions, the other in normal individuals, indicate that the normal process of retrieving words that denote concrete entities depends in part on multiple regions of the left cerebral hemisphere, located outside the classic language areas. Moreover, anatomically separable regions tend to process words for distinct kinds of items." Damsio, H., et al., *Nature* 380, no. 6574 (11 April 1996): 499–505; Caramazza, A., "The Brain's Dictionary," Ibid., 485–86.

33. Neural networks can learn much if not all of the structure of the Black-Scholes equations by simply sampling data that the equations generate. This holds some promise that neural networks can learn the structure of real financial derivatives processes that do not obey log-normal or other artificial assumptions: Hutchison, J. M., Lo, A. W., and Poggio, T., "A Nonparametric Approach to Pricing and Hedging Derivatives Securities via Learning Networks," *Journal of Finance* 49, no. 3 (1994): 851–89; Barucci, E., Landi, L., and Cherubini, U., "Computational Methods in Finance: Option Pricing," *IEEE Computational Science & Engineering* 3, no. 1 (Spring 1996): 66–80; Refenes, A.-P., N., Burgess, A. N., and Bentz, T., "Neural Networks in Financial Engineering: A Study in Methodology," *IEEE Transactions on Neural Networks* 8, no. 6 (November 1997): 1222–67.

11. IT FROM FIT

1. Gregory J. Chaitin is one of the founders of modern complexity theory. He was perhaps the first to define the complexity of data as the shortest program or algorithm that can reproduce the data. See his article "Randomness and Mathematical Proof," *Scientific American,* May 1975, 47–54, or his book *Information, Randomness, and Incompleteness: Papers on Algorithmic Information,* 2d edition, World Scientific, 1990. Chaitin made the epigraph quote during the meeting "The Limits to Scientific Knowledge" at the Santa Fe Institute, 24–26 May 1994: "Chaitin denounced real numbers as 'nonsense.' Their precision is a sham, given the noisiness, the fuzziness, of the world. 'Physicists know that every equation is a lie,' he declared." Horgan, J., *The End of Science: Facing the Limits of Knowledge in the Twilight of the Scientific Age,* Addison Wesley, 1996, p. 231.

2. Physicist Frank Meno exploits the fluid-dynamical structure of Maxwell's equations of electromagnetism to derive the mathematical structure of many physical processes. These processes arise from the local kinetics of a post-Kelvin gyrostatic ether or compressible fluid of spinning gyroscopes: "The aether, as a homogenous anisotropic compressible fluid, can sustain certain permanent dynamic states that are manifested in the form of electromagnetism, matter, and gravity." Meno, F. M., "Photons, Electrons, and Gravitation as Aether Dynamics," *Physics Essays* no. 2 (1995): 245–54. Meno further explores the fluid structure of Maxwell's equations in Meno, F. M., "Electromagnetics as Fluid Mechanics," *Physics Essays* 7, no. 4 (1994): 450–52. Meno first published his gyron theory in Meno, F. M., "A Planck-Length Atomistic Kinetic Model of Physical Reality," *Physics Essays* 4, no. 1 (1991): 94–104.

   This ether theory fails to model the curved space-time of general relativity and string theory. It views the gravitational force as "the net result of the impacts of transversely oriented gyrons that keep replacing the longitudinally oriented gyrons that have escaped from the vortices without colliding." It instead merely multiplies the usual Newtonian gravitational force by an exponential because "it is sensible to expect that the rate at which the gyrons will be flipped over is proportional to the number present": $F = -G \, m_1 m_2 / (r^2 \, e^{r/R})$ for some large scaling factor R that "must be very large so that this [exponential gyron] effect would be observable only at galactic distances." For a more modern view of ether see Wilczek, F., "The Persistence of Ether," *Physics Today,* vol. 52, no. 1 (January 1999): 11.

3. String theorists have to balance one of the most powerful math models in physics with its use of 10 dimensions when we have measured only the 4 of space and time: "A crucial question is what to do with the six extra spatial dimensions. The natural guess is that they curl up to form a six-dimensional space K that is sufficiently small not to have been observed. . . . Since string theory contains gravity it should determine space-time geometry dynamically, and so we must require that the 'background' geometry [of 10 dimensions] $M_4 \times K$ corresponds to a solution of the classical equations of motion. That is the classical statement sometimes called 'spontaneous compactification,' " Schwartz, J. H., "Superstring Unifications," chapter 15 in *300 Years of Gravitation,* ed. S. Hawking and W. Israel, Cambridge University Press, 1987, p. 670.

   Most string theorists now work with some form of the so-called heterotic (hybrid-vigor) string first put forth in Gross, D. J., Harvey, Martinec, E., and Rohm, R., "Heterotic String," *Physical Review Letters* 54, no. 6 (11 February 1985): 502–5. This model combines a bosonic string of 26 dimensions with a fermion string of 10 dimensions. This superspace in effect gives back the world's bosons and fermions as symmetry operations on the same high-dimensional substance. The simplest string models commence with an "action" S that describes how the string wiggles in some complex space-time:

$$S = \iint f(\sigma, \tau) d\sigma \, d\tau$$

for some complicated function f of the spacelike variable σ and the timelike variable τ. For details see Brink, L. and Henneaux, M., *Principles of String Theory,* Plenum Press, 1988.

   British and Finnish research teams found indirect evidence for the formation of superstrings in the early universe just a fraction of a second after the Big Bang exploded and then cooled. They supercooled liquid helium (^3He) and then slightly heated it and let it return to its cooler superfluid state where it has little or no friction or viscosity. There they observed stringlike vortices in the liquid helium: "The formation of cosmic strings during a symmetry-breaking phase transition shortly after the Big Bang is analogous to vortex creation in liquid helium following a rapid transition into the superfluid state." Bauerle, C., et al., "Laboratory Simulation of Cosmic String Formation in the Early Universe Using Superfluid ^3He," *Nature* 382, no. 6589 (25 July 1996): 332–34. The Finnish group found a like result: "These bubbles of normal liquid cool extremely rapidly, and we find that their transition back to the superfluid state is accompanied by the formation of a random network of vortices (the superfluid analogue of cosmic strings)." Ruutu, V. M. H., et al., "Vortex Formation in Neutron-Irradiated Superfluid ^3He as an Analogue of Cosmological Defect Formation," *Nature* 383, no. 6589 (25 July 1996): 334–36. Other research has used this superfluid as a test bed for studying the structure of cosmic strings: Bevan, T. D. C., et al., "Momentum Creation by Vortices in Superfluid ^3He as Model of Primordial Baryogenesis," *Nature* 386 (17 April 1997): 689–92.

For a popular account of recent disputes and advances in string theory see Duff, M. J., "The Theory Formerly Known as Strings," *Scientific American,* February 1998, 64–69: "There seem to be billions of different ways of crunching 10 dimensions down to four. So there are many competing predictions of how the real world works—in other words, no predictions at all." String theory may offer little hope for the old dream of traveling through wormholes in space-time: "The possibility of practically traversable static wormhole solutions in the string-inspired gravity appears to be rather slim." Nandi, N. K., and Alam, S. M. K., "Stringy Wormholes," *General Relativity and Gravitation* 30, no. 9 (September 1998): 1331–40.

4. Leonhard Euler was born in the Swiss town of Basel in 1707 and died blind but productive in 1783 in the high social circle of Catherine the Great in St. Petersburg, Russia. He was the star pupil of Johann Bernoulli. Among his many other deep contributions to mathematics that make up the 73 volumes of his monumental *Opera Omnia* he proved many series representations for *pi*. He published one of his most famous series representations in 1734. It shows how we can recover *pi* or π from the sum of inverse squared whole numbers:

$$\sum_{k=1}^{\infty} \frac{1}{k^2} = 1 + \frac{1}{4} + \frac{1}{9} + \frac{1}{16} + \cdots = \frac{\pi^2}{6}.$$

For a proof see Dunham, W., *Journey Through Genius: The Great Theorems of Mathematics,* Wiley, 1990, pp. 212–17. What we now call *Euler's formula* states that a complex exponential breaks into a cosine and sine function

$$e^{iz} = \cos z + i \sin z$$

for all real or complex numbers z and for the imaginary unit $i = \sqrt{-1}$. The value $z = \pi$ gives $\cos \pi = -1$ and $\sin \pi = 0$. Then Euler's formula reduces to his famous relation among the five mathematical "constants" e, i, π, 0, and 1 and the basic operations of addition and identity:

$$e^{i\pi} + 1 = 0.$$

Indian number theorist Srinivara Ramanujan was born in Madras in 1887 and died there in 1920. He arguably put forth more and deeper power-series expansion of *pi* than any human being before or since. He sent to Godfrey Hardy (1877–1947) a letter that contained without proof what we now call the two *Rogers-Ramanujan identities*:

$$1 + \sum_{k=1}^{\infty} \frac{x^{k^2}}{(1-x)(1-x^2)\cdots(1-x^n)} = \left[\prod_{m=1}^{\infty} (1 - x^{5m-4})(1 - x^{5m-1}) \right]^{-1}$$

$$1 + \sum_{k=1}^{\infty} \frac{x^{k(k+1)}}{(1-x)(1-x^2)\cdots(1-x^n)} = \left[\prod_{m=1}^{\infty} (1 - x^{5m-3})(1 - x^{5m-2}) \right]^{-1}.$$

These equations are typical constructs of Ramanujan's imagination.

5. John Wheeler's former Ph.D. student Kip Thorne tells how the name *black hole* came to be and came to spread: "Such was his [Wheeler's] search for a replacement for 'frozen star' / 'collapsed star.' Finally, in late 1967, he found the perfect name. . . . *Black hole* was Wheeler's new name. Within months it was adopted enthusiastically by relativity physicists, astrophysicists, and the general public, in East as well as in West—with one exception: In France, where the phrase *trou noir* (black hole) has obscene connotations, there was resistance for several years." Thorne, K. S., *Black Holes and Time Warps: Einstein's Outrageous Legacy,* W. W. Norton, 1994, pp. 256–57.

6. "Evidence for a massive black hole at the center of our [Milky Way] Galaxy has been accumulating for the past two decades. . . . Taken together, the [new] observations provide strong evidence for a central dark mass of $2.45 \pm 0.4 \times 10^6$ solar masses located within 0.015 parsecs of the compact radio source Sagittarius A$\star$." Eckart, A., and Genzel, R., "Observations of Stellar Proper Motions Near the Galactic Center," *Nature* 383, no. 6599 (3 October 1996): 415–17.

Mitchell Begelmen and Martin Rees argue that we now have solid evidence that black holes exist: "We now know that black holes are not mere theoretical constructs; they exist in profusion and account for many of the most spectacular astronomical discoveries of recent times." Begelmen, M. C., and Rees, M. J., *Gravity's Fatal Attraction: Black Holes in the Universe,* W. H. Freeman, 1996.

Stephen Hawking once bet against X-ray source Cygnus X-1 as a black hole but has since changed his mind: "There is now fairly good observational evidence to suggest that black holes of about this size [ten times the mass of the sun] exist in double-star systems such as the X-ray source known as Cygnus X-1." Hawking, S. W., *Black Holes and Baby Universes and Other Essays,* Bantam, 1993, p. 104.

Massive spinning black holes may explain cosmic jets and quasars and so may explain the vast energy emissions from active galaxies such as Centauras A (which lies 15 million light years from earth): "For this [process of matter falling into a gravitational field] to be the mechanism of the central engine of Centaurus A there would have to be a source of a very strong gravitational field at the center of the galaxy. Such a field could be provided by what is referred to as a collapsed object: a black hole with a mass of about a billion solar masses. If the collapsed object exists at the center of the galaxy, it is undoubtedly spinning. . . . The spin would cause nearby matter to precess around the [black-hole energy] engine and form the disk of gas known as an accretion disk. . . . The disk might have the shape of a torus with the black hole at its center. Some of the infalling gas would be collimated within the narrow confines of the black hole and driven back out along the rotation axis of the hole by the radiation pressure of emission from the inner radio jet." Burns, J. O, and Price, R. M., "Centaurus A: The Nearest Active Galaxy," *Scientific American,* November 1983. The Hubble Space Telescope has led to new evidence for a black hole in the nearby elliptical galaxy called M32. Van der Marel, R. P., de Zeeuw, P. T., Rix, H.-W., and Quinlan, G. D., "A Massive Black Hole at the Center of the Quiescent Galaxy M32," *Nature* 385 (13 February 1997): 610–12.

New evidence also suggests that both massive and small-star black holes can account for bright quasars and so-called microquasars in the Milky Way Galaxy: "On its way into the hole, the accreting matter (probably in the form of a rotating disk) heats up through viscous dissipation and so converts gravitational energy into radiation. Calculations show that in such accretion discs about 10% of the initial rest-mass energy is radiated away. This is the most efficient radiation source one can think of, nuclear fusion in stars being a distant second." Genzel, R., "How Black Holes Stay Black," *Nature* 391 (1 January 1998): 17–18. "The recent finding in our own Galaxy of microquasars, a class of objects that mimics—on scales millions of times smaller—the properties of quasars, has opened new perspectives for the astrophysics of black holes. These scaled-down versions of quasars are believed to be powered by spinning black holes but with masses up to a few tens of times that of the Sun." Mirabel, I. F., and Rodriguez, L. F., "Microquasars in our Galaxy," *Nature* (16 April 1998): 673–76.

7. Michell presented his dark-star result to the Royal Society of London on 27 November 1783: "On the Means of Discovering the Distance, Magnitude, Etc., of the Fixed Stars, in Consequence of the Dimunition of Their Light, in Case Such a Diminuition Should be Found to Take Place in Any of Them, and Such Other Data Should be Procured from

Observations, as Would be Further Necessary for That Purpose," *Philosophical Transactions of the Royal Society of London* 74 (1783).

Laplace published his essay "Proof of the Theorem that the Attractive Force of a Heavenly Body Could Be So Large that Light Could Not Flow Out of It" in 1799. For details see Thorne, K. S., *Black Holes and Time Warps: Einstein's Outrageous Legacy,* W. W. Norton, 1994, pp. 121–23, 594. Stephen Hawking and George Ellis present a modern translation of Laplace's dark-star essay in the appendix of their classic math text on black holes: Hawking, S. W., and Ellis, G. F. R., *The Large Scale Structure of Space-Time,* Cambridge University Press, 1973, pp. 365–68.

8. Albert Einstein arrived in 1915 at what we now call the Einstein field equation of general relativity:

$$R_{ab} - \frac{1}{2}\, Rg_{ab} = 8\pi T_{ab}$$

in the notation of Wald, R. M., *General Relativity,* University of Chicago Press, 1984, pp. 72–73. This compact equation states Einstein's theory of gravity in terms of a coupled system of nonlinear second-order partial derivatives for the components of the Lorentz metric g_{ab}. The left-hand side describes the curvature of space-time as a Riemann manifold in terms of the Riemann-based Ricci curvature tensor R_{ab} and its scalar curvature R (the trace of the Ricci tensor). The right-hand side describes the distribution of mass and energy in terms of the energy-momentum tensor T_{ab}.

A Schwarzschild star is static and spherically symmetric with radius r and null Ricci tensor: $R_{ab} = 0$. Then the Schwarzschild solution to Einstein's equation describes the metric or arc length ds^2 in terms of time t and in terms of the star's radius r and its mass m:

$$ds^2 = -\left(1 - \frac{2m}{r}\right)dt^2 - \frac{dr^2}{\left(1 - \dfrac{2m}{r}\right)} + r^2(d\theta^2 + \sin^2\theta\; d\phi^2)$$

in spherical coordinates. The first two terms on the right-hand side show that the curvature can "blow up" to infinity or produce a singularity when $r = 0$ or when $r = 2m$. These distances define the center and boundary of a Schwarzschild black hole. Most precise tests of general relativity have been tests of the Schwarzschild solution or some version of it.

Gravitational collapse of a spherical mass produces a Schwarzschild black hole in the space-time continuum. The zero singularity $r = 0$ is the proper singularity inside the black hole and denotes one of the most inscrutable objects in the universe. The singularity at $r = 2m$ is a mere "coordinate" singularity. It defines the Schwarzschild radius or the boundary of the black hole that enshrouds the physical singularity at $r = 0$. The Schwarzschild radius is approximately $3m\,/\,m_{sun}$ kilometers where m_{sun} is the mass of our sun (about 2×10^{33} grams). The Schwarzschild radius lies deep inside most spherical masses. The sun has a Schwarzschild radius of about 1 kilometer. Earth has a Schwarzschild radius of about 1 centimeter.

9. John Wheeler's former Ph.D. student Jacob Bekenstein describes how Wheeler posed to him the question of what happens when we throw matter in a black hole: "As a graduate student of Wheeler's at Princeton I found 'black holes have no hair' distressing for a reason he brought home to me in a 1971 conversation. The principle, he argued, allows a wicked creature—call it Wheeler's demon—to commit the perfect crime against the second law of thermodynamics. It only has to drop a package containing some entropy into a stationary black hole, thus decreasing the entropy in the part of universe visible from the exterior. The associated changes in M [mass], Q [charge], and L [spin or angular momentum] do not uniquely reveal how much entropy is then inside the hole, so an *exterior* observer with no inside information about the package can never be sure that the total entropy in the universe has not decreased. For him the second law is transcended—made irrelevant. It loses its predictive power, so that black holes seem to be outside the province of thermodynamics. This circumstance seemed disastrous, not only because it would deprive us of the use of model-free thermodynamic reasoning in investigating the bizarre black holes, but also because it could

be seen as throwing doubt on their very existence, even in principle." Bekenstein, J. D., "Black Hole Thermodynamics," *Physics Today,* January 1980, 24–31.

A more recent result questions the validity of the information paradox: "Recent progress in understanding black-hole micro-physics carries a simple message: pure quantum states do not form black holes. Rather, such states are likely to correspond to field configurations with 'singularities,' strong curvature regions, which are not clothed by horizons. Thus the information paradox results from an ill-posed question." Myers, R., "Pure States Don't Wear Black," *General Relativity and Gravitation* 29, no. 10 (October 1998): 1217–21.

10. "The area of the boundary of a black hole cannot decrease with time, and if two or more black holes merge to form a single black hole, the area of its boundary will be greater than the areas of the boundaries of the original black holes." Hawking, S. W., and Ellis, G. F. R., *The Large-Scale Structure of Space-Time,* Cambridge University Press, 1973, 318. Stephen Hawking first proved this theorem in 1971. See Hawking, S. W., "Gravitational Radiation from Colliding Black Holes," *Physics Review Letters* 26 (1971): 1344–46. Black holes also obey relations that have the same form as the first and third (and zeroth) laws of thermodynamics.

Hawking extended the area theorem and showed that a black hole's entropy is just one fourth the surface area of its event horizon in his landmark paper "Black Holes and Thermodynamics," *Physical Review D* 13, no. 2 (15 January 1976): 191–97. Here Hawking used quantum mechanics to ensure that the black hole contained only a finite number of internal configurations and used the "no hair" theorems to ensure that these states have the same probability. Then Shannon entropy gives the black hole entropy as just the logarithm of these equi-probable states: "If quantum effects were neglected, the number of different internal configurations would be infinite because one could form the black hole out of an indefinitely large number of small mass particles. . . . Let $\sigma\ dM\ dQ\ d^3\ J$ be the number of internal configurations or quantum states of a black hole [in some small volume]. . . . By the 'no hair' theorems one has no information about the internal state of the hole, and therefore all these configurations are equally probable [so $p_i = 1/\sigma$]. Thus the entropy S of the black hole is

$$S = - \sum_{i=1}^{\sigma} p_i \ln p_i = \ln \sigma."$$

There is some controversy over Hawking's view that throwing matter in a black hole causes an "information paradox." Physicist Leonard Susskind has argued that we can avoid any information loss if we view use string theory and view the world as a swarm of tiny vibrating strings with string pieces defining information bits: "A string is a minute object, $1/10^{20}$ the size of a proton. But as it falls into a black hole, its vibrations slow down, and more of them become visible. Mathematical studies done at Stanford by Amanda Peet, Thorlacius, Arthur Mezhlumian, and me have demonstrated the behavior of a string as its higher modes freeze out. The string spreads and grows, just as if it were being bombarded by particles and radiation in a very hot environment. In a relatively short time the string and all the information that it carries are smeared over the entire [event] horizon.

"This picture applies to all the material that ever fell into the black hole—because according to string theory, everything is ultimately made of strings. Each elementary string spreads and overlaps all the others until a dense tangle covers the horizon. Each minute segment of string, measuring 10^{-33} centimeter across, functions as a bit. Thus strings provide a means for the black hole's surface to hold the immense amount of information that fell during its birth and thereafter." Susskind, L., "Black Holes and the Information Paradox," *Scientific American,* April 1997, 52–57.

11. Some black holes depend on low-energy string states and permit a type of count of black-hole microstates. The count improves for black holes with large electrical charge. These results depend on the symmetry structure of the string theory. Some models fail to specify some key technical parameters such as the dimension of the cohomology of moduli spaces. Strominger, A., and Vafa, C., "Microscopic Origin of the Bekenstein-Hawking Entropy," *Physics Letters B* 379 (27 June 1996): 99–104.

12. Physicist Paul C. W. Davies derives his cosmic bit count from the squared-mass relation of black hole entropy: "One may obtain a natural measure of I the information capacity of the cosmos using the Hawking-Bekenstein formula for black hole entropy. If the entire universe were converted into a black hole, it would conceal a quantity of information I given by

$$I \approx \frac{Gm^2}{\hbar c}$$

where m is the mass of the observable universe (i.e., within the particle horizon). At the current epoch t_0, $I \approx 10^{120}$. At epoch t

$$I(t) \approx 10^{120}\left(\frac{t}{t_0}\right)^2$$

so that at the Planck time, $t_p \approx 10^{-43}s$, $I \approx 1$, as expected." Davies, P. C. W., "Why Is the Physical World So Comprehensible?" in *Complexity, Entropy, and the Physics of Information: Proceedings of the Santa Fe Institute in the Sciences of Complexity*, vol 8, ed. W. H. Zurek, Addison-Wesley, 1990, p. 68. The other terms in the equation are the constants of gravitation and light speed and the normalized Planck's constant.

13. "It is not unreasonable to estimate the entropy of the final crunch [final state of the universe] by using the Bekenstein-Hawking formula as though the whole universe had formed a black hole. This gives an entropy per baryon of 10^{43}, and the absolutely stupendous total, for the entire big crunch would be 10^{123}." Penrose, R., *The Emperor's New Mind: Concerning Computers, Minds, and the Laws of Physics*, Oxford University Press, 1989, p. 343. So Penrose and Davies differ in their estimates of the universe's bit count by a factor of a thousand. This difference would be huge in our daily affairs of finance and macroscopic energy conversion. It is reasonable given the ballpark nature of the assumptions and the fact that it occurs with "huge" numbers that exceed the first "googol" 10^{100}.

14. After Wheeler, J., "Time Today," *Physical Origins of Time Asymmetry*, editors Halliwell, J., Perez-Mercader, J., and Zurek, W., Cambridge University Press, 1994.

15. "I, like other searchers, attempt formulation after formulation of the central issues, and here present a wider overview, taking for a working hypothesis the most effective one that has survived winnowing: *It from bit.* Otherwise put, every *it*—every particle, every field of force, even the space-time continuum itself—derives its function, its meaning, its very existence entirely—even if in some contexts indirectly—from the apparatus-elicited answers to yes-or-not questions, binary choices, *bits.*" Wheeler, J. A., "Information, Physics, Quantum: The Search for Links," in *Complexity, Entropy, and the Physics of Information: Proceedings of the Santa Fe Institute in the Sciences of Complexity,* vol. 8, ed. W. H. Zurek, Addison-Wesley, 1990, p. 5. The it-from-bit quote in the chapter comes from page 3 of the same lead essay in the Sante Fe proceedings.

Wheeler goes on to cite three physical examples of it from bit. These are a black hole's thermodynamic entropy, the space-time field strength, and the magnetic flux: "Field strength or space-time curvature reveals itself through a shift of interference fringes, fringes that stand for nothing but a statistical pattern of yes-or-no registrations. . . . We deal with bits wholesale rather than retail when we run the fiducial current through the magnetometer coil, but the definition of fields founds itself no less decisively on bits."

Proceedings editor W. H. Zurek paints a like picture: "The specter of information is haunting science. Thermodynamics, much of the foundation of statistical mechanics, the quantum theory of measurement, the physics of computation, and many of the issues of the theory of dynamical systems, molecular biology, genetics, and computer science share information as a common theme." Zurek, W. H., "Complexity, Entropy, and the Physics of Information—A Manifesto," in *Complexity, Entropy, and the Physics of Information: Proceedings of the Santa Fe Institute in the Sciences of Complexity,* vol. 8, ed. W. H. Zurek, Addison-Wesley, 1990, p. vii.

Wheeler returned to his thesis of it from bit in a later work on the direction of time in "Time Today," *Physical Origins of Time Assymetry*, ed. J. J. Halliwell, J. Perez-Mercader, and W. Zurek, Cambridge University Press, 1994. See also Wilczek, F., "Getting Its from Bits," *Nature,* vol. 397 (28 January 1999): 303–6.

16. "It has been said that quantum field theory is the most accurate physical theory ever, being accurate to about one part in about 10^{11}. However, I would like to point out that general relativity has, in a certain clear sense, now been tested to be correct to one part in 10^{14} (and this accuracy has apparently been limited merely by the accuracy of clocks on Earth). I am speak-

ing of the Hulse-Taylor binary pulsar PST 1913 + 16, a pair of neutron stars orbiting each other, one of which is a pulsar. General relativity predicts that this orbit will slowly decay (and the period shorten) because energy is lost through the emission of gravitational waves. That has been observed, and the entire description of the motion agrees with general relativity (which I am taking to include Newtonian theory) to the remarkable accuracy, noted above, over an accumulated period of 20 years. The discoverers of this system have now rightly been awarded Nobel Prizes for their work." Penrose, R., "The Nature of Space and Time," *Scientific American,* July 1996, 61–62. The same material appears in Hawking, S., and Penrose, R., *The Nature of Space and Time,* Princeton University Press, 1996.

17. Note 17 in chapter 6 reviews the binary-set framework of probability theory that leads to the extremal outcomes $P(A \cap A^c) = P(\varnothing) = 0$ and $P(A \cup A^c) = P(X) = 1$ for *any* set event A in the sample space X. A fuzzy truth operator t: {*Statements*} $\rightarrow$ [0, 1] breaks just these two Aristotelian laws:

$$t(S \text{ and } {\sim}S) + t(S \text{ or } {\sim}S) = 1$$

if S is a statement in a multivalued logic. This theorem follows if pairwise minimum defines the conjunction or *and* operator and if pairwise maximum defines the disjunction or *or* operator and if $1 - x$ defines the negation or *not* operator. It also holds for continuum-many other *and* and *or* operators that transform as DeMorgan duals. This theorem gives Aristotle's "laws" of noncontradiction and excluded middle if and only if we assume that the statement S has a binary truth value.

A measure of set or event fuzziness $F(A)$ in [0, 1] behaves in a like way. Then $F(A) = 0$ iff A is a binary set (a unit-cube vertex point for discrete sets) and $F(A) = 1$ iff A is the maximally fuzzy point where $A = A^c$ (or A is cube midpoint in the discrete case). So $F(A) > 0$ or fuzziness or vagueness occurs in general and also in just those cases where $P(A \cap A^c) = P(\varnothing) = 0$. The natural cube measure of fuzziness has the ratio form $c(A \cap A^c) / c(A \cup A^c)$ for some counting measure c of a discrete fuzzy set A or an integrable continuous fuzzy set A. This ratio also gives the formal measure of equality between the set A and its complement. For details see chapters 1 and 7 of Kosko, B., *Neural Networks and Fuzzy Systems: A Dynamical Systems Approach to Machine Intelligence,* Prentice Hall, 1991. See also notes 16 and 17 in chapter 1 above.

18. Boltzmann's original H theorem says that $-H(v, t)$ grows in time for a 3-D ensemble of molecules with joint velocity vector v and with

$$H(v,t) = \int_{R^3} f(v,t) \ln f(v,t) dv$$

The density term $f(v, t)$ counts the gas molecules that have velocity v in the tiny gas volume dv. The entropy stays constant only for the Maxwell probability density. The H theorem result follows from the so-called Boltzmann equation of colliding dilute gas molecules that Boltzmann also derived in his 1872 memoir (now in English translation in Brush, S. G., "Kinetic Theory, Volume 2: Irreversible Processes," in *Selected Readings in Physics,* Pergamon Press, 1966). The negative of H has the same form as Claude Shannon's entropy of a continuous probability density function.

Many see the H theorem as the mathematical foundation of the second law of thermodynamics: "Gradually the term H theorem has acquired a much broader meaning, referring essentially to any statement about the approach to equilibrium which asserts that an appropriate quantity decreases (or increases) with time." Kac, M., and Ford, G. W., "H Theorem," *Encyclopedia of Physics,* ed. R. G. Lerner and G. L. Trigg, 2d edition, 1991.

19. Shannon published his MIT master's thesis on electrical circuits and Boolean logic as Shannon, C., E., "A Symbolic Analysis of Relay and Switching Circuits," *Transactions American Institute of Electrical Engineers* 58 (1938); reprinted in Shannon, C. E., *Claude Elwood Shannon: Collected Papers,* ed. N. J. A. Sloan and A. D. Wyner, IEEE Press, 1993. Shannon published the main results of his theory of information in Shannon, C. E., "A Mathematical Theory of Communication," *The Bell System Technical Journal* 27 (July): 379–23; (October 1948): 623–56.

Suppose $P = (p_1, \ldots, p_n)$ is a discrete probability density. So $p_i \geq 0$ and $\sum_{i=1}^{n} p_i = 1$. All such P are points in the fuzzy unit hypercube of n. They define the simplex in such a cube. Then the Shannon entropy $H(P)$ of P is just the expected value of the logarithmic information:

$$H(P) = \sum_{i=1}^{n} p_i \ln \frac{1}{p_i} = - \sum_{i=1}^{n} p_i \ln p_i$$

The minimum value $H(P) = 0$ occurs only when P is a bit vector with one 1 in it and with the other terms equal to 0 and thus it occurs when P describes a certain world.

The maximum value $H(P) = \ln n$ occurs only for the uniform distribution where each $p_i = 1/n$. This holds for black holes because we cannot say that any one of its n internal configurations is more probable than any other. Hence a black hole's entropy is proportional to the logarithm of the total number of states it contains or $\ln n$. The Shannon entropy of a continuous probability density function $p(x)$ replaces the discrete sum with an integral: $H(p) = -\int_{R^n} p(x) \ln p(x) \, dx$. For details and related results see Cover, T. M., and Thomas, J. A., *Elements of Information Theory*, Wiley, 1991.

Even "eternal" black holes obey this logarithmic structure of Shannon entropy as shown in note 10 above. These black holes are "not formed from matter collapse but are permanent features of solutions of the nonlinear Einstein equations. . . . An event horizon is not time symmetric." Matzner, R. A., et al., "Geometry of a Black Hole Collision," *Science* 270 (10 November 1995): 941–47.

20. Let F be a fuzzy set or fit vector in the fuzzy n-cube: $F = (f_1, \ldots, f_n) \in [0, 1]^n$. So F is the fuzzy subset of the space X of n objects: $F \subset X = \{x_1, \ldots, x_n\}$. Object x_i belongs to the fuzzy set F to the degree or fit value $f_i \in [0, 1]$ and does not belong to F to degree $1 - f_i$. So F has the fuzzy set complement $F^c = (1 - f_1, \ldots, 1 - f_n)$. The scalar fuzzy mutual entropy $M(F)$ combines the Kullback mutual entropy between F and F^c:

$$M(F) = H(F \,/\, F^c) - H(F^c \,/\, F) = \sum_{i=1}^{n} f_i \ln \frac{f_i}{1 - f_i} - \sum_{i=1}^{n} (1 - f_i) \ln \frac{1 - f_i}{f_i}$$

For details see Kosko, B., "Addition as Fuzzy Mutual Entropy," *Information Sciences* 73, no. 3 (1 October 1993): 273–84.

Note that here the fit vector F can be any point in the fuzzy cube. So its count $c(F)$ can be any value between 0 and n. It need not equal one. This extends both Shannon entropy and mutual entropy to the entire unit cube and not just the simplex embedded inside it (on which $c(\Gamma) = 1$ must hold). Probability vectors P lie in this simplex. But they face a problem if $n > 2$. Then the cube complement $P^c = (1 - p_1, \ldots, 1 - p_n)$ still lies in the cube but not in the simplex. The simplex is not closed under set complementation for $n > 2$. It is closed only when $n = 2$ and thus when $P = (p, 1 - p)$. The fuzzy cube itself is the maximal sigma algebra of fuzzy sets and is always closed under set complementation. In this sense fuzzy mutual entropy forces the analysis from the simplex to the cube.

Modern logicians have also recast the usual "orthomodular" logic of quantum mechanics as a multivalued Lukasiewicz or fuzzy logic in Pykacz, J., "Fuzzy Quantum Logics and Infinite-Valued Lukasiewicz Logic," *International Journal of Theoretical Physics* 33, no. 7 (1994): 1403–16; and Mesiar, R., "h-Fuzzy Quantum Logics," *International Journal of Theoretical Physics* 33, no. 7 (1994): 1417–25. Mathematician Stanley Gudder presents the standard orthomodular logic of quantum mechanics in terms of probability measures in Gudder, S. P., *Quantum Probability*, Academic Press, 1988.

21. A reaction-diffusion equation follows from Gauss's divergence theorem and a conservation assumption (such as conservation of energy). It has the form

$$\frac{\partial u}{\partial t} = f(u,x,t) - c \, \nabla^2 u$$

for the reaction forcing function f and a smooth potential $u:R^n \rightarrow R$ and some positive constant $c > 0$. The potential $u(x, t)$ gives rise to the gradient vector field ∇u (or the n-vector of partial spatial derivatives of u). The divergence operator

$$\nabla \cdot = \frac{\partial}{\partial x_1} + \cdots + \frac{\partial}{\partial x_n}$$

measures the compressibility of this gradient vector field or "fluid" ∇u and gives back the Laplacian (or the sum of second partial spatial derivatives) of the potential u:

$$\nabla \cdot \nabla u = \nabla^2 u = \frac{\partial^2 u}{\partial x_1^2} + \cdots + \frac{\partial^2 u}{\partial x_n^2}.$$

For a wide range of reaction–diffusion equations in all aspects of biology see Murray, J. D., *Mathematical Biology*, Springer-Verlag, 1989. The R–D equation for fuzzy mutual entropy assumes a null reaction function f.

22. Let $H{:}[0, 1]^n \rightarrow R$ define the extended Shannon entropy potential on the *entire* fuzzy cube $[0, 1]^n$ and not just the probability simplex embedded in it:

$$H(F) = - \sum_{i=1}^n f_i \ln f_i - \sum_{i=1}^n (1 - f_i) \ln (1 - f_i) = \hat{H}(F) + \hat{H}(F^c)$$

Then we ask a fundamental it-from-bit question: What gradient vector field F does the Shannon entropy potential H generate in the sense of $F = \nabla H(F)$? The answer is the field of fuzzy mutual entropy:

$$\mathbf{M}(F) = \left(-f_1 \ln \frac{f_1}{1 - f_1} + (1 - f_1) \ln \frac{1 - f_1}{f_1}, \ldots, -f_n \ln \frac{f_n}{1 - f_n} + (1 - f_n) \ln \frac{1 - f_n}{f_n} \right)$$

This compressible fluid has divergence

$$\nabla \cdot \mathbf{M}(F) = - \sum_{i=1}^n \frac{1}{f_i(1 - f_i)}$$

Then conservation of information leads to a wave or reaction-diffusion equation of the form

$$\frac{\partial H}{\partial t} = -c \, \nabla \cdot \mathbf{M} \quad \text{or} \quad \frac{\partial H}{\partial t} = -c \, \nabla^2 H = \sum_{i=1}^n \frac{1}{f_i(1 - f_i)}. \quad \text{So} \quad \frac{\partial H}{\partial t} > 0$$

holds and thus resembles the entropy increase of the H theorem or the second law of thermodynamics. The entropy always grows but the its rate of growth depends on the system's position in the fuzzy parameter space. The H velocity is infinite if and only if the fit vector F lies on the skin or boundary of the fuzzy cube. For then and only then does some fit value obey $f_k = 0$ or $f_k = 1$. So the entropy wave equation and a finite bound B on the wave velocity ($|\partial H/\partial t| \leq B < \infty$) imply that the system cannot achieve binary certainty. For then $0 < f_i < 1$ holds for all fit values f_i.

The Shannon wave fluctuates most slowly at the cube midpoint where there is least certainty. It fluctuates faster away from the midpoint and closer to the cube surface where the certainty or information is greatest. Again this analysis makes sense only if the uncertainty arguments F range over a unit hypercube. The wave view also assumes that the world does not just trace out a path in the fuzzy cube over any stretch of time. It assumes that all possible worlds are tracing out paths over any stretch of time because the wave equation holds on the entire cube. For details see chapter 12 "Fuzzy Cubes and Fuzzy Mutual Entropy," of Kosko, B., *Fuzzy Engineering*, Prentice Hall, 1996.

23. A fuzzy view replaces a binary point mass m of a particle with a topological cloud or fuzzy subset A of the space in question. The set $A \subset R^n$ has integrable set function $a_j{:}R^n \rightarrow [0,1]$ where fit value $a_j(x) \, dx$ measures the degree to which the mass exists in the infinitesimal space volume dx. Then the mass m is just the fuzzy count $c(A)$ of the cloud: $m = \int_{R^n} a_j(x)dx$. A like definition holds for the symmetric *negative* mass $\tilde{m}$ for the reflected mass cloud $\tilde{a}_j{:}R^n \rightarrow [0,-1]$. So $\tilde{m} = c(\tilde{A}) = -c(A)$. Then the fuzzy mass and negative-mass clouds still obey the conservation of linear momentum and kinetic energy

$$mv + \tilde{m}v = 0 \quad \text{and} \quad \frac{1}{2}mv^2 + \frac{1}{2}\tilde{m}v^2 = 0$$

in accord with the like binary point-mass results in Forward, R. L., "Negative Matter Propulsion," *Journal of Propulsion* 6, no. 1 (January 1990): 28–37. Sums of any number of fuzzy particles can replace those for just two in these conservation results. The two mass clouds define the two halves of a vague topological object. The two halves meet at the null set where $a(x) = \tilde{a}(x) = 0$ defines the vacuum state. The fuzzy cloud A can itself have a probabilistic uncertainty cloud about it that measures the likelihood $\Pr\{u \le a(x) \le v\}$ of existence values in infinitesimal regions dx.

The fuzzy-cloud view also holds if we identify the unit of kinetic energy $mv^2/2$ with the basic bit-like unit of logarithmic information $\ln 1/p = -\ln p$ for some event occurrence probability value p: $mv^2/2 = -\ln p$. This identification of energy with information gives back a Gaussian-like density function for the occurrence probability: $p \sim \exp(-mv^2/2)$. A finite mass m (cloud count $c(A)$) and finite velocity v give $p > 0$. This ensures that the local information $-\ln p$ is finite. A classical vacuum with $m = 0$ or $v = 0$ gives $p = 1$ and hence no local information. But quantum uncertainty forbids this binary extreme because it forbids a zero energy value. So even the vacuum contains information.

Quantum uncertainty should also forbid the extremal binary set functions $a_j:R^n \rightarrow \{0,1\}$ that underlie all quantum models of existence. Whether existence at the quantum level is binary or multivalent is an empirical and not a logical question. Only the fiat of the pen can force the binary extremal value $a_j(x) = 1$ when even the slightest deviation can result in some value in $0 < a_j(x) < 1$. In this sense existence itself may be fuzzy or vague.

24. Many thinkers have come to see the world as information or as some type of computer: "We could envisage a simulation that encompassed the whole of the Universe. Given such a possibility, some scientists argue that the Universe is in some sense a gigantic computational system. . . . So the question arises: Can the Universe be a computer if the laws of physics are reversible and computers are not?" Brown, J., "Is the Universe a Computer," *The New Scientist*, 14 July 1990, 37–39. The first mainstream popular book to deal with these issues was Barrow, J. D., and Tipler, F. J., *The Anthropic Cosmological Principle*, Oxford University Press, 1986.

Physicist John Barrow returns to this theme at the end of a later book: "The great unsolved puzzle for the future is to decide which is more fundamental: symmetry or computation. Is the universe a cosmic kaleidoscope or a cosmic computer, a pattern or a program? Or neither? The choice requires us to know whether the laws of physics constrain the ultimate capability of abstract computation. Do they limit its speed and scope? Or do the rules governing the process of computation control what laws of Nature are possible?" Barrow, J. D., *Theories of Everything: The Quest for Ultimate Explanation*, Oxford University Press, 1991, pp. 203–4.

Physicist Tom Stonier argues for a massless information particle called a phonon: "1. *An infon is a photon whose wavelength has been stretched to infinity. And conversely, 2. A photon is an infon traveling at the speed of light."* Stonier, T., *Information and the Internal Structure of the Universe*, Springer-Verlag, 1993. Stonier assumes that information I varies inversely with Boltzmann thermodynamic entropy S and so derives an exponential relationship between the two variables: $I = I_0 e^{-S/k}$ where k is Boltzmann's constant 3.2983×10^{-24} cal/deg.

Stanford physicist Leonard Susskind has used the Bekenstein-Hawking equations and a bit-based view to argue that we can project the world's 3-D information onto a 2-D lattice much as a planar hologram stores a 3-D image: "According to this [older] philosophy the world is about as rich in structure as a three-dimensional discrete lattice theory with a spacing of order the Planck length. In this paper I will follow 't Hooft and argue for a far more radical decrease in the number of degrees of freedom. Instead of a three-dimensional lattice, a full description of nature requires only a two-dimensional lattice at the spatial boundaries of the world. In a certain sense the world is two-dimensional and not three-dimensional as previously supposed. . . . The lattice is composed of binary pixels with spacing of order the Planck length. We may find it convenient to consider larger lattice spacing and have a larger number of configurations at each site but the net result should be one bit per Planck cell." Susskind, L., "The World as Hologram," *Journal of Mathematical Physics* 36, no. 11 (November 1995): 6377–96.

25. Physicist Frank Tipler uses assumptions about a closed universe and the Poincare recurrence theorems to argue for Nietzche's dream of an Eternal Recurrence: "The Omega Point has the physical power to resurrect all humans who have ever lived and grant them eternal life. In brief, the physical mechanism of individual resurrection is the emulation of each and every long-dead person—and their worlds—in the computers of the far future. . . . Remarkably, the argument boils down to a proof that we will be granted eternal life because it is probable that the Omega Point loves us! Thus the ultimate cause of eternal life for humans in the Omega Point Theory is exactly the same as it is in the Judeo-Christian-Islamic tradition: God's selfless love." Tipler, F. J., Bantam, 1994, p. 14. For critical responses from physicists see Stenger, V. J., "Scientist Nitwit Atheist Proves Existence of God," *Free Inquiry* 15, no. 2 (Spring 1995): 54–55; and Price, M. C., "Review of *The Physics of Immortality*," *Extropy: The Journal of Transhumanist Thought* 7, no. 1 (1995): 42–45.

Many scientists do not believe that the universe is closed and thus the existence of an Omega Point may be moot. There does not seem to be enough matter to support a gravitational recollapse of the expanding universe: "But while cosmologists have been detecting more and more mass as they survey the universe on larger scales, with few exceptions they are not finding more than about a third to half of the critical amount. . . . And no matter how much ordinary matter turns up, the nucleosynthesis calculations imply that the total can be no more than 5% and 15% of the critical density [needed for recollapse]. That leaves the fate of the universe in the hands of whatever nonbaryonic dark matter may exist in the form of clouds of exotic elementary particles left over from the Big Bang." Appenzeller, T., "Weighing the Universe," *Science* 272 (7 June 1996): 1426.

Further evidence for no Big-Crunch collapse comes from the study of distant supernova explosions: "The ultimate fate of the Universe, infinite expansion or a big crunch, can be determined by using the redshifts and distances of very distant supernovae to monitor changes in the expansion rate. We can now find large numbers of these distant supernovae and measure their redshifts and apparent brightnesses. . . . Spectra and photometry from the largest telescopes on the ground and in space show that this ancient supernova is strikingly similar to nearby, recent type *Ia* supernovae. When combined with previous measurements of nearer supernovae, these new measurements suggest that we may live in a low-mass-density universe." Perlmutter, S., et al., "Discovery of a Supernova Explosion at Half the Age of the Universe," *Nature* 381, no. 1 (January 1998): 51–54.

26. "All final singularities 'bounce' or tunnel to initial singularities of new universes at which point the dimensionless parameters of the standard models of particle physics and cosmology undergo *small* random changes. This speculative hypothesis, plus the conventional physics of gravitational collapse, together comprise a mechanism for natural selection, in which those choices of parameters that lead to universes that produce the most black holes during their lifetime are selected for. If our universe is a typical member of the ensemble that results from many generations of such reproducing universes then it follows that the parameters of our present universe are near a local maximum of the number of black holes produced per universe. Thus [and this is the main testable part] modifications of the parameters of particle physics and cosmology from their present values should tend to decrease the number of black holes in the universe." Smolin. L., "Did the Universe Evolve?" *Classical Quantum Gravity* 9 (1992): 173–91. Sociobiologist John Maynard Smith critiques this use of evolutionary concepts at the cosmic or super-cosmic level in "On the Likelihood of Habitable Worlds," *Nature* 384 (14 November 1996): 107. Smolin presents a popular account of his Darwinian theory of universal evolution in Smolin, L., *The Life of the Cosmos,* Oxford University Press, 1997.

13. SMART ART

1. "The art establishment by and large spurns the computer when it is used by artists to replicate traditional methods and materials. The feeling seems to be that artists who utilize computer software solely to produce and output a digital painting or other conventional work are neither battling the traditional mainstream nor advancing the field of visual art in any other way. For today's artist, using the computer as a tool requires far less skill than mastering the techniques of oil painting or charcoal sketching, for example, since with the click of

a mouse users can now automate what were once virtuoso artistic techniques." Miller, P., "Technology for Art's Sake," *IEEE Spectrum* 35, no. 7 (July 1998): 30–37.

2. "To many archaeologists, art—or symbolic representation as they prefer to call it—burst on the scene after 50,000 years ago, a time when modern humans are widely thought to have migrated out of Africa to the far corners of the globe. These scholars say the migrants brought with them an ability to manipulate symbols and make images that earlier humans had lacked. An explosion of art resulted, its epicenter in ice age Europe starting about 40,000 years ago when most anthropologists believe modern humans were replacing the earlier Neanderthal people. The new Europeans decorated their bodies with beads and pierced animal teeth, carved exquisite figurines from ivory and stone, and painted hauntingly lifelike animals on the walls of deep caves." Appenzeller, T., "Art: Evolution or Revolution?" *Science* 282 (20 November 1998): 1451–54.

3. Kosko, B., "Art For Computer's Sake," *IEEE Spectrum* 32, no. 5 (May 1995): 10–12.

4. "There is a melancholy fantasy, propounded a century and more ago by the psychologist Theodor Fechner and taken up by Kurt Lasswitz, Theodor Wolff, Jorge Luis Borges, George Gamow, and Willy Ley, of a complete library. The library is strictly complete, boasting as it does all possible books within certain rather reasonable limits. It admits no books in alien alphabets, nor any beyond the reasonable length say of the one you are now reading, but within those restrictions it boasts all possible books. . . . At 2,000 characters to the page we get 500,000 to the 250-page volume, so with say 80 capitals and smalls and other marks to choose from we arrive at the 500,000th power of 80 [or $80^{500,000}$ or roughly $10^{951,545}$] as the number of books in the library. . . . The miracle of the finite but universal library is a mere inflation of the miracle of binary notation: everything worth saying, and everything else as well, can be said with two characters. It is a letdown befitting the Wizard of Oz, but it has been a boon to computers." Quine, W. V. O., *Quiddities: An Intermittently Philosophical Dictionary,* Harvard University Press, 1987, pp. 225–27.

5. For the MRI analysis of musician brains see Schlaug, G., Gaencke, L., Huang, Y., and Steinmetz, H., *Science* 268 (1995): 699–701. This result led to the research on how music training affects verbal memory: "The left temporal area in musicians might have a better developed cognitive function than the right temporal lobe. Because verbal memory is mediated mainly by the left temporal lobe, and visual memory by the right, adults with music training should have better verbal, but not visual, memory than adults without such training. Here we show that adults who received music training before the age of 12 have a better memory for spoken words than those who did not. Music training in childhood may therefore have long-term positive effects on verbal memory." Chan, A. S., Ho, Y.-C., and Cheung, M.-C., "Music Training Improves Verbal Memory," *Nature* 396 (12 November 1998): 128.

6. "The knowledge gained in music is knowledge about musical objects—tones and intervals and phrases and harmonies and rhythms. It is also knowledge about how these objects are organized and associated. We learn to recognize musical objects in ever larger groupings, then learn to recognize when these groupings are transformed: what changes and what remains the same. We learn about musical mappings." Rothstein, E., *Emblems of Mind: The Inner Life of Music and Mathematics,* Avon, 1995, p. 125.

The media were quick to report the computer birth of one of infinitely many possible Mozart 42nd symphonies: "To extravagant flourishes of the media's trumpets and drums, Mozart's 42nd symphony was revealed earlier this year. But this was no treasure trove rediscovered in a Salzburg attic; it was the creation of a computer in Santa Cruz." "The Mozart of Santa Cruz," *The Economist,* 30 August 1997, 64.

7. Many modern artists have based their work on formal structures: "As a result of Saussure's work, language is viewed as a formal system. And all forms of communication—all semiotic systems, all vehicles for communicating ideas—are viewed as languages. Kandinsky imagined 'grammars' for visual composition. [Pierre] Boulez developed explicit grammars and music composition techniques that could be automated." Holtzman, S. R., *Digital Mantras: The Languages of Abstract and Virtual Worlds,* MIT Press, 1994.

8. Teuvo Kohonen developed his adaptive rule-based musical system as a type of neural associative memory: "It has to be emphasized that *we are not constructing any rules heuristically;* all the 'grammatical' productions are derived from examples, for which we initially took the well-known inventions of J. S. Bach. The rules of the grammar in fact describe what might

be called a *motive* in music; i.e., a rather short elementary sequence, say, three to six successive notes which do not yet form a theme. It turned out that the original motives convey the style and beauty to the results. Our procedure, however, should not be regarded as 'scrambled Bach.' The special grammar that I am describing below is in fact able to pick up from memory long sequences of motives of the same style." Kohonen, T., "A Self-Learning Musical Grammar, or 'Associative Memory of the Second Kind,' " *Proceedings of the 1989 International Joint Conference on Neural Networks (IJCNN-89)* 1, June 1989, 1–5. See also the edited volume *Music and Connectionism,* MIT Press, 1991.

9. Chaos drives musical variation and associative recall of Bach's Minuet in G Major in a feedback neural network in Kawashima, J., and Nagashima, T., "An Experiment on Arranging Music by a Chaos Neural Network," *Proceedings of the 3rd International Conference on Fuzzy Logic, Neural Networks, and Soft Computing (Iizuka-94),* August 1994, 429–30. Chaos varies musical themes in a feedforward neural-fuzzy system in Kang, H., J., and Park, M., "Music Autocomposition Using Chaotic Dynamics," *Proceedings of the 3rd International Conference on Fuzzy Logic, Neural Networks, and Soft Computing (Iizuka-94),* August 1994, 431–34.

10. Spread spectrum and other noise-based schemes can produce digital watermarks: "Like a traditional watermark on paper, which is only visible when held up to the light or chemically processed, a digital watermark is information hidden within a data set. It can only be detected by a computer program that has the same secret element as was used in embedding the watermark. Digital watermarking can be used to claim ownership, by embedding permanent author identifiers when the work is created, and then to track distribution, by adding a customer identifier when the work is copied. Other information such as user rights, contact addresses and key words can also be embedded, for a wide range of applications. And to remind the user of copyright, a public notice or logo can be embedded as a watermark in a publicly known location." Zhao, J., "Watermarking by Numbers," *Nature* 384, no. 6609 (12 December 1996): 514. See also Okerson, A., "Who Owns Digital Works?" *Scientific American,* July 1996, 80–84.

11. Lyall, S., "Book Notes: Program for a Best Seller," *New York Times,* 23 June 1993. Computer programmer Scott French wrote the novel-writing program HAL with thousands of rules and plenty of editing: " 'The most difficult thing was trying to analyze exactly what constitutes a writer's style,' he [French] says. 'I broke it up into several hundred things, ranging from frequency and type of sexual acts to sentence structure.' . . . I'd say it [HAL] came up with almost 100% of the theme and style. Often it came up with three adjectives in a row and I had to put a verb in there. But I didn't change its basic story line or themes. I didn't feel I had the right to do that because I would have violated Miss Susann's style.' " Boudreau, J., "A Romance Novel with Byte: Author Teams up with Computer to Write Book in Steamy Style of Jacqueline Susann," *Los Angeles Times,* 11 August 1993, p. E-6.

12. "The hardball decision to replace a reporter with software was easy enough to justify, Sportswriter, as the program is called, costs $100. 'We were laying out $1,500 a month for the sports reporter,' says Donald Zavadil, editor and owner of the 1,800-circulation weekly [*Humphrey Democrat*]. . . . Sportswriter is one of a new genre of software that can intelligently string words between facts. Pop a disk into a PC, type in a few numbers and comments, and presto—out come stories awash in clutch shots, runaway victories and at-the-buzzer heroics." Bulkeley, W. M., "Semi-Prose, Perhaps, But Sportswriting By Software is a Hit," *Wall Street Journal,* 29 March 1994, p. A-1.

13. "Using information on the repeat sales of identical prints, I estimate a semiannual index of prices for the period 1977–1992. From this index, the mean and standard deviation of the (semiannual) returns to modern prints are readily derived. The mean real return (annual rate) on the aggregate print portfolio is only 1.51%, well beneath the real returns on stocks, US government bonds, and even US Treasury bills. Further, the risk (as measured by the standard deviation of real returns) of investing in prints is comparable to the risk of investing in stocks or long-term bonds. One must conclude, based on the risk-return criterion, that investments in this segment of the art market compares unfavorably to an investment in traditional financial assets." Pesando, J. E., "Art as an Investment: The Market for Modern Prints," *American Economic Review* 83, no. 5 (December 1993): 1075–89.

14. Today's smart rooms or mood matchers act as simple probabilistic associative memories: "Each room contains several machines, none more powerful than a personal computer. These

units tackle different problems. For instance, if a smart room must analyze images, sounds, and gestures, we equip it with three computers, one for each type of interpretation. If greater capabilities are needed, we add more machines. Although the modules take on different tasks, they all rely on the same statistical method, known as maximum likelihood analysis: the computers compare incoming information with models they have stored in memory. They calculate the chance that each stored model describes the observed input and ultimately pick the closest match. By making such comparisons, our smart-room machines can answer a range of questions about their users, including who they are and sometimes even what they want." Pentland, A. P., "Smart Rooms," *Scientific American,* April 1996, 68–76.

15. Hayes-Roth, B., "Interactive Fiction: Character-Based Interactive Story Systems," *IEEE Intelligent Systems* (November 1998): 12–15.

16. Current digital actors define paths or points that move through a complex joint space: "Motion control is the heart of computer animation. In the case of a digital actor, it essentially consists in describing the evolution over time of the joint angles of a hierarchical structure called a skeleton," Thalmann, N. M., and Thalmann, D., "Digital Actors for Interactive Television," *Proceedings of the IEEE,* July 1995, 1022–31.

17. "What awaits is not oblivion but rather a future which, from our present vantage point, is best described by the words 'postbiological' or even 'supernatural.' It is a world in which the human race has been swept away by the tide of cultural change, usurped by its own artificial progeny. . . . Unleashed from the plodding pace of biological evolution, the children of our minds will be free to grow to confront immense and fundamental challenges in the larger universe. We humans will benefit for a time from their labors, but sooner or later, like natural children, they will seek their own fortunes while we, their aged parents, silently fade away." Moravec, H., *Mind Children: The Future of Robot and Human Intelligence,* Harvard University Press, 1988, p. 1.

14. SECRET AGENTS

1. "While we or our electronic alter egos are busily looking at Web sites, a good number of the owners and advertisers on these sites are looking right back. All those mouse clicks and keystrokes—which electronic sites we visit, how long we stay and where we go before and after—are not the ephemera they seem. Clickstreams, as they are called, enjoy a digital afterlife in commercial databases, where raw statistics about our on-line behavior are transformed into useful information and then warehoused for future application, sale, or barter. These are known as the clickstreams that keep on giving—to advertisers, mass marketers, and lucky venture capitalists." Eisenberg, A., "Privacy and Data Collection on the Net," *Scientific American,* March 1996, 120.

The DoubleClick Network is one such marketing agency: "The comic-strip hero Dilbert shills for a Madison Avenue spy network on the World Wide Web. So do Sonic the Hedgehog, Doonesbury, and the popular Alta Vista Internet search engine, among others. All are affiliates of the DoubleClick Network *(www.doubleclick.net),* an ad agency collecting dossiers on the millions of people who visit dozens of popular Web sites daily—and who may not be aware that someone's gathering private information. As soon as you visit a site like 'The Dilbert Zone,' the DoubleClick Network records into its private database whatever information it can get about you—your zip code, area code, the organization you work for, or the type of computer you use. DoubleClick says it doesn't collect your name and e-mail address. At the same time, DoubleClick can also start collection about actions you take online—including any name, word, or subject you type into Alta Vista's search engine." "Is Your Computer Spying on You?" *Consumer Reports,* May 1997, 6. The software market has been quick to offer employers software to let them watch their employees in cyberspace: "More and more, corporations are using software tools to track employees' use of the Internet at the office—and managers don't always like what they see. Products like SurfWatch Professional Edition, WebSense, LittleBrother and Elron Internet Manager enable companies to follow virtually every mouse click a worker makes across the Internet. They can track access to specific Web sites and, with some programs, calculate the corporate cost of Web-surfing slackers. Bosses can even retrieve the results of an employee's search through

Internet directories such as Yahoo! and Excite." Branscum, D., "bigbrother@the.office.com: Your Boss Can Track Every Click You Make," *Newsweek,* 27 April 1998, 78.

2. Studies have shown that home-based midwives perform low-risk birth deliveries with risk rates comparable to those of hospital-based physician deliveries: Rooks, J., et al., "Outcomes of Care in Birth Centers: The National Birth Center Study," *New England Journal of Medicine* 321 (1989): 1804. Durand, M. A., "The Safety of Home Birth: The Farm Study," *American Journal of Public Health* 82 (March 1992): 450–53; Hafner-Eaton, C., and Pearce, L., "Birth Choices, the Law, and Medicine: Balancing Indivisual Freedoms and Protection of the Public's Health," *Journal of Health Politics, Policy, and Law* 19 (Winter 1994): 815. Studies have also shown that chiropractors often provide as good or better treatment for lower-back pain than do physicians in hospitals and that patients tend to prefer chiropractic treatment for lower-back pain: Meade, T. W., et al., "Low Back Pain of Mechanical Origin: Randomized Comparison of Chiropractic and Hospital Patient Treatment," *British Medical Journal* 300 (1990): 1435; Cherkin, D., et al., "Patient Evaluations of Low Back Pain Care From Family Physicians and Chiropractors," *Western Journal of Medicine* 150 (1989): 351.

3. Eisenberg, D., et al., "Unconventional Medicine in the United States: Prevalence, Costs, and Patterns of Use," *New England Journal of Medicine* 328, no. 4 (1993): 246–52.

4. Over half of the 130 or so US medical schools are state owned: Mullan, F., et al., "Doctors, Dollars, and Determination: Making Physician Work-Force Policy," *Health Affairs,* 1993, 138–51.

A Cato Institute study concludes that the United States and other countries can lessen their many problems with health care if they deregulate the supply side of medicine: "What should government do if it is serious about cutting health spending and improving access to affordable health care? The first step should be to eliminate the anti-competitive barriers that restrict access to low-cost providers, namely licensure laws and federal reimbursement regulations. Americans should not be forced to substitute providers against their will; rather, they should be free to choose among all types of health care providers. Instead of imposing strict licensure laws that focus on entry into the market but do not guarantee quality control, states should hold professionals equally accountable for the quality of their outcomes. That will reduce the need for strict licensure laws and other regulations that are purported to protect the public at large." Blevins, S. A., "The Medical Monopoly: Protecting Consumers or Limiting Competition?" *Policy Analysis,* no. 246 (Cato Institute, Washington, D.C., 15 December 1995): 1–36.

5. There are now many conferences and workshops on intelligent agents held each year around the world. The proceedings of these conferences and workshops summarize the past year's research frontier in software agents. For background see both the paper Maes, P., "Agents that Reduce Work and Information Overload," *Communications of the ACM* 37, no. 7 (July 1994): 31–40; and the entire *Proceedings of the First International Conference on Autonomous Agents,* ACM Press, February 1997. For a list of current workshops and conferences see the Agent Society at www.agent.org. Readers can download the agent tutorial Green, S., et al., "Software Agents: A Review" at www.cs.tcd.ie/research_group/aig/iag/.

For examples of medical agents see Detmer, W. M., and Shortliffe, E. H., "Using the Internet to Improve Knowledge Diffusion in Medicine," *Communications of the ACM* 40, no. 8 (August 1997): 101–8; Larsson, J. E., and Hayes-Roth, B., "Guardian: An Intelligent Autonomous Agent for Medical Monitoring and Diagnosis," *IEEE Intelligent Systems,* February 1998, 58–64. For an overview of search agents see Morreale, P., "Agents on the Move," *IEEE Spectrum,* April 1998, 34–41. Advances in charge-coupled devices and other imaging techniques will increase Internet telemedicine and the potential for medical software agents: Hill, J. W., "Telepresence Technology in Medicine: Principles and Applications," *IEEE Proceedings: Special Issue on Virtual and Augmented Reality in Medicine* 86, no. 5 (March 1998): 569–80. Teleradiology and teledermatology are already established fields: "Radiology is ideal for telemedicine because, even using conventional film-based systems, the radiologist rarely sees the patient. In a sense, conventional radiology is a store-and-forward technique: The film stores the image, which is then forwarded to the radiologist to examine it, make a diagnosis and report to the primary physician [or agent]. Now thanks to digital technology, the Internet, image compression software, and e-mail, the radiologist's office can be almost anywhere in the world." Robinson, K., "Telemedicine: Technology Arrives but Barriers Remain," *Biophotonics International,* July 1998, 40–47.

6. Intelligent software agents have extended and revitalized the field of computer science and its overreliance on binary expert systems: "Autonomous or semiautonomous agents are the dominant AI device in virtual organizations, proving to be a paradigm apparently suited to multiple, heterogeneous database structures. Agents have been given various roles within virtual organizations ranging from purchasing to selling, and to facilitating communications with other agents." O'Leary, D. E., Kuokka, D., and Plant, R., "Artificial Intelligence and Virtual Organizations," *Communications of the ACM* 40, no. 1 (January 1997): 52–59; O'Leary, D., "The Internet, Intranets, and the AI Renaissance," *IEEE Computer Magazine,* January 1997, 71–78. Agent software has since found their way into corporate planning because they offer a way to let marketers and others test their ideas on a "synthetic public": "If all goes well, the actions of the 'adaptive agents' will so closely mimic human behavior that managers for the first time will be able to use them to test the impact of their decisions before implementing them in the real world." Byrne, J. A., "Virtual Management," *Business Week,* 21 September 1998, 80–82; Falchuk, B., and Karmouch, A., "Visual Modeling for Agent-Based Applications," *IEEE Computer Magazine,* December 1998, 31–38.

7. A generic associative memory maps data to data. A random-address memory maps addresses to data. A content addressable memory maps data to addresses. For details see Kohonen, T., *Content Addressable Memories,* Springer-Verlag, 1980; Kohonen, T., *Self-Organization and Associative Memory,* Springer-Verlag, 1985. For a general class of feedback nonlinear content addressable memories see Kosko, B., "Unsupervised Learning in Noise," *IEEE Transactions on Neural Networks* 1, no. 1 (March 1990): 44–57.

8. Researchers have applied a wide range of fuzzy techniques to document search and retrieval: Radecki, T., "Mathematical Model of Information Retrieval System Based on the Concept of Fuzzy Thesaurus," *Information Processing & Management* 12, no. 5 (1976): 313–18; Zemankova, M., "FILIP: A Fuzzy Intelligent Information System with Learning Capabilities," *Information Systems* 14, no. 6 (1989): 473–86; Nomoto, K., et al., "A Document Retrieval System Based on Citations Using Fuzzy Graphs," *Fuzzy Sets and Systems* 38 (1990): 207–22, de Mantaras, R. L., et al., "Knowledge Engineering for a Document Retrieval System," *Fuzzy Sets and Systems* 38 (1990). 223–40; Ogawa, Y., et al., "A Fuzzy Document Retrieval System Using the Keyword Connection Matrix and a Learning Method," *Fuzzy Sets and Systems* 39 (1991): 163–79; Wu, J. K., et al., "CORE: A Content-Based Retrieval Engine for Multimedia Information Systems," *Multimedia Systems* 3 (1995): 25–41; Gomes, R., Pacheco, R., Martins, A., Weber, R., and Barcia, R., "Product Pricing Decision Support Fuzzy Systems Through the Internet," *Proceedings of the 1998 IEEE International Conference on Fuzzy Systems (FUZZ-98),* vol. 2, May 1998, 1664–69.

These and other text-search techniques should someday lead to cheap and widespread "concept search": "Grand visions in 1960 led first to the development of text search from bibliographic databases to full-text retrieval. Next, research prototypes catalyzed the rise of document search from multimedia browsing across local-area networks to distributed search on the Internet. By 2010 the visions will be realized with concept search enabling semantic retrieval across large collections." Schatz, B. R., *Science* 275 (17 January 1997): 327–34. Fuzzy systems can also apply at the higher multimedia level of predicting where an agent in a movie will move from frame to frame: Kim, H. M., and Kosko, B., "Neural Fuzzy Motion Estimation and Compensation," *IEEE Transactions on Signal Processing* 45, no. 10 (October 1997): 2515–32.

9. The measure E of fuzzy equality between two fuzzy subsets A and B of the same space has the form $E(A, B) = Degree(A = B) = c(A \cap B)/c(A \cup B)$. See note 17 of chapter 1 for details and the reduction of fuzzy equality to fuzzy subsethood. The papers in note 12 below show how search data can allow an agent to adapt this fuzzy search metric in the search process.

10. A utility function $u:X \to R$ maps each object $o \in X$ to a real number $u(o)$. Economists have proven that under technical conditions we can replace an agent's subjective relative ranking of all objects in the space X with a utility function that is unique up to linear transformation. So we can replace the ordinal claim "I like object o_1 at least as much as I like object o_2" with the cardinal relation $u(o_1) \geq u(o_2)$ and vice versa: "Representation of a Preference

Ordering by a Numerical Function," in Debreu, G., *Mathematical Economics: Twenty Papers of Gerard Debreu,* Cambridge University Press, 1983, pp. 105–110.

11. Neural and other statistical systems have joined with small electronic sensors and chips to produce the first practical biosensors: "Electrical impulses from nerves and muscles can command computers directly, a method that aids people with physical disabilities." Lusted, H. S., and Knapp, R. B., *Scientific American,* October 1996, 82–87; Proctor, P. "Retinal Displays Portend Synthetic Vision, HUD Advances," *Aviation Week & Space Technology,* 15 July 1996, 58; Wildes, R. P., "Iris Recognition: An Emerging Biometric Technology," *Proceedings of the IEEE* 85, no. 9 (September 1997): 1348–63. For a privacy perspective on the emerging field of biometrics see Woodward, J. D., "Biometrics: Privacy's Foe or Privacy's Friend?" *Proceedings of the IEEE* 85, no. 9 (September 1997): 1480–92. Research in patient-specific cochlear implants and related signal processing algorithms suggest that neural or fuzzy or other "smart" techniques may assist auditory biosensors: Loizou, P. C., "Mimicking the Human Ear," *IEEE Signal Processing Magazine,* Fall 1998, 101–30. German scientists showed that fully paralyzed patients could use only their thoughts or slow cortical potentials (SCPs) to spell or type 2 characters per minute: "Our data indicate that patients who lack muscular control can learn to control variations in their SCPs sufficiently accurately to operate an electronics spelling device." Birbaumer, N., et al., "A Spelling Device for the Paralyzed," *Nature* 398, 25 March 1999, 297–98.

12. Neural fuzzy systems can both approximate a user's utility map over sampled search objects and can adapt the fuzzy measure of equality to help search for objects in new databases that match the utility rankings of objects in the training database: Mitaim, S., and Kosko, B., "Neural Fuzzy Agents that Learn Profiles and Search Databases," *Proceedings of the IEEE International Conference on Neural Networks (ICNN-97),* June 1997, 467–72; Mitaim, S., and Kosko, B., "Neural Fuzzy Agents for Profile Learning and Adaptive Object Matching," *Presence* 7, no. 6 (December 1998): 617–37.

13. Most pattern-recognition schemes convert an image into a feature vector. The feature vector has far fewer dimensions than does the image. For a typical color-based scheme see Caelli, T., and Reye, D., "On the Classification of Image Regions of Colour, Texture and Shape," *Pattern Recognition* 26, no. 4 (1993): 461–70.

14. Nineteenth-century English philosopher John Stuart Mill appears as both a neural-fuzzy software intelligent agent and a *dramatis persona* in a cyber-thriller novel: Kosko, B., *Nanotime,* Avon Books, 1997.

15. Lanier, J., "My Problems with Agents," *Wired,* November 1996, 157–60.

16. Minsky, M., *The Society of Mind,* Simon & Schuster, 1985.

17. "Which is the more profitable way to sell a company: an auction with no reserve price or an optimally structured negotiation with one less bidder? We show under reasonable assumptions that the auction is always preferable when bidders' signals are independent. . . . A simple competitive auction with $N + 1$ bidders will yield a seller more expected revenue than she could expect to earn by fully exploiting her monopoly selling position against N bidders." Bulow, J., and Klemperer, P., "Auctions Versus Negotiations," *American Economic Review* 86, no. 1 (March 1996): 180–94. This result can fail in the general case when there are "affiliated" bidders who do not act independently of one another. Digital marketplaces will favor ever more competition and hence auctions: "The dynamics of friction-free markets are not attractive for sells that had previously depended on geography or customer ignorance to insulate them from the low-cost sellers in the market." Bakos, Y., "The Emerging Role of Electronic Marketplaces on the Internet," *Communications of the ACM,* August 1998, 35–42. These markets may face competing state jurisdictions: Aalberts, R. J., Townsend, A. M., and Whitman, M. E., "The Threat of Long-Arm Jurisdiction to Electronic Commerce," *Communications of the ACM* 41, no. 12 (December 1998): 15–20.

18. Hanson, R., "Idea Futures," *Wired,* September 1995, 125.

19. The World Wide Web hosts the first trial idea-futures markets in ideas at http://if.arc.ab.ac/if.shtml and at http://www.ideafutures.com. The Web also hosts a fictitious stock market that lets users "invest" $2 worth of shares in upcoming Hollywood feature movies on the Hollywood Stock Exchange at www.hsx.com.

20. "The concrete proposal for the near future . . . is that the countries of the Common Market, preferably with the neutral countries of Europe (and possibly later the countries of North

America) mutually bind themselves by formal treaty not to place any obstacles in the way of the free dealing throughout their territories in one another's currencies (including gold coins) or of a similar free exercise of the banking business by any institution legally established in any of their territories. This would mean in the first instance the abolition of any kind of exchange control or regulation of the movement of money between these countries, as well as the full freedom to use any of the currencies for contracts and accounting." Hayek, F. A., *Denationalisation of Money: The Argument Refined,* 2d edition, Institute of Economic Affairs, 1978. Free banking has a long history in the United States and in other countries: Selgin, G. A., *The Theory of Freebanking: Money Supply Under Competitive Note Issue,* Rowman and Littlefield, 1983; White, L. H., *Competition and Currency: Essays on Free Banking and Money,* New York University Press, 1984; Selgin, G. A., and White, L. H., "How Would the Invisible Hand Handle Money?" *Journal of Economic Literature* 32 (December 1994): 1718–49.

Free banking flourished in many states before the Civil War and its return to a government monopoly on the money supply: "The argument for free banking is also very simple. If markets are generally better at allocating resources than governments are, then what is 'different' about 'money' and the industry that provides it, the banking industry, to lead one to conclude that money and banking are an exception to the general rule?" Dowd, K., *Laissez-Faire Banking,* Routledge, 1993. One argument put forth in favor of central banks or "natural monopolies" in banking is that banks enjoy increasing returns to scale from their reserve holdings: Some math models show that a bank's optimal reserve holdings grow with the square root of its liabilities and hence its optimal reserve-liability ratio falls. This explains why banks tend to merge at least until some form of diminishing returns sets in (which tends to happen quickly as the marginal return steadily falls with the inverse square root of its liabilities). This argument implies only that small banks have an incentive to grow larger. It does not imply that the banking industry cannot benefit from competition among even these larger banks. There is no evidence that a natural banking monopoly ever emerged in those countries (from the United States and Canada and Scotland to Switzerland and Australia and China) that have practiced freebanking.

21. "FBI: Advanced Communications Technologies Pose Wiretapping Challenges," Briefing Report to the Chairman, Subcommittee on Telecommunications and Finance, Committee on Energy and Commerce, House of Representatives; General Accounting Office; B-249358, July 1992, p. 2.

22. Quoted from interview in Adam, J. A., "Wanted: Wiretappable Equipment," *IEEE Institute* 16, no. 5 (October 1992): 7.

23. For an algorithmic overview of standard encryption schemes see Patterson, W., *Mathematical Cryptology for Computer Scientists and Mathematicians,* Rowman & Littlefield, 1987. For a popular view see Beth, T., "Confidential Communication on the Internet," *Scientific American,* December 1995, 88–91. Engineers and computers can find it hard to crypto-analyze coded messages that have not been encrypted: Gillman, D. W., Mohtashemi, M., and Rivest, R. L., "On Breaking a Huffman Code," *IEEE Transactions on Information Theory* 42, no. 3 (May 1996): 972–76. Encryption pioneer Philip Zimmermann gives an accessible overview of popular encryption schemes such as public-key encryption in Zimmermann, P. R., "Cryptography for the Internet," *Scientific American,* October 1998, 110–15.

24. Kosko, B., ". . . And Constitutional Rights Lag Behind," *Los Angeles Times,* 19 December 1994, op-ed page. Newspaper editors often impose their own titles on op-ed articles that others write. The original title of this op-ed essay was "Time for a Digital Rights Act."

25. "From street agents to the seat of judicial power in Washington, federal law enforcement officials are preparing a major push this coming year to broaden legal authority to wiretap telephones in criminal cases. Specifically, they want the right to put electronic bugs not only on fixed locations—such as an office or home telephone—but also on the mobile communication technology used by the suspects they are pursuing. . . . The Supreme Court declared all wiretapping illegal in 1967. A year later, Congress legalized the technique as long as the FBI obtained court approval. In the last eight years, no federal judge has turned down a proper request for federal eavesdropping." Serrano, R. A., "Agencies Seek Update in Wiretap Access," *Los Angeles Times,* 29 November 1996, p. A-1.

Encryption pioneer Ronald Rivest (who co-invented the popular RSA encryption algorithm) has challenged the argument that the state should control encryption to help it fight crime: "The government's concern is that the 'bad guys' will benefit from the new cryptographic technology. This is certainly possible—the sun shines on the evil as well as the good. But it is poor policy to clamp down indiscriminately on a technology merely because some criminals might be able to use it to their advantage. For example, any US citizen can freely buy a pair of gloves even though a burglar might use them to ransack a house without leaving fingerprints. Cryptology is a data-protection technology just as gloves are a hand-protection technology. Cryptography protects data from hackers, corporate spies and con artists, whereas gloves protect hands from cuts, scrapes, heat, cold and infection. The former can frustrate FBI wiretapping and the latter can thwart FBI fingerprint analysis. Cryptography and gloves are both dirt-cheap and widely available. In fact, you can download good cryptographic software from the Internet for less than the price of a good pair of gloves." Rivest, R. L., "The Case Against Regulating Encryption Technology," *Scientific American,* October 1998, 116–17.

26. "The agency that manages the nation's spy satellite program has accumulated unspent funds totaling more than $1 billion without informing its supervisors at the Pentagon and CIA or its overseers in Congress, according to Capitol Hill sources. . . . The 35-year-old agency, whose operations were so cloaked in secrecy that even its name [The National Reconnaissance Office] was classified until three years ago, supervises design, development, procurement and launching of satellites." Pincus, W., *Los Angeles Times,* 24 September 1995, p. A-29.

15. Heaven in a Chip

1. Zerin, E., "Karl Popper on God: The Lost Interview," *Skeptic* 6, no. 2 (1998): 46–49.
2. MIT artificial intelligence pioneer Marvin Minsky made this comment in December of 1995 during an address at an anti-aging conference in Las Vegas: Cheney, R., "The 3rd International Conference on Anti-Aging Medicine and Biomedical Technology: A Review," *Cryonics* 17, no. 1 (First Quarter 1996): 19–22.
3. Nobel laureates David Hubel and Torston Wiesel carried out their pioneering research on the kitten visual neural system at Harvard University: Hubel, D. H., and Wiesel, T. N., "Receptive Fields of Striate Cortex of Very Young, Visually Inexperienced Kittens," *Journal of Neurophysiology* 26 (1963): 994–1002; Wiesel, T. N., and Hubel, D. H., "Comparison of the Effects of the Unilateral and Bilateral Eye Closure on Cortical Unit Responses in Kittens," *Journal of Neurophysiology* 28 (1965): 1029–40. More recent research on visual neural networks has supported and extended many of these findings: Sinha, P., and Poggio, T., "Role of Learning in Three-Dimensional Form Perception," *Nature* 384 (5 December 1996): 460–63. Wylie, D. R. W., Bischof, W. F., and Frost, B. J., "Common Reference Frame for Neural Coding of Translational and Rotational Optic Flow," *Nature* 392 (19 March 1998): 278–81.
4. Adaptive resonance theory or ART describes the nonlinear math of how a sensory pattern excites a field of neurons and how this field evokes an expected pattern that then matches or mismatches the sensory pattern in a dynamic equilibrium or resonance. Boston University's Stephen Grossberg first developed the ART architecture in 1976 in the seminal paper Grossberg, S., "Adaptive Pattern Classification and Universal Recoding, II: Feedback, Expectation, Olfaction, and Illusions," *Biological Cybernetics* 23 (1976): 187–202. See also Grossberg, S., *Studies of Mind and Brain,* Reidel, 1982.
5. "Cells from multicellular organisms self-destruct when they are no longer needed or have become damaged. They do this by activating genetically controlled cell suicide machinery that leads to programmed cell death (PCD). To survive, all cells from multicellular animals depend on the constant repression of this suicide program by signals from other cells. . . . Why has evolution favored such a system of cell destruction? PCD allows constant selection for the fittest cell in the colony, optimal adaptation of cell numbers to the environment, and tight regulation of the cell cycle and cell differentiation." Ameisen, J. C., "The Origin of Programmed Cell Death," *Science* 272 (31 May 1996): 1278–79.
6. For reviews of recent results in cell-based theories of aging see Lithgow, G. J., and Kirkwood, T. B. L., "Mechanisms and Evolution of Aging," *Science* 273 (5 July 1996): 80–81; Irmler, M., et al., "Inhibition of Death Receptor Signals by Cellular FLIP," *Nature* 388 (10 July 1997): 190–95; Barinaga, M., "Death by Dozens of Cuts," *Nature* 280 (3 April 1998): 32–34.

Further research suggests that some cells will still age or experience "senescence" despite telomeric attempts to give them immortality: Weinberg, R. A., "Bumps on the Road to Immortality," *Nature* 396 (5 November 1998): 23–24.

7. Sociologist Franz Oppenheimer casts the origin of the state in stark terms of conquest and exploitation: "The moment when the first conqueror spared his victim in order to permanently exploit him in productive work, was of incomparable historical importance. It gave birth to nation and state, to right and the higher economics, with all the developments and ramifications which have grown and which will hereafter grow out of them." Oppenheimer, F., Viking Press, 1975, p. 27.

Economists Martin McGuire and Mancur Olson have presented a like economic view of the state as a monopoly on power: "Consider the interests of the leader of a group of roving bandits in an anarchic environment. In such an environment, there is little incentive to invest or produce and therefore not much to steal. If the bandit leader can seize and hold a given territory, it will pay him to limit the rate of his theft in that domain and to provide a peaceful order and other public goods. By making it clear that he will take only a given percentage of output—that is, by making himself a settled ruler with a given rate of tax theft—he leaves his victims with an incentive to produce. By providing a peaceful order and other public goods, he makes his subjects more productive. Out of the increase in output that results from limiting his rate of theft and from providing public goods, he obtains more resources for his own purposes than from roving banditry. This rational monopolization of theft also leaves the bandit's subjects better off: they obtain the increase in income not taken in taxes. The bandit leader's incentive to forgo confiscatory taxation and to provide public goods is due to his 'encompassing interest' in the conquered domain." McGuire, M. C., and Olson Jr., M., "The Economics of Autocracy and Majority Rule: The Invisible Hand and the Use of Force," *Journal of Economic Literature* 34 (March 1996): 72–96.

8. Most theists or believers base their belief in God on faith. Those who wish to go beyond faith and base their belief on reason or formal argument would do well to examine the vast philosophical and logical literature that disputes such claims. Perhaps the definitive book on the subject is Boston University philosopher Michael Martin's *Atheism: A Philosophical Justification,* Temple University Press, 1990. See also Flew, A., "The Presumption of Atheism," in *God, Freedom and Immortality,* Prometheus Books, 1984. Radcliffe scholar Wendy Kaminer presents a more personal view of the near absence of nontheism in modern culture: "What's striking about American intellectuals today, liberal or conservative alike, is not their Voltairean skepticism but their deference to belief and their utter failure to criticize, much less satirize, America's romance with God. They've abandoned the tradition of caustic secularism that once provided refuge for the faithless: people 'are all insane,' Mark Twain remarked in *Letters from Earth.* 'Man is a marvelous curiosity . . . he thinks he is the Creator's pet . . . he even believes the Creator loves him; has a passion for him; sits up nights to admire him; yes and watch over him and keep him out of trouble. He prays to Him and thinks He listens. Isn't it a quaint idea.' No prominent liberal thinker writes like that anymore." Kaminer, W., "The Last Taboo: Why America Needs Atheism," *The New Republic,* 14 October 1996, 24–32.

9. Pragmatist philosopher William James argued that our innate will to believe could decide or at least incline belief in matters of doubt: "The thesis I defend is, briefly stated, this: Our passional nature not only lawfully may, but must, decide an option between propositions, whenever it is a genuine option that cannot by its nature be decided on intellectual grounds; for to say, under such circumstances, 'Do not decide, but leave the question open,' is itself a passional decision—just like deciding yes or no—and is attended with the same risk of losing the truth." James, W., *The Will to Believe,* first published in 1896, reprinted in the anthology *William James: Writings 1878–1899,* Viking Press, 1992, p. 464.

10. Most math models of how synapses learn use some form of Hebbian learning:

$$m_{ij} = -m_{ij} + S_i(x_i) S_j(y_j)$$

where m_{ij} measures a synapse's strength or efficacy in the link from the ith pre-synaptic neuron to the jth post-synaptic neuron. The ith neuron uses a bounded monotone increasing (or S-shaped) signal function S_i to convert an arbitrary real-valued activation value x_i into the signal that lies in interval $[0, 1]$ or $[-1, 1]$. The passive decay term $-m_{ij}$ forces exponential forgetting on all such learning laws. The additive structure of this simple differential equation dictates the exponential structure of its solution:

$$\dot{m}_{ij}(t) = m_{ij}(0)e^{-t} + \int_0^t S_i(x_i(s))\,S_j(y_j(s))e^{s-t}\,ds$$

Psychologist Donald Hebb proposed a verbal version of this correlation learning law in Hebb, D. O., *The Organization of Behavior,* Wiley, 1949. Nobel laureate economist Friedrich Hayek independently proposed a verbal version at about the same time (in terms of German publication) in *The Sensory Order: An Inquiry in the Theoretical Foundations of Psychology,* University of Chicago Press, 1952. Neural theorist Stephen Grossberg may have published one of the first formal mathematical analyses of this and related unsupervised synaptic learning laws in "On Learning and Energy-Entropy Dependence in Recurrent and Nonrecurrent Signed Networks," *Journal of Statistical Physics* 1 (1969): 319–50.

Many biochemical factors govern how fast a real synapse learns or forgets neural patterns of sensation: "Sensory experiences leave their mark on the brain by altering the effectiveness of synapses between neurons. Based on how active they are during a sensory experience, some synapses on a neuron grow stronger while others grow weaker. And the pattern of synaptic change represents a memory of the experience. Changes in synaptic strength are due to modifications of existing synaptic proteins, by phosphorylation and dephosphorylation, for example." Bear, M. F., "How Do Memories Leave Their Mark?" *Nature* 385 (6 February 1997): 481–82.

11. Earlier versions of some of these ideas appear in Kosko, B., "Heaven in a Chip," *Datamation,* 15 February 1994, 12–13, reprinted in *Free Inquiry,* Fall 1994, 37–38; and "Chipping Away at Your Brain," *Datamation,* 15 April 1994, 96–97.

12. The brain may store as many as 10^{28} bits or as few as a few trillion or quadrillions. Most analyses arrive at a number far smaller than 10^{28} bits by factoring in the vast redundancy in neural coding and focusing on synapses rather than brain atoms as the units of computation. Ralph Merkle of Xerox Palo Alto Research Park uses the biophysics of a "Ranvier operation" to arrive at the brain's bit count of 10^{18} bits and a processing rate of 10^{16} bits per second: "When a single voltage-activated sodium channel opens, it has a conductance of about 15 picosiemens. (A siemen is the reciprocal of an ohm, and is also called a 'mho.') In myelinated nerve cells there are roughly 60,000 channels at each node of Ranvier (and nowhere else). The total charge that crosses the membrane at one node in one millisecond can thus be computed: about 5.4×10^{-11} coulombs (over 300 million ions per node). The energy dissipated is just the charge times the voltage or 3.2×10^{-12} joules. If we view this one millimeter jump [along an axon] as a 'basic operation' then we can easily compute the maximum number of such 'Ranvier ops' the brain can perform each second: 3.1×10^{12}. . . . If propagating a nerve impulse a distance of 1 millimeter requires about 3.2×10^{-12} joules and the total energy dissipated by the brain is about 10 watts, then nerve impulses in your brain can collectively travel at most 3.1×10^{12} millimeters per second. By estimating the distance between synapses we can in turn estimate how many synapses operations per second your brain can do. This estimate is three to four orders of magnitude smaller than an estimate based simply on counting synapses and multiplying by the average firing rate, and similar to an estimate based on functional estimates of retinal computational power. It seems reasonable to conclude that the human brain has a 'raw' computational power towards the low end of the range between 10^{12} and 10^{16} 'operations' per second. . . . Roughly, uploading will need a computer with a memory about 10^{18} bits, able to do around 10^{16} 'operations' per second. A computer of this capacity should fit comfortably into a cubic centimeter in the early 21st century." Merkle, R. C., "Uploading: Transferring Consciousness from Brain to Computer," *Extropy* 5, no. 1 (Fall 1993): 5–8. Computer scientist Hans Moravec arrives at comparable retina-based brain estimates in Moravec, H., *Mind Children,* Harvard University Press, 1988. Russian cryonicist Mikhail Soloviev builds on these results in one of many proposed nanotech-based cell repair algorithms: Soloviev, M. V., "A Cell Repair Algorithm," *Cryonics* 19, no. 1 (First Quarter 1998): 22–27. Uploading also has a home page: http://sunsite.unc.edu/jstrout/uploading/MUHomePage.html

13. Intel founder and chairman emeritus Gordon Moore believes some form of Moore's Law will still hold for a few years: "The complexity of the microprocessor as measured by the number of transistors on a chip has been doubling every 18 months or so since the original 4004 [Intel chip of 1971]. This exponential growth has helped increase performance and shrink the cost of computing. As recently as 1991, a PC based on the Intel 486 processor cost about $225 per million instructions per second (MIPS) of performance. Today a desktop computer using the Pentium Pro processor delivers dramatically increased performance at a cost of only about $7 per MIPS. There seems to be no reason why this general trend should not continue.

"Before the semiconductor industry reaches any fundamental limits, we should be able to make economical processors with well over a billion transistors performing at speeds far above today's. Another thousand-fold increase in computing power seems to me a conservative estimate." Moore, G. E., "The Microprocessor: Engine of the Technology Revolution," *Communications of the ACM* 40, no. 2 (February 1997): 112–14. See also Moore, G. E., "Cramming More Components onto Integrated Circuits," *Proceedings of the IEEE* 86, no. 1 (January 1998): 82–85.

This exponential growth in chip transistor density may depend on the shift from light etching or optical lithography to X-ray lithography: "Current state-of-the-art systems use UV light with a wavelength of 248 nanometers, which allows them to write features of just 0.25 microns. Industry scientists expect that by the year 2001, they'll be able to mass-produce features of 0.18 micrometers by reducing the wavelength of the UV light used to 193 nanometers. And many expect that, with a few more tweaks to the system, optical lithography should be able to produce features of 0.13 micrometers by about 2004. But past that optical lithography fails says Franco Cerrina, who heads the Center for X-ray Lithography at the University of Wisconsin, Madison. The problem, he explains, is that at wavelengths shorter than 193 nanometers, conventional quartz lenses absorb rather than refract, or bend, light. If a lithographic successor isn't found by then, the number of transistors on computer memory chips could top out at about 4 billion, compared to today's standard of 16 million. Perhaps the strongest candidate for succeeding optical lithography is its close cousin X-ray lithography. Typical X-ray systems fire photons with a wavelength of just 4 nanometers and so can write far sharper features than their UV counterparts can." Service, R. E., "Can Chip Devices Keep Shrinking?" *Science* 274 (13 December 1996): 1834–36. See also Geppert, L., "Semiconductor Lithography for the Next Millennium," *Spectrum,* April 1996, 33–38; Hutcheson, G. D., and Hutcheson, J. D., "Technology and Economics in the Semiconductor Industry," *Scientific American,* January 1996, 54–62; Hirschman, K. D., Tsybeskov, L., Duttagupta, S. P., and Fauchet, P. M., "Silicon-based Visible Light-emitting Devices Integrated into Microelectronic Circuits," *Nature* 384 (28 November 1996): 338–41; Klein, D. L., et al., "A Single-Electron Transistor Made From a Cadmium Selenide Nanocrystal," *Nature* 389 (16 October 1997): 699–701; Penzias, A. A., "The Next Fifty Years: Some Likely Impacts of Solid-State Technology," *Proceedings of the IEEE* 86, no. 1 (January 1998): 289–90.

14. Much research in nanotechnology has focused on how to build or grow tiny molecular assemblers or nanobots with carbon nanotubes or other materials. Related computational research has focused on how to combat energy dissipation and thermal noise in the design of nanoprocessors: "The historical trend in computer systems is to pack ever more logic gates into ever smaller volumes. This trend can only be continued if the energy dissipation per logic operation also continues to decline. The potential packing densities that nanoelectronic and molecular logic devices should be able to achieve will only be realized if the energy dissipated per logic operation can be reduced to extremely small values. Projections of current trends in energy dissipation per gate operation suggest that the kT 'barrier' [kT measures the thermal energy of an atom where k is Boltzmann's constant and T is temperature in kelvins] will become significant with 10–20 years. This barrier can be overcome by using reversible logic. Reversible logic will be valuable well before the kT barrier is reached. Even though not inherently required when the energy dissipation per logic operation is greater than kT, reversible designs can more easily reduce energy dissipation than irreversible designs even when the actual energy dissipation is orders of magnitude greater than kT. . . . Breaking the kT barrier is feasible in principle, and will eventually be necessary if we are to continue the

dramatic improvements in computer hardware performance and packing densities that we have seen during the last several decades. The ultimate limit for electronic devices will be reached when we are able to fabricate atomically precise logic elements that are thermodynamically reversible and use single electrons to represent information. This is likely to occur sometime early in the 21st century." Merkle, R. C., "Reversible Electronic Logic Using Switches," *Nanotechnology* 4 (1993): 21–40.

Half of Merkle's prediction has already come to pass. Researchers at the Nanostructure Laboratory of the University of Minnesota have built a single electron metal-oxide semiconductor memory (SEMM) device at room temperature: "To increase the storage density of semiconductor memories, the size of each memory cell must be reduced. A smaller memory cell also leads to faster speeds and lower power consumption. . . . The SEMM presented here is orders of magnitude smaller than the conventional floating gate MOS memory, has properties that conventional memories do not have, and is a major step forward in taking advantage of single electron effects to build ultrasmall and ultrahigh density transistor memories." Guo, L., Leobandung, E., and Chou, S. Y., "A Silicon-Electron Transfer Memory Operating at Room Temperature," *Science* 275 (31 January 1997): 649–51.

Quantum computers offer a formal math framework in which to design computational algorithms and devices on nano-scales and even on smaller scales. See DiVencenzo, D. P., "Quantum Computation," *Science* 270 (13 October 1995): 255–61; Lloyd, S., "Quantum-Mechanical Computers," *Scientific American*, October 1995, 140–45; Lloyd, S., "Universal Quantum Simulators," *Science* 273 (23 August 1996): 1073–78; Orlov, A. O., et al., "Realization of a Functional Cell for Quantum-Dot Cellular Automata," *Science* 277 (15 August 1997): 928–30; Jones, J. A., "Fast Searches with Nuclear Magnetic Resonance Computers," *Science* 280 (10 April 1998): 229.

15. Philosopher Daniel Dennett reviews and criticizes the main theories of mind in Dennett, D. C., *Consciousness Explained*, Little, Brown, 1992. Philosopher John Searle explores theories of mind from a different point of view in Searle, J. R., *The Rediscovery of Mind*, MIT Press, 1992. Searle reviews and criticizes Dennett's more anticonsciousness views in the two-part essay in Searle, J. R., "The Mystery of Consciousness," *New York Review of Books*, 2 November 1995, 60–66, and 16 November 1995, 54–61. Dennett responds with detailed letters to the editor in *New York Review of Books*, 21 December 1995, 83–85. Philosophers Paul and Patricia S. Churchland present "eliminative materialist" theories of mind from the perspective of modern neural network theory in Churchland, P., *A Neurocomputational Perspective: The Nature of Mind and the Structure of Science*, MIT Press, 1993; Churchland, P. S., and Sejnowski, T. J., *The Computational Brain*, MIT Press, 1992; More, M., "Thinking About Thinking: An Interview with Paul Churchland," *Wired*, December 1996, 252–53.

16. Oxford sociobiologist Richard Dawkins has been one of the most forceful advocates of the view that natural selection and random mutation is blind and that it does not in principle involve godlike or humanlike goals. It simply favors those "gene machines" that breed before they die: "We all, without a single exception, inherit all our genes from an unbroken line of successful ancestors. The world becomes full of organisms that have what it takes to become ancestors. That, in a sentence, is Darwinism. . . . It is tempting to think that when ancestors did successful things, the genes they passed on to their children were, as a result, upgraded relative to the genes they had received from their parents. Something about their success rubbed off on their genes, and that is why their descendants are so good at flying, swimming, courting. Wrong, utterly wrong! Genes do not improve in the using, they are just passed on, unchanged except for very rare random errors. It is not success that makes good genes. It is good genes that make success, and nothing an individual does during its lifetime has any effect whatever upon its genes. Those individuals born with good genes are the most likely to grow up to become successful ancestors; therefore good genes are more likely than bad to get passed on to the future. Each generation is a filter, a sieve: good genes tend to fall through the sieve into the next generation; bad genes tend to end up in bodies that die young or without reproducing." Dawkins, R., *River Out of Eden: A Darwinian View of Life*, Basic Books, 1995. See also Dawkins, R., *The Blind Watchmaker*, Norton, 1986; Dennett, D. C., *Darwin's Dangerous Idea: Evolution and the Meaning of Life*, Touchstone, 1995.

Paleontologist Stephen Gould states the issue in blunter terms: "There is no progress in evolution. The fact of evolutionary change through time doesn't represent progress as we know it. Progress is not inevitable. Much of evolution is downward in terms of morphological complexity rather than upward. We're not marching toward some greater thing." Gould, S. J., in chapter 2 of Brockman, J., *The Third Culture*, Touchstone, 1995, p. 52.

17. The "new growth theory" in modern economics focuses on how an economy or market grows from the inside out (from endogenous factors) rather from factors like technology outside the economy (exogenous factors). U.C. Berkeley economist Paul Romer has taken the lead in showing how the bit-based world of ideas drives growth in advanced economies: "New growth theorists now start by dividing the world into two fundamentally different types of productive inputs that can be called 'ideas' and 'things.' Ideas are nonrival goods that could be stored in a bit string. Things are rival goods with mass (or energy). With ideas and things [bits and atoms] one can explain how economic growth works. Nonrival ideas can be used to rearrange things, for example, when one follows a recipe and transforms noxious olives into tasty and healthy olive oil. Economic growth arises from the discovery of new recipes and the transformation of things from low to high value configurations. This slightly different initial cut leads to insights that do not follow from the neoclassical model. It emphasizes that ideas are goods that are produced and distributed just as other goods are." Romer, P. M., "Why, Indeed, in America? Theory, History, and the Origins of Modern Economic Growth," *American Economic Review* 86, no. 2 (May 1996): 202–6. See also Romer, P. M., "Growth Based on Increasing Returns Due to Specialization," *American Economic Review* 77, no. 2 (May 1987): 56–62; Romer, P. M., "The Origins of Endogenous Growth," *Journal of Economic Perspectives* 8, no. 1 (Winter 1994): 3–22; Solow, R. M., "Perspectives on Growth Theory," *Journal of Economic Perspectives* 8, no. 1 (Winter 1994): 45–54.

18. "Every available measure of piety, including frequency of prayer, belief in God, and confidence in religion, is greater in countries with numerous competing churches than in countries dominated by a single established church, and these relationships remain strong even after controlling for income, education, or urbanization. It is also true that *within* each country the average level of religious belief and participation is consistently lower in the established churches, which enjoy the financial and regulatory support of the state, than among the small denominations operating at the competitive fringe of the country's religious market." Iannaccone, L. R., "Introduction to the Economics of Religion," *Journal of Economic Literature* 36 (September 1998): 1465–96.

Acknowledgments

Several fine minds helped refine the manuscript that evolved into this book. Special thanks go to Eamon Dolan, Janet Goldstein, Rick Kot, Daniel McNeill, and especially Laura Wood and Douglas Pepper of Harmony Books.

Index

About the Author

Dr. Bart Kosko holds degrees in philosophy, economics, mathematics, and electrical engineering. He is an award-winning composer and has published several works of fiction and nonfiction. He is on the faculty of the electrical engineering department at the University of Southern California and has chaired or co-chaired many international conferences on fuzzy systems and neural networks. He is an elected governor of the International Neural Network Society. He lives in the Los Angeles area.